Mathematics Level 3

Mathematics Level 3

ERIC WALKER, MA(Cantab.)

Formerly Head of the Mathematics Department and Deputy Headmaster of Sir Roger Manwood's School, Sandwich; later at the South Kent College of Technology, Folkestone.

HOLT, RINEHART AND WINSTON

LONDON · NEW YORK · SYDNEY · TORONTO

Holt, Rinehart and Winston Ltd: 1 St Anne's Road,
Eastbourne, East Sussex BN21 3UN

British Library Cataloguing in Publication Data

Walker, Eric
Mathematics level 3.—(Holt technician texts)
1. Shop mathematics
I. Title
510′.246 TJ1165

ISBN 0-03-910355-2

Typeset by Macmillan India Ltd, Bangalore.
Printed in Great Britain by Mackays of Chatham Ltd, Chatham, Kent.

Last digit is print no: 9 8 7 6 5 4 3 2 1

Introduction

In a number of parts of the syllabus at Level 3 it is an advantage to be aware of certain trigonometric identities. In Chapter 21 it is essential to know them. These identities are readily proved by vector methods, yet vector methods are not included in the syllabus. Certain traditional methods which have otherwise been adopted to prove the identities are very lengthy. It is hoped that the following treatment is easy to understand and will serve the intended purpose. As an essential preliminary a formula is required for the perpendicular from a point to a line.

The perpendicular distance from a point to a line

In Fig. 1 suppose the line BM has an equation

$$y = mx + c$$

$$\text{or} \quad y - mx - c = 0 \qquad 1$$

P is the point (x_1, y_1). Suppose that the line perpendicular to BM from P is PM, and that the line AP is parallel to BM meeting Oy at A. Then the equation of AP is of the form:

$$y = mx + k \qquad 2$$

where k is to be determined. Because AP passes through $P(x_1, y_1)$:

$$y_1 = mx_1 + k \qquad 3$$

$$\text{i.e.} \quad k = y_1 - mx_1 \qquad 4$$

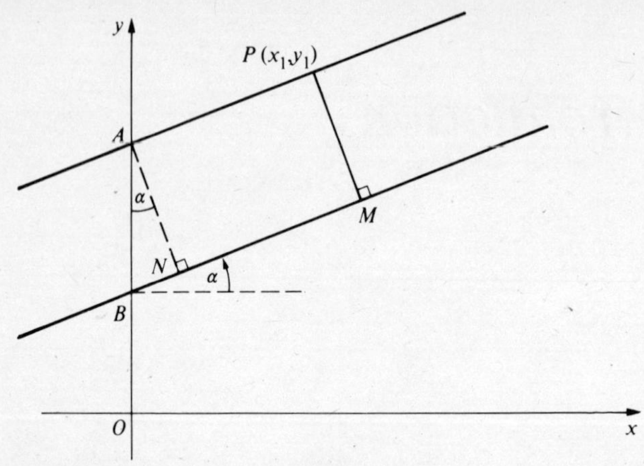

Figure 1

Then equation *2* may be written:

$$y = mx + y_1 - mx_1 \qquad 5$$

From the equation of BM, i.e. equation *1*, the intercept is c. Therefore the co-ordinates of B are $(0, c)$. Similarly, the co-ordinates of A, from equation *5*, are:

$$(0, y_1 - mx_1)$$

Then:

$$AB = OA - OB = y_1 - mx_1 - c$$

From Fig. 1:

$$PM = AN = AB \cos \alpha = (y_1 - mx_1 - c) \cos \alpha$$

$$= \frac{y_1 - mx_1 - c}{\sec \alpha}$$

$$= \frac{y_1 - mx_1 - c}{\sqrt{1 + m^2}} \qquad 6$$

where $\tan \alpha = m$. Therefore:

$$\sec^2 \alpha = 1 + \tan^2 \alpha = 1 + m^2$$

$$\text{i.e.} \quad \sec \alpha = \pm\sqrt{1 + m^2}$$

In equation *6* the issue of the alternative signs for $\sec \alpha$ is ignored because, in Fig. 2, the sign of the perpendicular is taken to be positive.

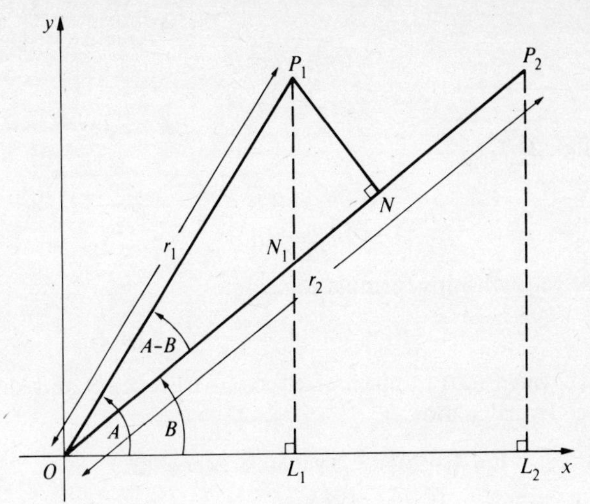

Figure 2

Trigonometric identities

In Fig. 2, suppose $P_1(x_1, y_1)$, $P_2(x_2, y_2)$ are any points in the plane such that $OP_1 = r_1$, $OP_2 = r_2$, and $P_1\hat{O}x = A$, $P_2\hat{O}x = B$, and angle A is greater than angle B. The equation of OP_2 is:

$$y = \frac{y_2}{x_2}.x, \text{ of gradient } y_2/x_2$$

$$\text{i.e.} \quad x_2y - y_2x = 0 \qquad 7$$

By formula *6* the value of P_1N_1, the perpendicular from P_1 to OP_2, is:

$$\frac{x_2y_1 - y_2x_1}{\sqrt{x_2^2 + y_2^2}} = \frac{x_2y_1 - y_2x_1}{r_2}$$

From triangle OP_1N:

$$P_1N = OP_1 \sin(P_1\hat{O}N) = r_1 \sin(A - B)$$

Therefore:

$$\sin(A - B) = \frac{x_2y_1 - y_2x_1}{r_1r_2}$$

$$= \left(\frac{y_1}{r_1}\right)\left(\frac{x_2}{r_2}\right) - \left(\frac{x_1}{r_1}\right)\left(\frac{y_2}{r_2}\right) \qquad 8$$

From triangle OP_1L_1:

$$x_1/r_1 = \cos A$$
$$y_1/r_1 = \sin A$$

From triangle OP_2L_2:

$$x_2/r_2 = \cos B$$
$$y_2/r_2 = \sin B$$

Using these relationships formula *8* becomes:

$$\sin(A-B) = \sin A \cos B - \cos A \sin B \qquad 9$$

In formula *9* put $(-B)$ in place of $B[\cos(-B) = \cos B$ and $\sin(-B) = -\sin B$, e.g., by calculator:

$$\cos(-20°) = 0.9396926 = \cos 20°$$
$$\sin(-20°) = -0.3420201 = -\sin 20°]$$

LHS becomes:

$$\sin A - (-B) = \sin(A+B)$$

RHS becomes:

$$\sin A . \cos(-B) - \cos A . \sin(-B)$$
$$= \sin A . \cos B - \cos A . (-\sin B)$$
$$= \sin A . \cos B + \cos A . \sin B$$

Therefore:

$$\sin(A+B) = \sin A . \cos B + \cos A . \sin B \qquad 10$$

In formula *9* put $\left(\frac{\pi}{2} - A\right)$ instead of A. LHS becomes:

$$\sin\left(\frac{\pi}{2} - A - B\right) = \sin\left[\frac{\pi}{2} - (A+B)\right] = \cos(A+B)$$

RHS becomes:

$$\sin\left(\frac{\pi}{2} - A\right)\cos B - \cos\left(\frac{\pi}{2} - A\right)\sin B = \cos A \cos B - \sin A . \sin B$$

Therefore:

$$\cos(A+B) = \cos A \cos B - \sin A \sin B \qquad 11$$

In formula *11* put $(-B)$ instead of B. LHS becomes:

$$\cos A + (-B) = \cos(A-B)$$

RHS becomes:

$$\cos A \cos(-B) - \sin A \sin(-B) = \cos A \cos B - \sin A(-\sin B)$$
$$= \cos A \cos B + \sin A \sin B$$

Therefore:

$$\cos(A - B) = \cos A \cos B + \sin A \sin B \qquad 12$$

From formula *11*, by putting $A = B = x$:

$$\cos 2x = \cos^2 x - \sin^2 x \qquad 13$$
$$= (1 - \sin^2 x) - \sin^2 x$$
$$= 1 - 2\sin^2 x$$

Rearranging:

$$\sin^2 x = \tfrac{1}{2}(1 - \cos 2x) \qquad 14$$

From formula *13*:

$$\cos 2x = \cos^2 x - (1 - \cos^2 x)$$
$$= 2\cos^2 x - 1$$

Rearranging:

$$\cos^2 x = \tfrac{1}{2}(1 + \cos 2x) \qquad 15$$

9 + *10* produces:

$$\sin(A + B) + \sin(A - B) = 2 \sin A \cos B \qquad 16$$

11 + *12* produces:

$$\cos(A + B) + \cos(A - B) = 2 \cos A \cos B \qquad 17$$

12 − *11* produces:

$$\cos(A - B) - \cos(A + B) = 2 \sin A \cdot \sin B \qquad 18$$

To Liza

Contents

Unit Reference

Unit AB10: Chapter 13
Unit AC12: Chapter 18
Unit BB8: Chapter 14
Unit BB9: Chapter 15
Unit BB10: Chapter 13
Unit BC7: Chapter 16
Unit BC8: Chapter 17
Unit BD12: Chapter 12
Unit CA11: Chapter 1
Unit CA12: Chapter 2
Unit CA13: Chapter 3
Unit CA14: Chapter 4
Unit CB12: Chapter 5
Unit CB14: Chapter 6
Unit CD13: Chapter 7
Unit CD14: Chapter 8

1

The Derivatives of Exponential, Logarithmic and Trigonometric Functions

1.1 The derivative of e^x

The function e^x may be defined in a number of different ways, all of them consistent with one another. At the moment the following definition suits our purposes.

Definition: e^x is that function of x whose derivative with reference to x (w.r.t. x) is itself. Fig. 1.1 is a sketch of the curve $y = e^x$.

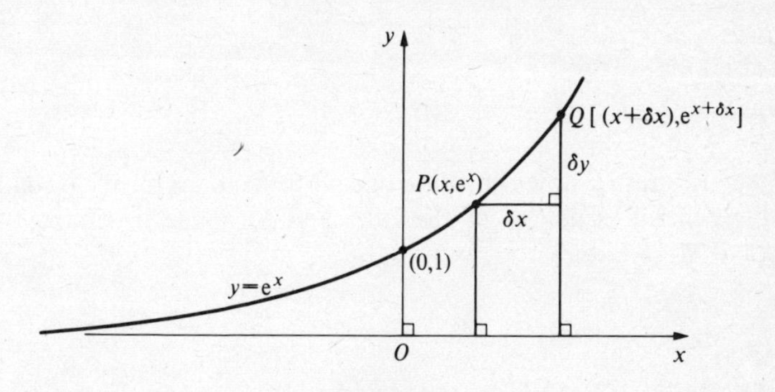

Figure 1.1

This means that where $y = e^x$, $\dfrac{dy}{dx} = e^x$. In other words:

$$\frac{d}{dx}(e^x) = e^x \qquad 1.1$$

It is a simple matter to check that this definition agrees with standard calculations based on e. From Fig. 1.1, the points P and Q on the curve are such that $P \equiv (x, e^x)$ and $Q \equiv [(x+\delta x), e^{x+\delta x}]$. Then:

$$\frac{\delta y}{\delta x} = \frac{e^{x+\delta x} - e^x}{\cancel{x} + \delta x - \cancel{x}} = \frac{e^x(e^{\delta x} - 1)}{\delta x}$$

$$= e^x \frac{(e^{\delta x} - 1)}{\delta x}$$

$\frac{dy}{dx}$ exists only if $\frac{\delta y}{\delta x} \to$ a limit as $\delta x \to 0$. Suppose δx is extremely small, e.g. 0.00001, then, by calculator:

$$e^{\delta x} = e^{0.00001} \approx 1.00001$$

$$\left(\frac{e^{\delta x} - 1}{\delta x}\right) \approx \frac{1.00001 - 1}{0.00001} = \frac{0.00001}{0.00001} = 1$$

$$e^x\left(\frac{e^{\delta x} - 1}{\delta x}\right) = e^x \times 1 = e^x$$

At this stage it will be assumed that formula *1.1* may be extended as follows. Where a is a constant and $y = e^{ax}$, then:

$$\frac{dy}{dx} = a \, . \, e^{ax}$$

Otherwise
$$\frac{d}{dx}(e^{ax}) = a \, . \, e^{ax} \qquad 1.2$$

Formulae *1.1* and *1.2* now constitute two more standard forms to combine with the general principles of the rules and formulae in Chapter 9 of *Analytical Mathematics 2.*

1.2 The derivative of ln x

From *1.1* the derivative of another special function can be obtained. The relationship $y = e^x$ makes y the subject. This relationship may be inverted to give $x = \ln y$, when x is the subject. By definition, in section 1.1:

$$\frac{dy}{dx} = e^x = y$$

Therefore, inverting:

$$\frac{dx}{dy} = \frac{1}{y}$$

In other words, when:

$$x = \ln y \tag{i}$$

then:

$$\frac{dx}{dy} = \frac{1}{y} \tag{ii}$$

Suppose now, in (i) and (ii), x and y are interchanged. Then (i) becomes:

$$y = \ln x \qquad \textit{1.3}$$

and (ii) becomes:

$$\frac{dy}{dx} = \frac{1}{x} \qquad \textit{1.4}$$

1.3 and *1.4* mean that:

$$\frac{d}{dx}(\ln x) = \frac{1}{x} \qquad \textit{1.5}$$

1.5 is another standard form which is to be added to the list of derivatives of special functions.

1.3 The derivatives of sin *ax* and cos *ax*

In Chapter 9 of *Analytical Mathematics 2* it was shown that:

$$\frac{d}{dx}(\sin x) = \cos x$$

and that:

$$\frac{d}{dx}(\cos x) = -\sin x$$

It will be assumed that these may be extended to the following rules:

$$\frac{d}{dx}(\sin ax) = a\cos ax \qquad \textit{1.6}$$

$$\frac{d}{dx}(\cos ax) = -a\sin ax \qquad \textit{1.7}$$

Since we shall frequently be using this list of standard forms, together with the linear operator rules of Chapter 8 of *Analytical Mathematics 2*, it is advisable to have this list readily available.

$$\frac{d}{dx}(f+g+h+\ldots) = \frac{df}{dx}+\frac{dg}{dx}+\frac{dh}{dx}+\ldots \qquad \text{(i)}$$

$$\frac{d}{dx}[a.f(x)] = a.\frac{d}{dx}[f(x)]\text{: } a \text{ is constant} \qquad \text{(ii)}$$

$$\frac{d}{dx}(x^n) = n.x^{n-1} \qquad \text{(iii)}$$

$$\frac{d}{dx}(e^{ax}) = a.e^{ax} \qquad \text{(iv)}$$

$$\frac{d}{dx}(\ln x) = 1/x \qquad \text{(v)}$$

$$\frac{d}{dx}(\sin ax) = a\cos ax\text{: } a \text{ is constant} \qquad \text{(vi)}$$

$$\frac{d}{dx}(\cos ax) = -a\sin ax\text{: } a \text{ is constant} \qquad \text{(vii)}$$

Rules (i) and (ii) are general principles, the rules which indicate that d/dx is a linear operator. Rules (iii), (iv), (v), (vi) and (vii) are the rules for special functions.

By combining the general principles with those of the standard forms a fairly wide variety of function may be differentiated. In fact we are now in a position to determine the derivatives of functions which are obtained from the listed special functions in the following manner:

1. The constant multiple of any of the five functions, e.g.

$$Ae^{ax},\ B\sin px,\ C\ln x,\ Dx^n,\ E\cos qx$$

2. The sum of two or more terms which are constant multiples of the special functions, e.g.

$$Ae^{ax}+B\ln x+C\sin bx+D\cos px+Ex^n$$

where A, B, C, D, E, a, b, p and n are constants.

Examples

Determine the derived functions of the following expressions w.r.t. the independent variable.

1. $y = Ae^{ax} + B\sin bx + C\ln px$, where A, B, C, a, b and p are constants.
   Step 1 Write $y = Ae^{ax} + B\sin bx + C\ln p + C\ln x$
   Step 2

$$\frac{dy}{dx} = \frac{d}{dx}(Ae^{ax} + B\sin bx + C\ln p + C\ln x)$$

$$= \frac{d}{dx}(Ae^{ax}) + \frac{d}{dx}(B\sin bx) + \frac{d}{dx}(C\ln p) + \frac{d}{dx}(C\ln x) \qquad \text{(by (i))}$$

$$= A\frac{d}{dx}(e^{ax}) + B\frac{d}{dx}(\sin bx) + C\frac{d}{dx}(\ln p) + C\frac{d}{dx}(\ln x) \qquad \text{(by (ii))}$$

$$= A\,.\,ae^{ax} + B\,.\,b\cos bx + 0 + C\,.\,\frac{1}{x} \qquad \text{(by (iv), (v) and (vi))}$$

2. $y = Ae^{ax} + Be^{bx} + Ce^{cx}$, where A, B, C, a, b and c are constants.

$$\frac{dy}{dx} = \frac{d}{dx}(Ae^{ax} + Be^{bx} + Ce^{cx})$$

$$= \frac{d}{dx}(Ae^{ax}) + \frac{d}{dx}(Be^{bx}) + \frac{d}{dx}(Ce^{cx}) \qquad \text{(by (i))}$$

$$= A\frac{d}{dx}(e^{ax}) + B\frac{d}{dx}(e^{bx}) + C\frac{d}{dx}(e^{cx}) \qquad \text{(by (ii))}$$

$$= A\,.\,ae^{ax} + B\,.\,be^{bx} + C\,.\,ce^{cx} \qquad \text{(by (iv))}$$

3. $V = 8\sin u - 7\sin 2u + 11\cos 3u - \frac{1}{4}\cos 4u$

$$\frac{dV}{du} = \frac{d}{du}(8\sin u - 7\sin 2u + 11\cos 3u - \tfrac{1}{4}\cos 4u)$$

$$= \frac{d}{du}(8\sin u) - \frac{d}{du}(7\sin 2u) + \frac{d}{du}(11\cos 3u) - \frac{d}{du}(\tfrac{1}{4}\cos 4u) \qquad \text{(by (i))}$$

$$= 8\frac{d}{du}(\sin u) - 7\frac{d}{du}(\sin 2u) + 11\frac{d}{du}(\cos 3u) - \frac{1}{4}\frac{d}{du}(\cos 4u) \qquad \text{(by (ii))}$$

$$= 8\cos u - 7(2\cos 2u) + 11(-3\sin 3u) - \tfrac{1}{4}(-4\sin 4u) \qquad \text{(by (vi) and (vii))}$$

$$= 8\cos u - 14\cos 2u - 33\sin 3u + \sin 4u$$

Exercise 1.1

Determine the derived functions of the following expressions w.r.t. the independent variable.

1. e^{2x}
2. e^{5x}
3. $e^{3x/5}$
4. $e^{-\frac{1}{4}x}$
5. $5.e^{2x}$
6. $6.e^{-\frac{1}{3}x}$
7. $-11e^{-4x}$
8. $\frac{2}{3}\cdot\frac{1}{e^{2x}}$
9. $-\frac{3}{4e^{-\frac{1}{4}x}}$
10. $\sin 2x$
11. $\sin(3x/5)$
12. $\sin\left(\frac{\pi x}{2}\right)$
13. $\cos 4x$
14. $\cos(-3x)$
15. $\cos(\frac{1}{4}x)$
16. $\cos\left(\frac{\pi x}{4}\right)$
17. $\frac{2}{3}\sin 3x$
18. $\frac{4}{3}\cos 4x$
19. $-\frac{16}{9}\sin(3x/2)$
20. $\frac{2}{3}\cos\left(\frac{\pi x}{6}\right)$
21. $\ln x$
22. $\ln(3x)$
23. $2\ln x$
24. $\ln x^2$ (Hint: use question 23)
25. $3\ln x$
26. $\ln x^3$
27. $\ln x^{\frac{1}{2}}$
28. $\ln\sqrt{x}$
29. $5\ln(16.x^{\frac{3}{5}})$

30. $\ln x + \ln x^2$
31. $2\ln x + 3\ln x^2$
32. $3e^x + 5e^{2x} + 6e^{3x}$
33. $\dfrac{e^x + e^{2x}}{e^x}$ (Hint: simplify before differentiating)
34. $4e^x(3e^x - 5e^{2x})$
35. $(2 + 3e^x)^2$
36. $2 \sin x - 3 \sin 2x + 5 \cos x + 8 \cos 2x$
37. $\dfrac{2}{5} \cos\left(\dfrac{3x}{5}\right) - \dfrac{4}{5} \sin\left(\dfrac{2x}{5}\right)$
38. $4e^{-x} + 7 \cos \frac{1}{4}x - \frac{3}{5}\sqrt{x}$
39. $\dfrac{4}{3} \sin\left(\dfrac{3x}{2}\right) + \dfrac{1}{4}.e^{-\frac{1}{3}x} - \dfrac{6}{5x^2}$
40. $5\ln x^2 - \frac{3}{4}\ln(16x) + \frac{5}{6}e^{2x/3} - 5x^3\sqrt{x}$

1.4 The values of the derivatives for the given values of the independent variables

The following examples illustrate the application of the methods above in the calculation of the values of the derivatives at special points of curves.

Examples

Evaluate the derivatives of the following functions at the points indicated.

1. $y = 5e^{-x} - \frac{2}{3}\ln x$, at $x = 2$

$$dy/dx = 5(-e^{-x}) - \frac{2}{3}\left(\frac{1}{x}\right)$$

$$= -5e^{-x} - \frac{2}{3x}$$

When $x = 2$, $dy/dx = -5e^{-2} - \frac{2}{6} = -5e^{-2} - \frac{1}{3}$. For many purposes the answer in this form is sufficient, but where an approximate answer in decimal form is required:

$$dy/dx = -5 \times 0.1353353 - \tfrac{1}{3} = -0.676676 - \tfrac{1}{3}$$

$$= -1.0100097 \approx -1.01$$

2. $y = 8\cos\left(\frac{3x}{5}\right) - \frac{3}{4}\sin\left(\frac{\pi x}{4}\right) + \frac{2}{3}x^{-\frac{3}{4}}$, at $x = 3$

$$dy/dx = 8\left(-\frac{3}{5}\sin\frac{3x}{5}\right) - \frac{3}{4}\left(\frac{\pi}{4}\cos\frac{\pi x}{4}\right) + \frac{2}{3}\left(-\frac{3}{4}x^{-\frac{7}{4}}\right)$$

$$= -\frac{24}{5}\sin\frac{3x}{5} - \frac{3\pi}{16}\cos\frac{\pi x}{4} - \frac{1}{2}x^{-\frac{7}{4}}$$

When $x = 3$:

$$dy/dx = -\frac{24}{5}\sin\frac{9}{5} - \frac{3\pi}{16}\cos\frac{3\pi}{4} - \frac{1}{2}(3^{-\frac{7}{4}})$$ (Note that the angle must be in radians)

$$= -\tfrac{24}{5}(0.9738476) + 0.4165203 - 0.0731152$$
$$= -4.6744686 + 0.4165203 - 0.0731152$$
$$= -4.3310636 \approx -4.33$$

Exercise 1.2

Evaluate the derivatives of the following functions w.r.t. the independent variables at the points indicated.

1. $y = e^x + \ln x$: $x = -1$
2. $y = 3e^{-x} + 5e^{-2x}$: $x = 2$
3. $y = \ln x$: $x = 4$
4. $y = 6\ln 2x$: $x = 1$
5. $y = \sin 2x$: $x = \pi/8$
6. $y = -2\cos 2x$: $x = \pi/4$
7. $y = 4\sin x - 3\sin 2x$: $x = \pi/2$
8. $y = 3\cos x - 5\cos 2x$: $x = \pi/2$
9. $y = e^{-x} + 3\cos x$: $x = \pi/6$
10. $y = \ln\left(\frac{3x}{2}\right) - \frac{4}{5}\sin x$: $x = \pi/4$
11. $y = 2 + x - 3x^2 + 4e^{-x}$: $x = 1$
12. $y = 3e^x(2 + e^{-x})$: $x = 1/2$
13. $y = -\frac{2}{5}\ln x^{\frac{1}{2}} + 2.5e^{-0.34x}$: $x = 5.2$
14. $y = 3.65\cos 2.6x - 4.52\sin(-3.7x)$: $x = -1.9$
15. $y = \frac{4.06}{2.75}e^{-1.54x} - 5.43\ln\left(\frac{4}{3x}\right)$: $x = 5.72$

2

The Derivatives of Sums, Products, Quotients and Functions of Functions

The laws listed in Chapter 1 enable us to differentiate only a limited number of functions. To extend the range of function that can be differentiated, more general rules must be obtained.

2.1 The derivatives of sums

The differentiation of sums is, in fact, covered by rules (i) and (ii) of the previous chapter.

Examples

1. $\frac{d}{dx}(5x^3 - 9x^2 + 11x - 4 + 6/x)$

$$= \frac{d}{dx}(5x^3) - \frac{d}{dx}(9x^2) + \frac{d}{dx}(11x) - \frac{d}{dx}(4) + \frac{d}{dx}(6/x) \qquad \text{(by (i))}$$

$$= 5\frac{d}{dx}(x^3) - 9\frac{d}{dx}(x^2) + 11\frac{d}{dx}(x) - 4\frac{d}{dx}(1) + 6\frac{d}{dx}(x^{-1}) \quad \text{(by (ii))}$$

$$= 5.3x^2 - 9.2x + 11.1 - 4.0 + 6.(-1)x^{-2} \qquad \text{(by (iii))}$$

$$= 15x^2 - 18x + 11 - 6/x^2$$

Note: $\frac{d}{dx}(1) = \frac{d}{dx}(x^0) = 0$, by (iii); or $\frac{d}{dx}(1) = \frac{d}{dx}(\text{constant}) = 0$.

2. $\frac{d}{dx}(6e^{3x} - 4e^x + 7e^{-x})$

$$= \frac{d}{dx}(6e^{3x}) + \frac{d}{dx}(-4e^x) + \frac{d}{dx}(7e^{-x}) \qquad \text{(by (i))}$$

$$= 6\frac{d}{dx}(e^{3x}) - 4\frac{d}{dx}(e^x) + 7\frac{d}{dx}(e^{-x}) \qquad \text{(by (ii))}$$

$$= 6 \times 3e^{3x} - 4 \times e^x + 7 \times (-e^{-x}) \qquad \text{(by (iv))}$$

$$= 18e^{3x} - 4e^x - 7e^{-x}$$

3. $\frac{d}{dx}(7\sin 2x + 8\sin x - 3\cos 2x + 5\cos x)$

$$= \frac{d}{dx}(7\sin 2x) + \frac{d}{dx}(8\sin x) - \frac{d}{dx}(3\cos 2x) + \frac{d}{dx}(5\cos x) \qquad \text{(by (i))}$$

$$= 7 \times 2\cos 2x + 8\cos x - 3 \times -2\sin 2x + 5 \times -\sin x$$

(by (ii), (vi) and (vii))

$$= 14\cos 2x + 8\cos x + 6\sin 2x - 5\sin x$$

4. $\frac{d}{dx}(6x^{-3} + 4e^{2x} - 3e^{-3x} + 2\sin 3x - 6\cos 2x + 3\ln x)$

$$= \frac{d}{dx}(6x^{-3}) + \frac{d}{dx}(4e^{2x}) - 3\frac{d}{dx}(e^{-3x})$$
$$+ 2\frac{d}{dx}(\sin 3x) - 6\frac{d}{dx}(\cos 2x) + 3\frac{d}{dx}(\ln x)$$

$$= 6 \times -3x^{-4} + 4 \times 2e^{2x} - 3 \times -3e^{-3x}$$
$$+ 2 \times 3\cos 3x - 6 \times -2\sin 2x + 3 \times \frac{1}{x}$$

$$= -18/x^4 + 8e^{2x} + 9e^{-3x} + 6\cos 3x + 12\sin 2x + 3/x$$

Exercise 2.1

Differentiate the following functions w.r.t. x:

1. $4x^2 + 3x + 2$
2. $5x^3 - 6x - 11/x$

3. $7x^2 + 2 - 3/x^2$
4. $2e^x + 3 + 4e^{-x}$
5. $7e^{2x} - 8e^x + 4$
6. $10 \sin x + 4 \cos x - 2x$
7. $3 \sin 2x - 4 \sin 3x + \cos 2x + 3x^2$
8. $\frac{1}{2}e^{2x} + \frac{1}{4}e^x - 3/x^2$
9. $2\ln x + 3e^x - 2x^3$
10. $5 \sin 4x - 6 \cos 2x + \frac{3}{4}e^{-x} - \frac{3x^2}{4}$

2.2 The derivative of a function of a function

As a preliminary to the statement of the law for the derivative of a function of a function we will make sure we understand the very basic ideas of differentiation on which the law depends. Fig. 2.1 illustrates the steps taken to find dy/dx when $y = f(x)$. P and Q are points close together on the curve. δx is the small displacement along Ox and δy is the corresponding displacement, small, along Oy. $\delta y/\delta x$ is the gradient of the line PQ, and dy/dx is the value of $\delta y/\delta x$ as $\delta x \to 0$. Then dy/dx is the gradient of the curve at P.

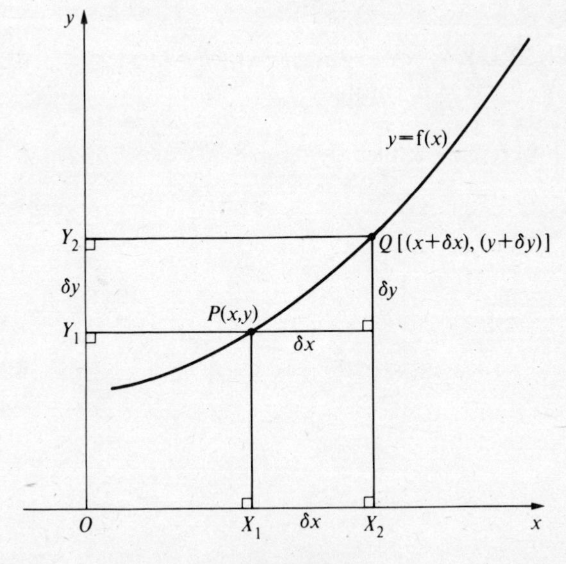

Figure 2.1

Fig. 2.2, which shows the mapping of x onto $f(x)$, illustrates that $\delta y/\delta x$ is the average magnification of that mapping for the interval X_1X_2. Then dy/dx is also the magnification of the mapping at X_1. It is this magnification aspect of the derivative of a function of a function which will be used to interpret the derivative of a function of a function.

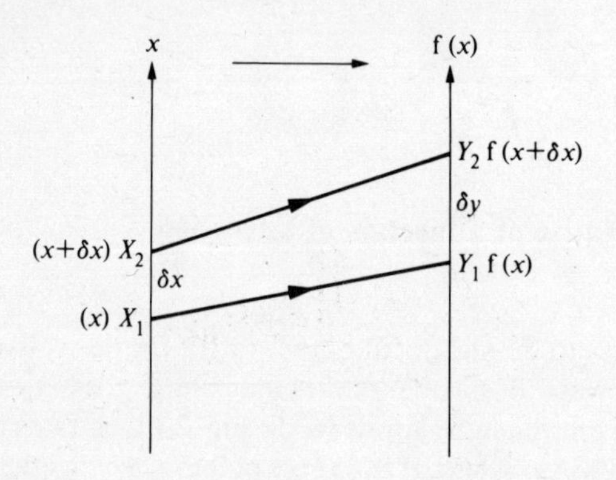

Figure 2.2

Examples

1. $y = ax + b$
   From Fig. 2.3, $Y_1Y_2 = \delta y = a(x+\delta x) + b - (ax+b) = a.\delta x$
   This gives $\delta y/\delta x = a$, constant.
   Then $dy/dx = a$
   If $ax+b$ were $3x+2$, then $dy/dx = 3$

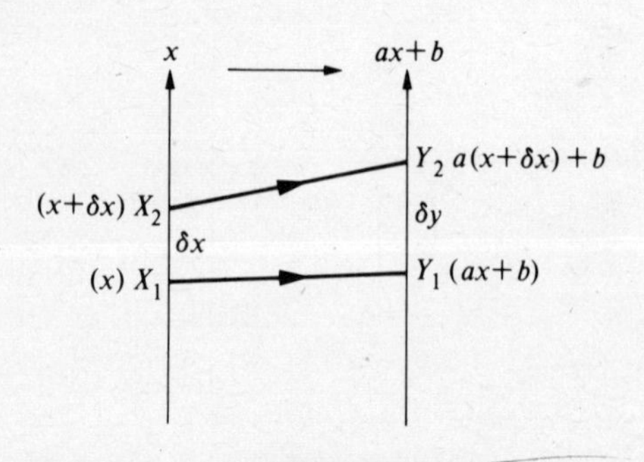

Figure 2.3

2. $y = x^2$
From Fig. 2.4,
$Y_1 Y_2 = \delta y = (x + \delta x)^2 - x^2 = x^2 + 2x \,.\, \delta x + \delta x^2 - x^2 = 2x \,.\, \delta x + \delta x^2$
$\delta y / \delta x = 2x + \delta x$
Then $\mathrm{d}y/\mathrm{d}x = 2x$

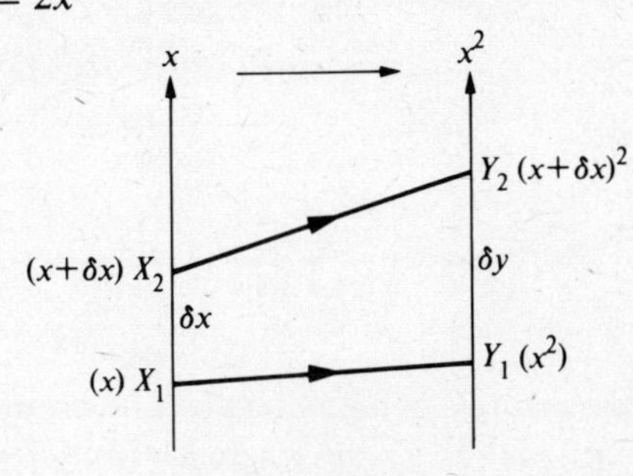

Figure 2.4

3. $y = b\mathrm{e}^{ax}$
From Fig. 2.5, $Y_1 Y_2 = \delta y = b\mathrm{e}^{a(x+\delta x)} - b\mathrm{e}^{ax} = b\mathrm{e}^{ax}(\mathrm{e}^{\delta x} - 1)$

$$\delta y / \delta x = \frac{b\mathrm{e}^{ax}(\mathrm{e}^{\delta x} - 1)}{\delta x} \approx \frac{b\mathrm{e}^{ax} \,.\, \delta x}{\delta x}, \text{ from Chapter 1.}$$

Then $\mathrm{d}y/\mathrm{d}x = b\mathrm{e}^{ax}$

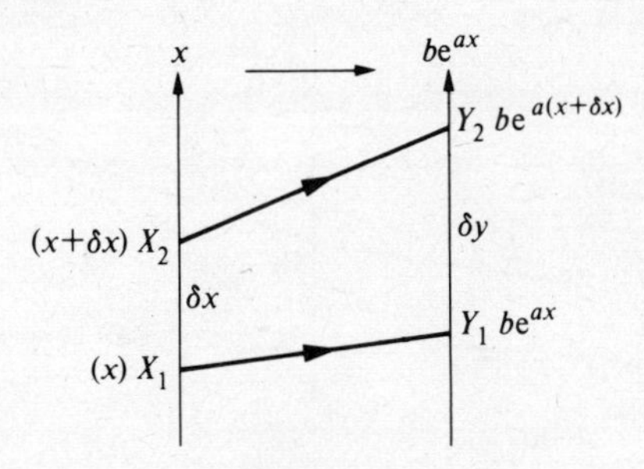

Figure 2.5

4. $y = (3x + 2)^2 = 9x^2 + 12x + 4$
From Fig. 2.6:

$$\begin{aligned} Y_1 Y_2 = \delta y &= 9(x + \delta x)^2 + 12(x + \delta x) + 4 - 9x^2 - 12x - 4 \\ &= 9x^2 + 18x \,.\, \delta x + 9\delta x^2 + 12x + 12\delta x + 4 - 9x^2 - 12x - 4 \\ &= 18x \,.\, \delta x + 12x + 9\delta x^2 \end{aligned}$$

$$\delta y / \delta x = 18x + 12 + 9 \,.\, \delta x$$

Then $\mathrm{d}y/\mathrm{d}x = 18x + 12 = 6(3x + 2)$

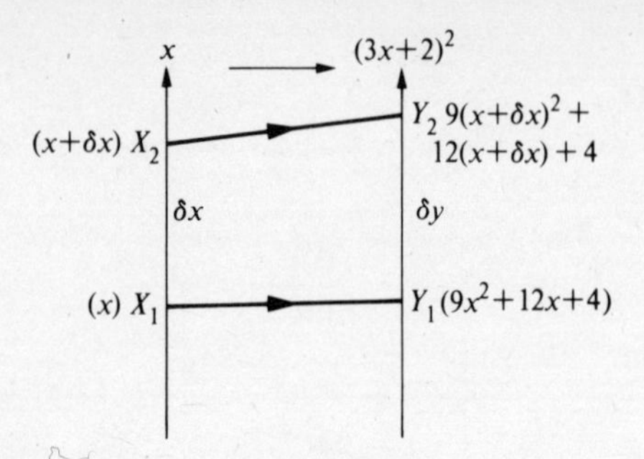

Figure 2.6

Example 4 is a simple example of a function of a function. For instance, we may interpret $y = (3x+2)^2$ as $y = u^2$ where $u = 3x+2$; in other words, a two-stage mapping.

Stage 1 $x \rightarrow 3x+2$ (or u)
Stage 2 $u \rightarrow u^2$ (or y)

Fig. 2.7 represents this diagrammatically. From Fig. 2.7:

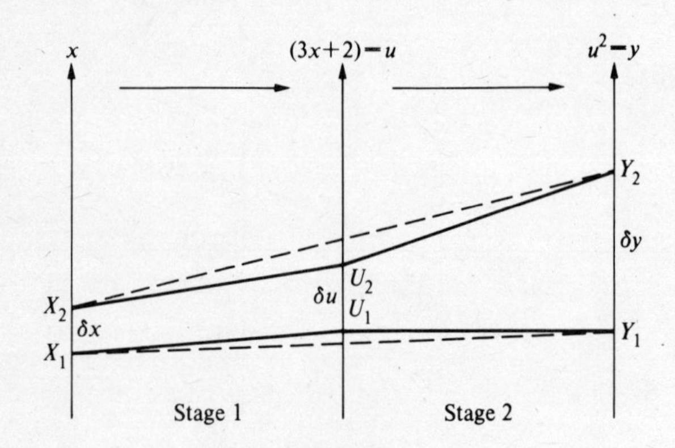

Figure 2.7

$$\text{Stage 1 magnification} = \frac{U_1U_2}{X_1X_2}$$

$$\text{Stage 2 magnification} = \frac{Y_1Y_2}{U_1U_2}$$

The overall magnification is $\dfrac{Y_1Y_2}{X_1X_2}$

And this is equal to $\frac{Y_1Y_2}{U_1U_2} \times \frac{U_1U_2}{X_1X_2}$ or $\frac{\delta y}{\delta x} = \frac{\delta y}{\delta u} \cdot \frac{\delta u}{\delta x}$

As $\delta x \to 0$, this produces:

$$\frac{dy}{dx} = \frac{dy}{du} \cdot \frac{du}{dx} \qquad 2.1$$

Example 4 may be regarded as a combined mapping of examples 1 and 2.

Example 1 gave $y = 3x + 2$, $dy/dx = 3$. In it replace y by u to produce $u = 3x + 2$, $du/dx = 3$.

Example 2 gave $y = x^2$, $dy/dx = 2x$. In it replace x by u to produce $y = u^2$, $dy/du = 2u$.

Combining the two results gives:

$$\frac{du}{dx} \times \frac{dy}{du} = 3 \times 2u = 6u = 6(3x + 2)$$

This result agrees with the answer above to example 4. In fact rule *2.1* applies generally to a function of a function whether it is algebraic or not.

Examples

Apply formula *2.1* to the following functions:

1. $y = \sin ax$
   Write $y = \sin u$ and $u = ax$
   Then $dy/du = \cos u$ and $du/dx = a$

   By law *2.1* $\frac{dy}{dx} = \frac{dy}{du} \times \frac{du}{dx} = \cos u \, . \, a = a \cos ax$

   This is a formula previously assumed to be true.
2. $y = e^{ax}$
   Write $y = e^u$ and $u = ax$
   Then $dy/du = e^u$ and $du/dx = a$

   By law *2.1* $\frac{dy}{dx} = \frac{dy}{du} \times \frac{du}{dx} = e^u \times a = a \, . \, e^{ax}$

   That is another formula which we have assumed before.

3. $y = e^{ax^2+bx+c}$
   Write $y = e^u$ and $u = ax^2+bx+c$
   Then $dy/du = e^u = e^{ax^2+bx+c}$ and $du/dx = 2ax+b$
   By law *2.1* $\dfrac{dy}{dx} = \dfrac{dy}{du} \times \dfrac{du}{dx} = e^u \times (2ax+b) = (2ax+b)e^{ax^2+bx+c}$
4. $y = A.e^{p \sin ax}$, where A, p and a are constants.
   Write $y = Ae^u$ and $u = p \sin ax$
   Then $dy/du = Ae^u = Ae^{p \sin ax}$ and $du/dx = p \times a \cos ax = ap \cos ax$

   $\dfrac{dy}{dx} = Ae^{p \sin ax}.ap \cos ax = Aap.e^{p \sin ax}.\cos ax$
5. $y = 3\cos(x^2-3x)$
   Write $y = 3\cos u$ and $u = x^2-3x$
   $dy/du = -3\sin u$ and $du/dx = 2x-3$
   $\dfrac{dy}{dx} = -3\sin u \times (2x-3) = -3(2x-3)\sin(x^2-3x)$
6. $y = a\sin^2 px$
   Write $y = au^2$ and $u = \sin px$
   $dy/du = a \times 2u$ and $du/dx = p\cos px$

   $\dfrac{dy}{dx} = a \times 2u \times p\cos px = 2ap \sin px \times \cos px$

Exercise 2.2

Apply rule *2.1* to differentiate the following functions:

1. $y = (2x+3)^2$
2. $y = (2x+3)^3$
3. $(5x+1)^2$
4. $y = (5x+1)^3$
5. $(2x-1)^{-1}$
6. $y = (3x+2)^{-2}$
7. $y = (ax+b)^9$
8. $y = (x^2-3x+5)^2$
9. $y = (x^2-3x+6)^3$
10. $y = 3e^{4x+1}$
11. $y = 8e^{\frac{1}{4}x^2}$

12. $y = -e^{-\sin 2x}$
13. $x = Ae^{5\cos 3t}$: differentiate w.r.t. t
14. $z = 7\ln(x^2 + 5x - 1)$
15. $x = a\ln(\cos pt)$
16. $y = (3e^x + 4)^2$

The general interpretation of rule *2.1* is, where:

$$y = g\{f(x)\}$$

i.e. $y = g(u)$ and $u = f(x)$, then:

$$\frac{dy}{dx} = \frac{dy}{du} \times \frac{du}{dx}$$

This rule may be extended to more complicated chains of functions. Where y is a function of a function of a function:

$$\frac{dy}{dx} = \frac{dy}{du} \times \frac{du}{dv} \times \frac{dv}{dx} \qquad 2.2$$

2.3 The derivative of a product

An example of a function which is a product is:

$$y = e^{ax} \times \cos bx$$

or $y = u \times v$, where $u = e^{ax}$ and $v = \cos bx$. Before we look at the problem in general we will analyse some simple problems applied to the above idea of a product.

Examples

1. Differentiate w.r.t. x $y = x^6$ (a) first in the normal way and (b) then as a product.
   (a) $dy/dx = 6x^5$ (by (iii))
   (b) Write $y = x^6 = x^4 \times x^2 = u \times v$, where $u = x^4$ and $v = x^2$
   Then $du/dx = 4x^3$ and $dv/dx = 2x$
   Now calculate $u \cdot \frac{dv}{dx} + v \cdot \frac{du}{dx} = x^4(2x) + x^2(4x^3) = 2x^5 + 4x^5 = 6x^5$

2. Differentiate w.r.t. x $y = (3x+2)(5x^2-7)$
   (a) Write $y = 15x^3 + 10x^2 - 21x - 14$
   Then $dy/dx = 45x^2 + 20x - 21$
   (b) Put $u = 3x+2$ and $v = 5x^2 - 7$
   Then $du/dx = 3$ and $dv/dx = 10x$

$$u \cdot \frac{dv}{dx} + v \cdot \frac{du}{dx} = (3x+2)10x + (5x^2-7)3$$

$$= 30x^2 + 20x + 15x^2 - 21$$

$$= 45x^2 + 20x - 21$$

In general, where $y = u \times v$, and u and v are both functions of x, then:

$$\frac{dy}{dx} = u \cdot \frac{dv}{dx} + v \cdot \frac{du}{dx}$$

To prove this, Figs 2.8, 2.9 and 2.10 are required. Fig. 2.8 represents a sketch of the graph of $u = u(x)$. The arc PP^1 involves increments δx, δu in x and u respectively. Then:

$$\delta u/\delta x = \frac{(u+\delta u) - u}{(x+\delta x) - x}$$

As $\delta x \to 0$, $\delta u \to 0$ and $\delta u/\delta x \to du/dv$ because we assume that u is differentiable.

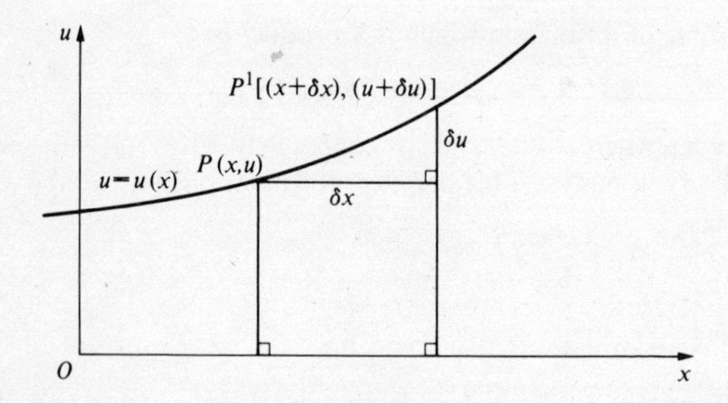

Figure 2.8

From Fig. 2.9:

$$\delta v/\delta x = \frac{(v+\delta v) - v}{(x+\delta x) - x}$$

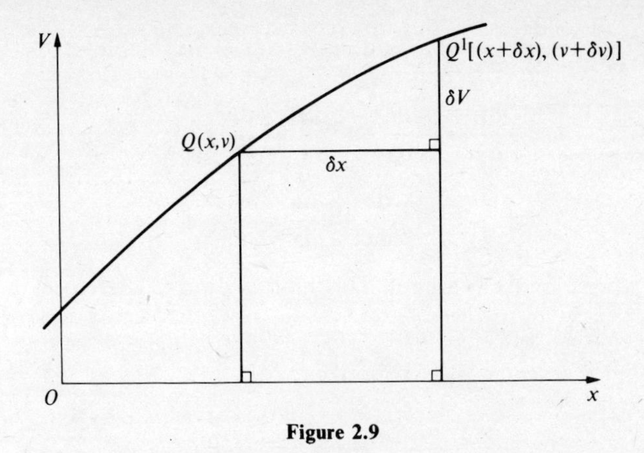

Figure 2.9

As $\delta x \to 0$, $\delta v \to 0$ and $\delta v/\delta x \to \mathrm{d}v/\mathrm{d}x$ because v is differentiable.
From Fig. 2.10:

$$\frac{\delta y}{\delta x} = \frac{(u+\delta u)(v+\delta v)-uv}{(x+\delta x)-x}$$

$$= \frac{uv + u\delta v + v\delta u + \delta u \,.\, \delta v - uv}{\delta x}$$

$$= u\frac{\delta v}{\delta x} + v\frac{\delta u}{\delta x} + \frac{\delta u \,.\, \delta v}{\delta x}$$

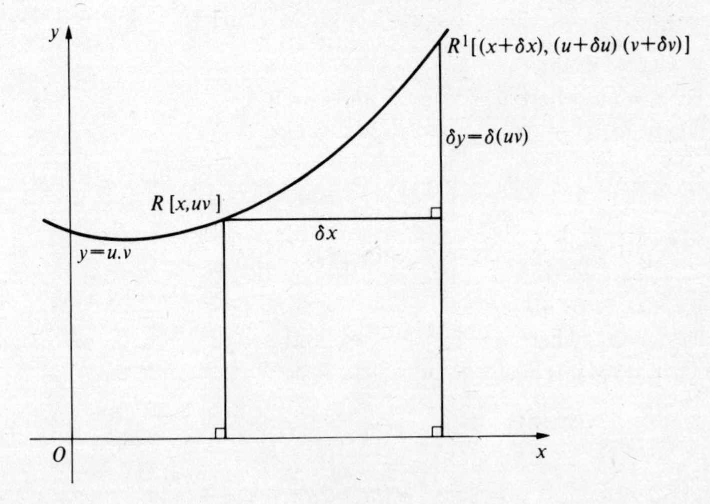

Figure 2.10

As $\delta x \to 0$, $\dfrac{\delta v}{\delta x} \to \dfrac{dv}{dx}$, $\dfrac{\delta u}{\delta x} \to \dfrac{du}{dx}$ and $\delta u \to 0$, $\delta v \to 0$. Therefore:

$$\frac{\delta y}{\delta x} \to u \cdot \frac{dv}{dx} + v \cdot \frac{du}{dx} + \left(\frac{du}{dx}\right) \times 0$$

$$= u\frac{dv}{dx} + v\frac{du}{dx}$$

In other words, $\delta y/\delta x \to$ a limit. This limit is dy/dx. Therefore:

$$\frac{dy}{dx} = u\frac{dv}{dx} + v\frac{du}{dx} \qquad 2.3$$

Examples

Using *2.3* find the derivatives of the following functions.

1. $y = e^{ax} \sin bx$
   i.e. $y = uv$, where $u = e^{ax}$ and $v = \sin bx$
   Then $du/dx = ae^{ax}$ and $dv/dx = b \cos bx$
   By *2.3*:

$$\frac{dy}{dx} = u\frac{dv}{dx} + v\frac{du}{dx}$$

$$= e^{ax} . b \cos bx + \sin bx . ae^{ax}$$

$$= e^{ax}(b \cos bx + a \sin bx)$$

2. $y = (x^3 - 5x)\ln x$
   i.e. $y = uv$, where $u = x^3 - 5x$ and $v = \ln x$
   Then $du/dx = 3x^2 - 5$ and $dv/dx = 1/x$

$$\frac{dy}{dx} = (x^3 - 5x)\frac{1}{x} + \ln x(3x^2 - 5)$$

$$= (x^2 - 5) + (3x^2 - 5)\ln x$$

3. $y = (ax^2 + bx + c)e^{px}$
   i.e. $y = uv$, where $u = ax^2 + bx + c$ and $v = e^{px}$
   Then $du/dx = 2ax + b$ and $dv/dx = pe^{px}$

$$\frac{dy}{dx} = (ax^2 + bx + c) . pe^{px} + e^{px}(2ax + b)$$

$$= [apx^2 + (bp + 2a)x + (pc + b)]e^{px}$$

4. If $y = u \times v \times w$, where u, v and w are functions of x which are

differentiable, then *2.3* may be extended to:

$$\frac{dy}{dx} = vw\frac{du}{dx} + wu\frac{dv}{dx} + uv\frac{dw}{dx} \qquad 2.4$$

$y = (x^2 + 5x + 2) . e^{-4x} . \sin\frac{3}{4}x$,
i.e. $y = uvw$, where $u = x^2 + 5x + 2$ and $v = e^{-4x}$ and $w = \sin\frac{3}{4}x$
Then $du/dx = 2x + 5$, $dv/dx = -4e^{-4x}$ and $dw/dx = \frac{3}{4}\cos\frac{3}{4}x$
By *2.4*:

$$\begin{aligned}\frac{dy}{dx} &= e^{-4x} . \sin\tfrac{3}{4}x . (2x + 5) + (x^2 + 5x + 2) . \sin\tfrac{3}{4}x . (-4e^{-4x}) \\ &\quad + (x^2 + 5x + 2)e^{-4x}(\tfrac{3}{4}\cos\tfrac{3}{4}x) \\ &= e^{-4x}[(2x + 5)\sin\tfrac{3}{4}x - 4(x^2 + 5x + 2)\sin\tfrac{3}{4}x \\ &\quad + \tfrac{3}{4}(x^2 + 5x + 2)\cos\tfrac{3}{4}x] \\ &= e^{-4x}[\tfrac{3}{4}(x^2 + 5x + 2)\cos\tfrac{3}{4}x - (4x^2 + 18x + 3)\sin\tfrac{3}{4}x]\end{aligned}$$

Exercise 2.3

Differentiate the following:

1. $e^{ax}\cos bx$
2. $(x^2 + 3)\ln x$
3. $(x^2 - 2x + 3)e^{2x}$
4. $\cos 2x . \ln x$
5. $4\sin 3x . \cos 2x$
6. $\left(x + \frac{1}{x}\right) . \cos 4x$
7. $e^{-3x} . \sin 2x$
8. $(2x + 5) . e^x . \cos x$
9. $(3x + 1) . \ln x . \sin x$
10. $e^{-\frac{1}{4}x} . \sin\frac{2}{3}x . \ln x$

2.4 The derivative of a quotient

A quotient is of the form $y = \dfrac{u}{v}$, where u and v are each functions of x. We take each to be differentiable. The formula for the derivative is:

$$\frac{dy}{dx} = \frac{v(du/dx) - u(dv/dx)}{v^2} \qquad 2.5$$

Example

Demonstrate that formula *2.5* gives the correct answer for the following functions:

$$y = e^{x^2 - 3x}$$

Rewrite: $y = \dfrac{e^{x^2}}{e^{3x}} = u/v$, where $u = e^{x^2}$ and $v = e^{3x}$

$du/dx = 2x \,.\, e^{x^2}$ (function of function); $dv/dx = 3e^{3x}$, by (iv)
Formula *2.5* gives:

$$\frac{dy}{dx} = \frac{e^{3x}(2xe^{x^2}) - e^{x^2}(3e^{3x})}{(e^{3x})^2}$$

$$= \frac{2xe^{x^2+3x} - 3e^{x^2+3x}}{e^{6x}} = \frac{(2x-3)e^{x^2+3x}}{e^{6x}}$$

$$= (2x-3)e^{x^2+3x-6x}$$

$$= (2x-3)e^{x^2-3x}$$

However, $y = e^{x^2-3x}$ is itself a function of a function, where $y = e^w$ and $w = x^2 - 3x$
$dy/dw = e^w$ and $dw/dx = 2x - 3$
Then:

$$\frac{dy}{dx} = \frac{dy}{dw} \times \frac{dw}{dx} = e^w(2x-3)$$

$$= (2x-3)e^{x^2-3x}$$

The above example bears out formula *2.5*. To prove that formula we shall refer to Figs 2.8, 2.9 and 2.11.
From Fig. 2.11:

$$\delta y = \frac{u+\delta u}{v+\delta v} - \frac{u}{v}$$

$$= \frac{(u+\delta u)v - u(v+\delta v)}{v(v+\delta v)}$$

$$= \frac{\cancel{uv} + v\,.\,\delta u - \cancel{uv} - u\,.\,\delta v}{v(v+\delta v)}$$

$$\frac{\delta y}{\delta x} = \frac{v\cdot\dfrac{\delta u}{\delta x} - u\cdot\dfrac{\delta v}{\delta x}}{v(v+\delta v)}$$

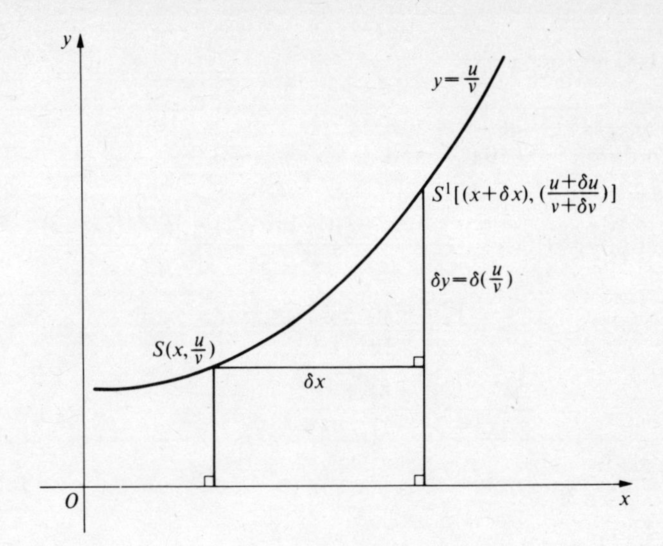

Figure 2.11

From Fig. 2.8, $\delta u/\delta x \to \mathrm{d}u/\mathrm{d}x$ as $\delta x \to 0$. From Fig. 2.9, $\delta v/\delta x \to \mathrm{d}v/\mathrm{d}x$ as $\delta x \to 0$, and $\delta v \to 0$ as $\delta x \to 0$. Therefore:

$$\frac{\mathrm{d}y}{\mathrm{d}x} = \frac{v(\mathrm{d}u/\mathrm{d}x) - u(\mathrm{d}v/\mathrm{d}x)}{v^2}$$

Examples

Differentiate the following:

1. $y = \dfrac{2x+5}{3x-4}$

   i.e. $y = \dfrac{u}{v}$, where $u = 2x+5$ and $v = 3x-4$
   Then $\mathrm{d}u/\mathrm{d}x = 2$ and $\mathrm{d}v/\mathrm{d}x = 3$
   By formula *2.5*:

$$\frac{\mathrm{d}y}{\mathrm{d}x} = \frac{(3x-4).2-(2x+5).3}{(3x-4)^2}$$

$$= \frac{6x-8-6x-15}{(3x-4)^2} = -\frac{23}{(3x-4)^2}$$

2. $y = f(x) = \dfrac{5.e^{-2x}}{(x^2 + 3x - 1)}$

Put $u = 5.e^{-2x}$ and $v = x^2 + 3x - 1$
Then $du/dx = -10e^{-2x}$ and $dv/dx = 2x + 3$
By *2.5*:

$$\frac{dy}{dx} = f'(x) = \frac{(x^2 + 3x - 1)(-10e^{-2x}) - 5e^{-2x}(2x+3)}{(x^2 + 3x - 1)^2}$$

$$= \frac{-5e^{-2x}(2x^2 + 6x - 2 + 2x + 3)}{(x^2 + 3x - 1)^2}$$

$$= \frac{-5e^{-2x}(2x^2 + 8x + 1)}{(x^2 + 3x - 1)^2}$$

3. $f(t) = \dfrac{\ln t}{2 \sin 3t}$

Put $u = \ln t$ and $v = 2 \sin 3t$
$du/dt = 1/t$ and $dv/dt = 6 \cos 3t$

$$f'(t) = \frac{2 \sin 3t(1/t) - \ln t(6 \cos 3t)}{(2 \sin 3t)^2}$$

$$= \frac{(2/t) \sin 3t - 6 \cos 3t \, . \ln t}{(2 \sin 3t)^2}$$

$$= \frac{\sin 3t - 3t \cos 3t \, . \ln t}{2t \, . \sin^2 3t}$$

Exercise 2.4

Differentiate w.r.t. the independent variables the following functions:

1. $\dfrac{3x+1}{4x-7}$

2. $\dfrac{\sin x}{2x+1}$

3. $\dfrac{\cos 2x}{x^2 - 4}$

4. $\dfrac{a \cos bx}{px + q}$

5. $\dfrac{e^x}{\sin 3x}$

6. $\dfrac{\ln x}{\frac{1}{4}x^2+4}$

7. $\dfrac{x+1}{x+2}$

8. $\dfrac{x+1/x}{x-1/x}$

9. $\dfrac{3x+5}{x^2+4x+2}$

10. $\dfrac{e^x-1}{e^x+1}$

11. $\dfrac{e^{\frac{x}{2}}-e^{-\frac{x}{2}}}{e^{\frac{x}{2}}+e^{-\frac{x}{2}}}$

12. $\dfrac{e^{-2t}}{t+1}$

2.5 Miscellaneous differentiation

By combining the rules in the previous sections a much wider variety of functions may be differentiated, as the following examples illustrate.

Examples

Differentiate the following functions w.r.t. the independent variables.

1. $f(x) = \tan ax$

   Write $f(x) = \dfrac{\sin ax}{\cos ax}$

   Put $u = \sin ax$ and $v = \cos ax$
   So $du/dx = a \cos ax$ and $dv/dx = -a \sin ax$

Using formula *2.5*:

$$f'(x) = \frac{\cos ax(a \cos ax) - \sin ax(-a \sin ax)}{\cos^2 ax}$$

$$= \frac{a \cos^2 ax + a \sin^2 ax}{\cos^2 ax}$$

$$= \frac{a}{\cos^2 ax} = a \sec^2 ax \qquad 2.6$$

2. $y = \operatorname{cosec} ax$
   Write $y = 1/\sin ax = (\sin ax)^{-1}$
   Then $y = u^{-1}$ and $u = \sin ax$
   So $dy/du = -u^{-2}$ and $du/dx = a \cos ax$
   By formula *2.1*:

$$dy/dx = -\frac{1}{u^2} \times a \cos ax$$

$$= -\frac{a \cos ax}{\sin^2 ax} = -a\left(\frac{\cos ax}{\sin ax}\right)\left(\frac{1}{\sin ax}\right)$$

$$\frac{d}{dx}(\operatorname{cosec} ax) = -a \cot ax \,.\, \operatorname{cosec} ax \qquad 2.7$$

3. $x = \sec^2 at$

   Write $x = \dfrac{1}{\cos^2 at} = (\cos at)^{-2}$

   Put $x = u^{-2}$, where $u = \cos at$

   $dx/du = -2u^{-3} = -\dfrac{2}{u^3}$ and $du/dt = -a \sin at$

   Then:

$$dx/dt = -\frac{2}{u^3} \times -a \sin at = \frac{2a \sin at}{\cos^3 at}$$

$$= 2a\left(\frac{\sin at}{\cos at}\right) \times \left(\frac{1}{\cos^2 at}\right)$$

$$= 2a \tan at \,.\, \sec^2 at$$

4. $y = \tan(at^2 + bt + c)$
   Put $y = \tan u$, where $u = at^2 + bt + c$
   $dy/du = \sec^2 u$ (by *2.6*) and $du/dt = 2at + b$

$$dy/dt = \sec^2 u \times (2at + b)$$

$$= (2at + b) \,.\, \sec^2(at^2 + bt + c)$$

5. $x = a.e^{b \sin pt}$
   Put $x = ae^u$, where $u = b \sin pt$
   $dx/du = ae^u$ and $du/dt = bp \cos pt$
   Then $dx/dt = abpe^u \cos pt = abp \cos pt . e^{b \sin pt}$

6. $y = \ln \sqrt{x^2 + 2/\sqrt{x}}$
   Put $y = \ln u$, $u = \sqrt{v}$ and $v = x^2 + 2/\sqrt{x}$
   $dy/du = 1/u$, $du/dv = \frac{1}{2}v^{-\frac{1}{2}}$ and $dv/dx = 2x - 1/x^{\frac{3}{2}}$

$$\frac{dy}{dx} = \frac{1}{u} \times \frac{1}{2}v^{-\frac{1}{2}} \times (2x - 1/x^{\frac{3}{2}})$$

$$= \frac{1}{\sqrt{x^2 + 2/\sqrt{x}}} \times \frac{1}{2\sqrt{x^2 + 2/\sqrt{x}}} \times (2x - 1/x^{\frac{3}{2}})$$

$$= \frac{(2x - 1/x^{\frac{3}{2}})}{2(x^2 + 2/\sqrt{x})}$$

Exercise 2.5

Differentiate the following functions w.r.t. the independent variable:

1. $\sec at$ (Hint: use a method similar to example 2 above.)
2. $\cot at$ (Hint: use a method similar to example 1 above.)
3. $\sin^2 (\frac{1}{2}x)$
4. $\cos^2 (\frac{3x}{4})$
5. $\tan^2 (at)$
6. $e^{\sin at}$
7. $e^{\cos at}$
8. $\ln (1 + \cos 2t)$
9. $\ln (1 - \cos 2t)$
10. $\sin (2t^2 + 1)$
11. $\frac{1}{3} \cos at + \frac{1}{4} \sin 2at$
12. $(3x + 4) \tan 2x$
13. $\dfrac{\sec 3x}{1 + 5x}$
14. $\dfrac{\tan x + 1}{\sec x}$
15. $e^{\tan x}$
16. $\ln (\sin x)$
17. $\ln (\cos ax)$

18. $\ln(\tan ax)$
19. $\ln(\operatorname{cosec} x)$
20. $\ln(\cot ax)$
21. $\ln(\sec x)$
22. $\sqrt{3x^2+5x-1}$
23. $\sqrt{1+\tan^2 x}$
24. $\sqrt{1+e^{ax}}$
25. $\sqrt{\ln(3x)}$
26. $\sqrt{\dfrac{1-x}{1+x}}$
27. $\sqrt{\dfrac{1-\cos x}{1+\cos x}}$
28. $\sqrt{\dfrac{1-e^{-t}}{1+e^{-t}}}$
29. $\ln\sqrt{(x+1)}$
30. $\ln(1+2x)^3$
31. $\dfrac{3}{\sqrt{x}}\sin 2x$
32. $\ln\left(\dfrac{3}{x+2}\right)$

Note:

$$\frac{d}{dx}(\sec ax) = a \sec ax \,.\, \tan ax \qquad 2.8$$

$$\frac{d}{dx}(\cot ax) = -a \operatorname{cosec}^2 ax \qquad 2.9$$

These may be added to the list of derivatives of special functions.

2.6 Evaluation of derivatives at special points

Examples

1. Where $f(x) = e^{-2x} \,.\, \sin 3x$, determine the values of $f'(x)$ when $x = 0, 1, \pi/2$.
   $f(x) = uv$, where $u = e^{-2x}$ and $v = \sin 3x$

Then $du/dx = -2e^{-2x}$ and $dv/dx = 3\cos 3x$

$$f'(x) = e^{-2x}.3\cos 3x + \sin 3x.(-2e^{-2x})$$
$$= e^{-2x}(3\cos 3x - 2\sin 3x)$$

When $x = 0$, $f'(0) = e^0(3\cos 0 - 2\sin 0) = 1 \times 3 = 3$
When $x = 1$:

$$f'(1) = e^{-2}(3\cos 3 - 2\sin 3) \text{ (Note that the angle is in radians)}$$
$$= 0.1353353(-2.9699775 - 0.28224)$$
$$= 0.1353353 \times -3.2522175$$
$$= -0.4401398 \approx -0.4401$$

When $x = \pi/2$:

$$f'(\pi/2) = e^{-\pi}(3\cos 3\pi/2 - 2\sin 3\pi/2)$$
$$= e^{-\pi}(0 + 2)$$
$$= 0.0432139 \times 2 = 0.0864278 \approx 0.0864$$

2. Given that $x = \dfrac{\tan 2t + 1}{\sec 2t}$, determine the values of dx/dt when $t = 0, 1, 2.5$.

Sometimes it is more convenient to rearrange the function before commencing differentiation, for example:

$$x = \frac{\tan 2t}{\sec 2t} + \frac{1}{\sec 2t} = \sin 2t + \cos 2t$$

Then $dx/dt = 2\cos 2t - 2\sin 2t$
When $t = 0$, $dx/dt = 2(\cos 0 - \sin 0) = 2(1 - 0) = 2$
When $t = 1$:

$$dx/dt = 2(\cos 2 - \sin 2) = 2(-0.4161468 - 0.9092974)$$
$$= 2 \times -1.3254443 = -2.6508885 \approx -2.6509$$

When $t = 2.5$:

$$dx/dt = 2(\cos 5 - \sin 5)$$
$$= 2(0.2836622 + 0.9589243)$$
$$= 2 \times 1.2425865 = 2.4851729 \approx 2.485$$

Exercise 2.6

Evaluate the derivatives w.r.t. the independent variables for the given values, for the following functions:

1. $e^{-3x} . \cos 2x : x = 0, 1, \pi/2$
2. $e^{-0.01x} \sin 4x$: $x = 0, \pi/2, \pi$
3. $\sqrt{\dfrac{1-t}{1+t}}$: $t = \frac{1}{4}, \frac{1}{2}$
4. $\sqrt{\dfrac{1-\cos x}{1+\cos x}}$: $x = 0, \pi/2, \pi$
5. $\ln\sqrt{x^2+1}$: $x = 0, 1, 4$
6. $e^{2x} . \ln\left(\dfrac{x}{2}\right)$: $x = \frac{1}{2}, 1, 10$
7. $(x+5)e^{\frac{1}{4}x}$: $x = 2, 3.5$
8. $3 \sin 2x . \ln(2x)$: $x = \frac{1}{2}, 1, 2$
9. $\dfrac{3x}{x^2+1}$: $x = 0, 1, 5.5$
10. Calculate the gradient of the curve $y = (x+1)e^{-2x}$ at the point (0, 1).
11. Determine the gradient of the curve

    $y = (x + 1/x) \tan\left(\dfrac{\pi}{4}x\right)$ at the point (1, 2).

3

Second Derivatives

3.1 Determination of second derivatives

The symbol $\frac{dy}{dx}$ may be interpreted in various ways. It is the $\lim \frac{\delta y}{\delta x}$ as $\delta x \to 0$. It is the function which determines the gradient of a curve. It is also that function which determines the magnification of a mapping. Where we regard it as $\frac{d}{dx}(y)$, it is also an operator on y, i.e. on $f(x)$. Now $\frac{d}{dx}(\sin ax) = a \cos ax$. This is still a function of x and so the operation $\frac{d}{dx}$ may be carried out again. This second operation, following the first, may be written:

$$\frac{d}{dx}\left[\frac{d}{dx}(\sin ax)\right] = \frac{d}{dx}(a \cos ax) = a \times -a \sin ax$$
$$= -a^2 \sin ax$$

The combined operation $\frac{d}{dx}\left(\frac{d}{dx}\right)$ needs to be condensed. It is written $\frac{d^2}{dx^2}$.

In other words, $\frac{d^2y}{dx^2}$ means:

$$\frac{d}{dx}\left[\frac{d}{dx}(y)\right] = \frac{d}{dx}\left(\frac{dy}{dx}\right)$$

It is called the second derivative. Further:

$$\frac{d^2x}{dt^2} = \frac{d}{dt}\left[\frac{d}{dt}(x)\right] = \frac{d}{dt}\left(\frac{dx}{dt}\right)$$

Examples

Determine the second derivatives of the following functions w.r.t. the independent variables.

1. $15x\sqrt{x} - 7/\sqrt{x}$
   Write $y = 15x^{\frac{3}{2}} - 7x^{-\frac{1}{2}}$

$$\frac{dy}{dx} = 15 \times \frac{3}{2}x^{\frac{1}{2}} - 7 \times -\frac{1}{2}x^{-\frac{3}{2}}$$

$$= \frac{45}{2}x^{\frac{1}{2}} + \frac{7}{2}x^{-\frac{3}{2}}$$

$$\frac{d^2y}{dx^2} = \frac{45}{2} \times \frac{1}{2}x^{-\frac{1}{2}} + \frac{7}{2} \times \left(-\frac{3}{2}x^{-\frac{5}{2}}\right)$$

$$= \frac{45}{4}x^{-\frac{1}{2}} - \frac{21}{4}x^{-\frac{5}{2}}$$

or $\dfrac{45}{4\sqrt{x}} - \dfrac{21}{4x^2\sqrt{x}}$

2. $e^{ax} \sin bx$

   Put $y = e^{ax} . \sin bx$

$$\frac{dy}{dx} = e^{ax} . b\cos bx + ae^{ax} . \sin bx \qquad \text{(by } 2.3\text{)}$$

$$\frac{d^2y}{dx^2} = e^{ax}(-b^2 \sin bx + ab\cos bx) + ae^{ax}(b\cos bx + a\sin bx)$$

$$= e^{ax}[(a^2 - b^2)\sin bx + 2ab\cos bx]$$

3. $\cot^2(3x+2)$
   Write $f(x) = \cot^2(3x+2)$
   Put $f = u^2$, $u = \cot v$ and $v = 3x + 2$
   Then $df/du = 2u$, $du/dv = -\operatorname{cosec}^2 v$ and $dv/dx = 3$

$$\frac{df}{dx} = -6u . \operatorname{cosec}^2 v = -6\cot(3x+2) . \operatorname{cosec}^2(3x+2) = -6pq,$$

where $p = \cot(3x+2)$, $q = r^2$ and $r = \operatorname{cosec}(3x+2)$
Then $dp/dx = -3\operatorname{cosec}^2(3x+2)$, $dq/dr = 2r$, $dr/dx = -3\cot(3x+2)\operatorname{cosec}(3x+2)$

$$\frac{d^2f}{dx^2} = f''(x) = -6p.\frac{dq}{dx} + \left(-6\frac{dp}{dx}\right).q$$

$$= -6\cot(3x+2)[-6\cot(3x+2).\operatorname{cosec}^2(3x+2)]$$
$$+ \operatorname{cosec}^2(3x+2)[18\operatorname{cosec}^2(3x+2)]$$
$$= -36\cot^2(3x+2)\operatorname{cosec}^2(3x+2) + 18\operatorname{cosec}^4(3x+2)$$
$$= 18\operatorname{cosec}^2(3x+2)[\operatorname{cosec}^2(3x+2) - 2\cot^2(3x+2)]$$
$$= 18\operatorname{cosec}^2(3x+2)[1 - \cot^2(3x+2)]$$

4. Where $y = ae^{3x} + be^{-2x}$ show that $\dfrac{d^2y}{dx^2} - \dfrac{dy}{dx} - 6y = 0$

Instead of determining dy/dx, d^2y/dx^2 and then substituting into the LHS of the equation on the right above, a quicker method is the following:

$$y = ae^{3x} + be^{-2x} \quad \text{(i)}$$
$$dy/dx = 3ae^{3x} - 2be^{-2x} \quad \text{(ii)}$$

Eliminate e^{-2x} from (i) and (ii). 2(i) + (ii) gives:.

$$dy/dx + 2y = 5ae^{3x} \quad \text{(iii)}$$

Differentiate (iii) w.r.t. x:

$$d^2y/dx^2 + 2(dy/dx) = 15ae^{3x} \quad \text{(iv)}$$

Substitute for $5ae^{3x}$ from (iii) in (iv):

$$d^2y/dx^2 + 2(dy/dx) = 3[(dy/dx) + 2y]$$

$$\text{i.e.} \quad \frac{d^2y}{dx^2} - \frac{dy}{dx} - 6y = 0$$

This is a technique which is useful in determining differential equations for a family of curves.

5. Where $z = y^2$, show: (a) $dz/dx = 2y(dy/dx)$; and
(b) $d^2z/dx^2 = 2y(d^2y/dx^2) + 2(dy/dx)^2$
Here we regard both z and y as functions of x.
So $z = y^2$ (a function of y) and y is a function of x.
That is, z is a function of a function of x.

Then: $\quad dz/dx = (dz/dy)(dy/dx)$

But: $\quad dz/dy = 2y$

Therefore: $$dz/dx = 2y.\frac{dy}{dx} \qquad \text{(a)}$$

In (i), dz/dx, $2y$ and dy/dx are all functions of x. So $2y.\frac{dy}{dx}$ is a product.

Differentiate (a) w.r.t. x:

$$\frac{d^2z}{dx^2} = 2y.\frac{d^2y}{dx^2} + \frac{dy}{dx}.\frac{d}{dx}(2y)$$

$$= 2y.\frac{d^2y}{dx^2} + 2\left(\frac{dy}{dx}\right)^2 \qquad \text{(b)}$$

6. Where $x = a\cos pt + b\sin pt$ find an equation relating x and d^2x/dt^2

$$x = a\cos pt + b\sin pt \qquad \text{(i)}$$

Differentiate (i) w.r.t. t:

$$\frac{dx}{dt} = -ap\sin pt + bp\cos pt \qquad \text{(ii)}$$

$$\frac{d^2x}{dt^2} = -ap^2\cos pt - bp^2\sin pt$$

$$= -p^2(a\cos pt + b\sin pt)$$

$$= -p^2x$$

Therefore:

$$\frac{d^2x}{dt^2} + p^2x = 0$$

This is a differential equation commonly encountered.

Exercise 3.1

Determine the second derivatives of the following functions w.r.t. the independent variables.

1. $y = 3x^2 - 5x + 7$
2. $y = 3x + 1/x$
3. $y = 4 - 3/x + 7/x^2$
4. $f(x) = 4x^2 + 8 + 4/x^2$
5. $f(x) = (2x - 3/x)^2$
6. $x = (2t^2 + 3t - 1)^3$
7. $x = a\sin 2t$

8. $x = b\cos 3t$
9. $A\cos at + B\sin at$
10. $f(x) = Ae^{3x}$
11. $f(x) = P.e^{-2x}$
12. $f(x) = Ae^{\frac{1}{2}x} + Be^{-\frac{1}{2}x}$
13. $x = Ae^{-t}\sin 2t$
14. $x = Ae^{-t}\cos 3t + Be^{-2t}\sin 3t$
15. $y = \cot(2x+1)$
16. $y = 3\ln x$
17. $y = 3\ln x^2$
18. $y = (2x-3)\ln x$
19. $y = (4x^2+1)^5$
20. $y = (x^2+3x)\sin 2x$
21. $y = \sqrt{x} + 1/\sqrt{x}$
22. $y = (3x-1)\tan x$
23. $y = 2\sec 3x + 3\operatorname{cosec} 2x$
24. Given $x = Ae^{2t} - Be^{-2t}$ prove $d^2x/dt^2 - 4x = 0$
25. Given $x = Pe^{t} + Qe^{-2t}$ prove that $d^2x/dt^2 + dx/dt - 2x = 0$
26. Given $x = a\cos 2t + b\sin 2t$ prove that $d^2x/dt^2 + 4x = 0$
27. Given $xy = \sin ax$ prove $x.\dfrac{d^2y}{dx^2} + 2.\dfrac{dy}{dx} + a^2xy = 0$
28. Given $x\sin pt = a\cos pt$ prove $\sin pt.\dfrac{d^2x}{dt^2} + 2p.\cos pt.\dfrac{dx}{dt} = 0$

3.2 Evaluation of second derivatives at given points

Examples

1. Determine the value of d^2y/dx^2 when $x = -2, -1, -\frac{1}{2}, 0, 1, 2$, where $y = 2x^2 - 3x - 2$.

$$dy/dx = 4x - 3$$
$$d^2y/dx^2 = 4$$

d^2y/dx^2 is constant and takes the value 4 for all values of x.

2. Where $y = 11 + 6x - 4x^2$ determine the value of d^2y/dx^2 when $x = -4, -2, 0, 1\frac{1}{2}, 5\frac{1}{4}$.

$$dy/dx = 6 - 8x$$
$$d^2y/dx^2 = -8$$

Again, d^2y/dx^2 is constant and takes the value -8 for all values of x.

3. Where $y = ax^2 + bx + c$:

$$\mathrm{d}y/\mathrm{d}x = 2ax + b$$

$\mathrm{d}^2y/\mathrm{d}x^2 = 2a$ = constant for all values of x.

4. Where $y = ax^3 + bx^2 + cx + d$:

$$\mathrm{d}y/\mathrm{d}x = 3ax^2 + 2bx + c$$
$$\mathrm{d}^2y/\mathrm{d}x^2 = 6ax + 2b$$

When $x = 0$, $\mathrm{d}^2y/\mathrm{d}x^2 = 2b$.
When $x = 1$, $\mathrm{d}^2y/\mathrm{d}x^2 = 6a + 2b$.
When $x = -1$, $\mathrm{d}^2y/\mathrm{d}x^2 = -6a + 2b$.

5. Given that $x = \dfrac{4+3t}{1-2t}$ determine the values of $\mathrm{d}^2x/\mathrm{d}t^2$

when $t = 3/4$, 1, 4.

$$\frac{\mathrm{d}x}{\mathrm{d}t} = \frac{(1-2t)3-(4+3t)(-2)}{(1-2t)^2} = \frac{3-6t+8+6t}{(1-2t)^2} = \frac{11}{(1-2t)^2}$$
$$= -11(1-2t)^{-2}$$

i.e. a function of a function.

$$\mathrm{d}^2x/\mathrm{d}t^2 = -11 \times -2(1-2t)^{-3} \times (-2)$$
$$= -\frac{44}{(1-2t)^3}$$

When $t = 3/4$, $\mathrm{d}^2x/\mathrm{d}t^2 = -\dfrac{44}{(-0.5)^3} = 352.$

When $t = 1$, $\mathrm{d}^2x/\mathrm{d}t^2 = 44.$

When $t = 4$, $\mathrm{d}^2x/\mathrm{d}t^2 = -\dfrac{44}{(-7)^3} = 0.1282799 \approx 0.128.$

6. Given that $x = 7\mathrm{e}^{-2t}\cos 3t + 11\mathrm{e}^{-2t}\sin 3t$, determine the values of $\mathrm{d}^2x/\mathrm{d}t^2$ when $t = 0, \pi/2, 3\pi/4$.

$$\mathrm{d}x/\mathrm{d}t = -14\mathrm{e}^{-2t}\cos 3t - 21\mathrm{e}^{-2t}\sin 3t - 22\mathrm{e}^{-2t}\sin 3t + 33\mathrm{e}^{-2t}\cos 3t$$
$$= 19\mathrm{e}^{-2t}\cos 3t - 43\mathrm{e}^{-2t}\sin 3t$$
$$\mathrm{d}^2x/\mathrm{d}t^2 = -38\mathrm{e}^{-2t}\cos 3t - 57\mathrm{e}^{-2t}\sin 3t + 86\mathrm{e}^{-2t}\sin 3t - 129\mathrm{e}^{-2t}\cos 3t$$
$$= -167\mathrm{e}^{-2t}\cos 3t + 29\mathrm{e}^{-2t}\sin 3t$$

When $t = 0$ $\mathrm{d}^2x/\mathrm{d}t^2 = -167 + 0 = -167.$

When $t = \frac{1}{2}\pi$ $d^2x/dt^2 = -167 \times 0.0432139 \times 0 + 29 \times 0.0432139 \times (-1) = -1.2532031 \approx -1.253$

When $t = 3\pi/4$:

$$d^2x/dt^2 = -167 \times 0.0089833 \times 0.7071068 + 29 \times 0.0089833 \times 0.7071068$$
$$= -1.0608084 + 0.1842124$$
$$= -0.876596 \approx -0.877$$

Exercise 3.2

Evaluate the second derivatives of the following functions w.r.t. the independent variables at the given values:

1. $y = 5x^2 - 3x - 7$ at $x = 0, 2$
2. $y = (4x - 1)^2$ at $x = 1, 5$
3. $y = 5x - 6/x$ at $x = 1, 2, 3.5$
4. $y = 7x^3 + 9x^2 - 12x - 3$ at $x = 0, 7.2$
5. $y = 3x^4 + 2x^2 - 7 + 2/x^2$ at $x = 0.5, 5$
6. $x = \dfrac{3 + 2t}{1 + t}$ at $t = 0, 1, 4$
7. $x = 2e^t - 3e^{-t}$ at $t = 0, 1, 10$
8. $x = 7.3e^{-t} - 4.7e^{-2t}$ at $t = 0, 1, 2.5$
9. $y = x + 1/\sqrt{x}$ at $x = 4, 5.2$
10. $x = 6 \cos t - 3 \sin \frac{1}{2}t$ at $t = 0, \pi, 3\pi/4$
11. $y = \operatorname{cosec}(x + 4)$ at $x = 0, 1$
12. $x = e^{-2t}($[illegible]$\cos t + 3 \cos 2t)$ at $t = 0, \pi/2, \pi$
13. $y = (2x + 5) \tan 2x$ at $x = 0, \pi/8, \pi/4$
14. $y = \ln(3x + 2)$ at $x = 1, 1.2, 1.4$

3.3 Motion along a line. Velocity and acceleration

In Fig. 3.1 suppose that P is a variable point, and that O is a fixed point on a line, and that OP (the displacement of P from O) $= x$. As P varies with the time, t, so x will vary. The velocity of $P(v)$ is dx/dt (by definition, the rate of

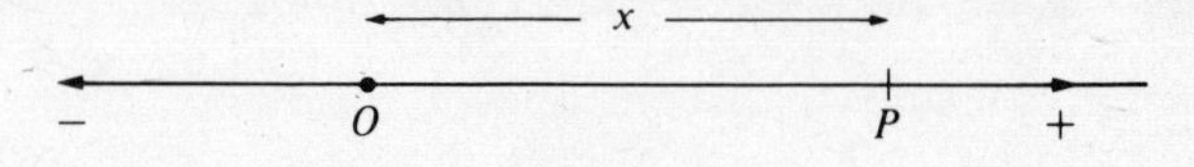

Figure 3.1

change of position). By definition the acceleration (a) is the rate of change of the velocity, i.e.

$$dv/dt = \frac{d}{dt}\left(\frac{dx}{dt}\right) = \frac{d^2x}{dt^2}$$

Where v and a are positive they are measured in the direction of x positive. Where x can be expressed as a function of t it is usually possible to express both v and a as functions of t.

Examples

The following problems presuppose motion of the type represented in Fig. 3.1.

1. When $x = 3t^2 - 4t + 7$
   Then $v = dx/dt = 6t - 4$
   And $a = dv/dt = d^2x/dt^2 = 6$
   Then a is constant $= 6$, for all t.
   When $t = 0$, $v = -4$; when $t = 1$, $v = 2$; when $t = 2$, $v = 8$
2. When $x = 2t^3 - 7t^2 + 11t - 8$
   Then $v = dx/dt = 6t^2 - 14t + 11$
   And $a = dv/dt = d^2x/dt^2 = 12t - 14$
   When $t = 0$, $v = 11$, $a = -14$
   When $t = 1$, $v = 6 - 14 + 11 = 3$, $a = -2$
   When $t = 2$, $v = 24 - 28 + 11 = 7$, $a = 24 - 14 = 10$
3. Where $x = 10 \sin 2\pi t$
   Then $v = dx/dt = 10 \times 2\pi \cos 2\pi t = 20\pi \cos 2\pi t$
   And $a = dv/dt = d^2x/dt^2 = 20\pi \times -2\pi \sin 2\pi t = -40\pi^2 \sin 2\pi t$
   When $t = 0$, $v = 20\pi \times 1 = 20\pi$, $a = -40\pi^2 \times 0 = 0$

   When $t = 1/8$, $v = 20\pi \cos \pi/4 = 20\pi \times \frac{1}{\sqrt{2}} = 10\sqrt{2}\pi$, $a = -40\pi^2 \times \frac{1}{\sqrt{2}} = -20\sqrt{2}.\pi^2$

   When $t = \frac{1}{4}$, $v = 20\pi \times 0 = 0$, $a = -40\pi^2 \times 1 = -40\pi^2$
   When $t = \frac{1}{2}$, $v = 20\pi \times -1 = -20\pi$, $a = -40\pi^2 \times 0 = 0$
   When $t = \frac{3}{4}$, $v = 20\pi \times 0 = 0$, $a = -40\pi^2 \times -1 = 40\pi^2$

   When $t = \frac{7}{8}$, $v = 20\pi \times \frac{1}{\sqrt{2}} = 10\sqrt{2}\pi$, $a = -40\pi^2 \times -\frac{1}{\sqrt{2}} = 20\sqrt{2}.\pi^2$

   When $t = 1$, $v = 20\pi \times 1 = 20\pi$, $a = -40\pi \times 0 = 0$

4. Where $x = 3e^{\frac{1}{2}t}$

   Then $v = dx/dt = \frac{3}{2}e^{\frac{1}{2}t}$

   And $a = dv/dt = d^2x/dt^2 = \frac{3}{4}e^{\frac{1}{2}t}$

   When $t = 0$, $v = \frac{3}{2} \times 1 = 3/2$, $a = \frac{3}{4} \times 1 = \frac{3}{4}$

   When $t = 1$, $v = \frac{3}{2}e^{\frac{1}{2}} = 2.4730819 \approx 2.47$, $a = \frac{3}{4}e^{\frac{1}{2}} = 1.236541 \approx 1.24$

5. Where $x = 5e^{-t} \cos \pi t$ (i)

   Then

   $v = dx/dt = -5e^{-t}\cos \pi t - 5\pi e^{-t}\sin \pi t = -5e^{-t}(\cos \pi t + \pi \sin \pi t)$

   And $a = dv/dt = 5e^{-t}(\cos \pi t + \pi \sin \pi t) - 5e^{-t}(-\pi \sin \pi t + \pi^2 \cos \pi t)$

   $= 5(1 - \pi^2)e^{-t}\cos \pi t + 10\pi e^{-t}\sin \pi t$

   When

   $t = 0$, $v = -5 \times (1 + 0) = -5$, $a = 5(1 - \pi^2) \times 1 \times 1 + 0 = 5(1 - \pi^2)$

   When $t = 1$, $v = -5 \times e^{-1}(-1 + 0) = 5e^{-1}$,

   $a = 5(1 - \pi^2)e^{-1} \times -1 + 0 = 5(\pi^2 - 1)e^{-1}$

Sometimes, in such calculations, the working may be simplified as follows.
Differentiate (i):

$$v = -5e^{-t}\cos \pi t - 5\pi e^{-t}\sin \pi t$$

$$= -x - 5\pi e^{-t}\sin \pi t, \text{ substituting from (i)} \qquad \text{(ii)}$$

Differentiate (ii):

$$dv/dt = -dx/dt - 5\pi^2 e^{-t}\cos \pi t$$

$$a = -v - \pi^2 x, \text{ substituting from (i)} \qquad \text{(iii)}$$

When $t = 0$:

from (i), $x = 5$
from (ii), $v = -5 - 0 = -5$
from (iii), $a = 5(1 - \pi^2)$

When $t = 1$:

from (i), $x = -5e^{-1}$
from (ii), $v = 5e^{-1} - 0 = 5e^{-1}$
from (iii), $a = -5e^{-1} + 5.\pi^2 e^{-1} = 5(\pi^2 - 1)e^{-1}$

Exercise 3.3

Where x represents the displacement of a moving point along a line from a fixed point in that line determine the velocity and acceleration at the times stated.

1. $x = 14t^2 - 11t + 6$: $t = 0$
2. $x = 3 + 8t + 6t^2$: $t = 0, 1$
3. $x = 17 - 9t - 5t^2$: $t = 0, 2$
4. $x = t^3 + t^2 - 4t - 5$: $t = 0, 2, 4$
5. $x = 6 - 8t + 11t^2 - 4t^3$: $t = 0, 1, 2, 3$
6. $x = t^4 - 2t^3 + 3t^2 - 6t + 9$: $t = 0, 2, 4$
7. $x = \frac{1}{12}t^4 - \frac{2}{3}t^3 + \frac{3}{2}t^2 - 6t + 8$: $t = 0, 1, 2, 3$
8. $x = 2e^t$: $t = 0, 1, 2$
9. $x = 5e^{-t}$: $t = 0, 1, 2$
10. $x = 3e^t + 4e^{-t}$: $t = 0, 1, 2$
11. $x = 2 \sin 2\pi t$: $t = 0, 1/8, 1/4$
12. $x = 3 \sin \frac{\pi}{2} t - 4 \cos \frac{\pi}{2} t$: $t = 0, \frac{1}{2}, 1$
13. $x = 10e^{-t} \sin \frac{\pi}{2} t$: $t = 0, \frac{1}{2}, 1$
14. $x = (t^2 + 3t - 2) \ln t$: $t = 1, 1.5, 3.5$

4

Turning Points of Quadratic and Cubic Functions

Figs 4.1, 4.2, 4.3 and 4.4 represent sketches of quadratic functions. L, M, N and R are called *turning points*. They are also *stationary points*. The values of y at such points are called turning values or stationary values. Fig. 4.5 is the mapping equivalent of Fig. 4.1 and this explains why the term 'turning point' is used. The arrows indicate that as x increases from $-\infty$ to 2 y decreases from ∞ to -1, and that as x increases from 2 to $+\infty$, y increases from -1 to $+\infty$. In other words, y turns round at the point -1. This is characteristic of all four graphs. Each possesses a point where the y value turns round as x increases from $-\infty$ to $+\infty$.

Where the graph of a function is drawn accurately it is an easy matter to locate any turning or stationary points. But drawing a graph is time-consuming, so we need an alternative method.

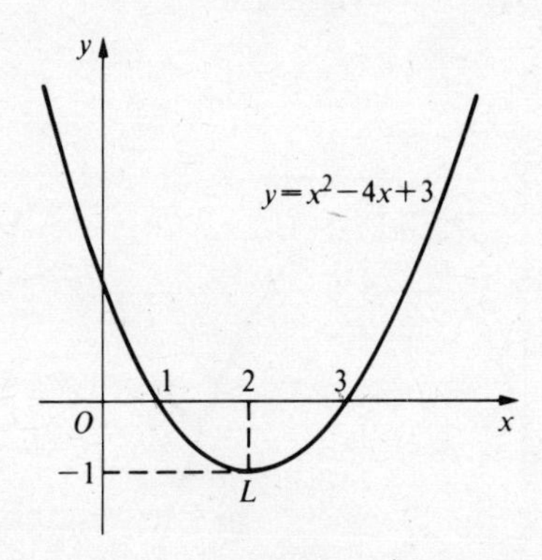

Figure 4.1

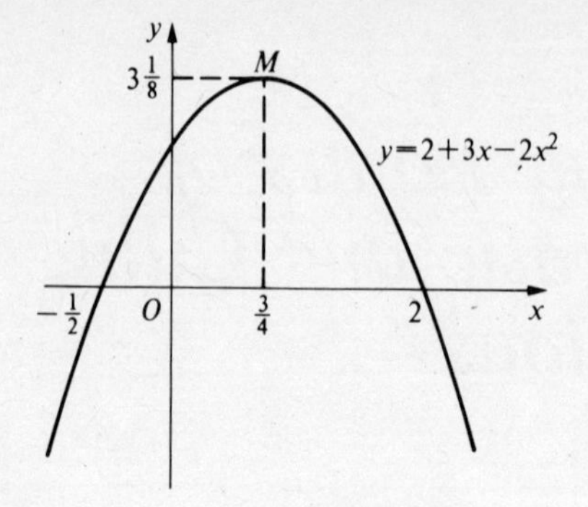

Figure 4.2

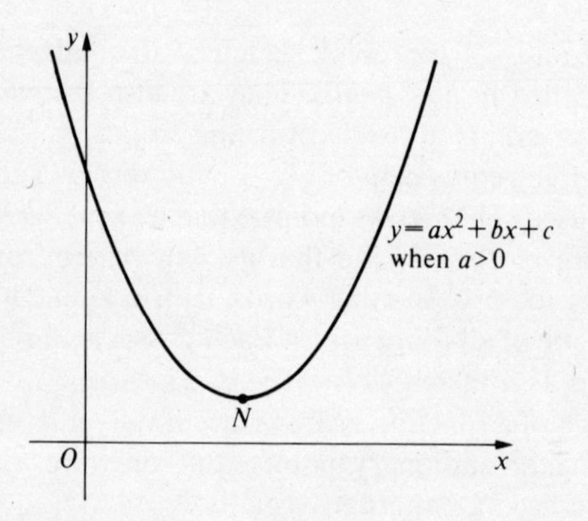

Figure 4.3

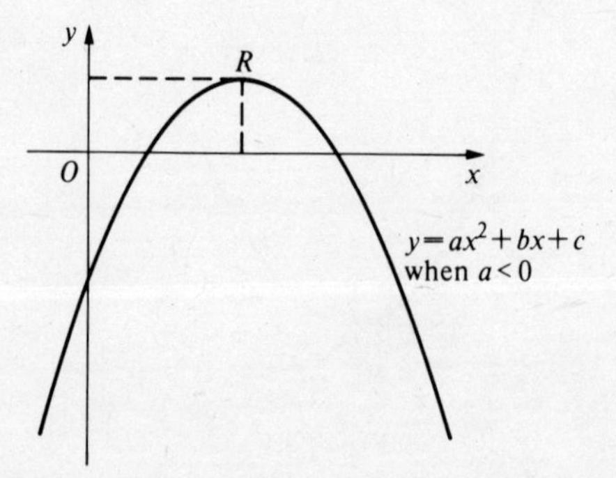

Figure 4.4

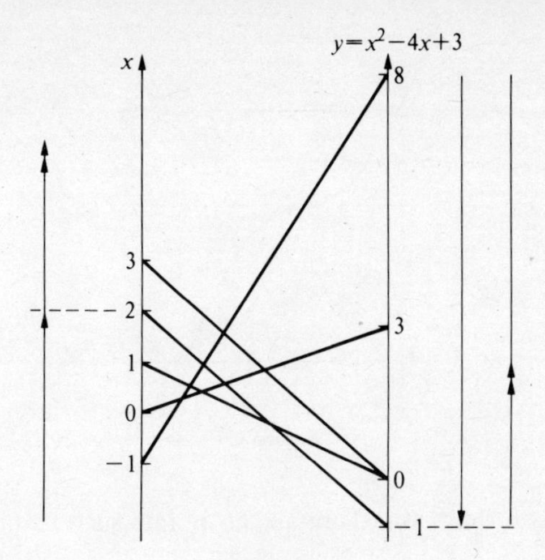

Figure 4.5

4.1 The definition of turning and stationary points

In fact not all stationary points are turning points. Figs 4.6 and 4.7 illustrate different kinds of stationary points. At all stationary points and at all turning points the gradient of the curve (dy/dx) is zero.

Definition

At a point of a curve, $y = f(x)$, where $f'(x) = 0$, the point in question is a stationary point. (If $f(x)$ is a quadratic function it will also be a turning

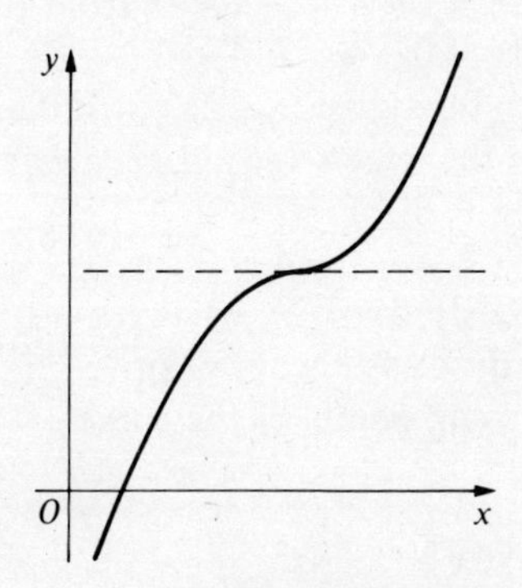

Figure 4.6

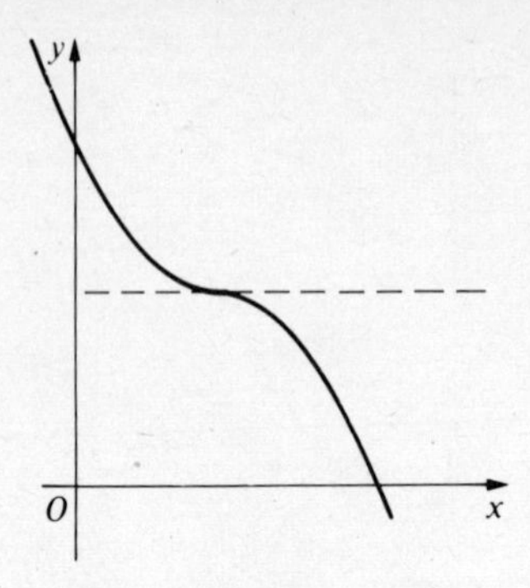

Figure 4.7

point because, in that case, there are no points such as those in Figs 4.6 and 4.7.)

Examples

1. Determine any turning points of the curve:

$$y = x^2 - 4x + 3 \quad \text{(Fig. 4.1)} \tag{i}$$

Then: $$dy/dx = 2x - 4 \tag{ii}$$

When $dy/dx = 0$, $2x - 4 = 0$, i.e. $x = 2$.
To find the value of y at this point substitute $x = 2$ in (i).
When $x = 2$, $y = 4 - 8 + 3 = -1$
The turning point is $(2, -1)$. See Fig. 4.1.
For future use we shall calculate dy/dx at a few different places on the curve.
When $x = 1$, $dy/dx = 2 - 4 = -2 < 0$, from (ii)
When $x = 0$, $dy/dx = 0 - 4 = -4 < 0$
When $x = -p(p > 0)$, $dy/dx = -2p - 4 < 0$
In other words, to the left of L dy/dx is always negative.
When $x = 3$, $dy/dx = 6 - 4 = 2 > 0$
When $x = 4$, $dy/dx = 8 - 4 = 4 > 0$
When $x = q(q > 2)$, $dy/dx = 2q - 4 = 2(q - 2) > 0$
To the right of L dy/dx is always positive.

2. Determine any turning points of the curve:

$$y = 2 + 3x - 2x^2 \quad \text{(Fig. 4.2)} \tag{iii}$$

Then: $$dy/dx = 3 - 4x \tag{iv}$$

When $dy/dx = 0$, $4x - 3 = 0$, i.e. $x = 3/4$ (v)

To find the value of y at this point substitute from (v) in (iii).
When $x = 3/4$, $y = 2 + 9/4 - 2 \times (9/16) = 2 + 2\frac{1}{4} - 1\frac{1}{8} = 3\frac{1}{8}$
The turning point is $(\frac{3}{4}, 3\frac{1}{8})$. See Fig. 4.2.
For future use we shall calculate dy/dx at a few different points on the curve.
When $x = 0$, $dy/dx = 3 - 0 = 3 > 0$, from (iv)
When $x = -\frac{1}{2}$, $dy/dx = 3 + 2 = 5 > 0$
When $x = -p\,(p > 0)$, $dy/dx = 3 + 4p > 0$
To the left of M dy/dx is always positive.
When $x = 1$, $dy/dx = 3 - 4 = -1 < 0$
When $x = 2$, $dy/dx = 3 - 8 = -5 < 0$
When $x = q\,(q > 3/4)$, $dy/dx = 3 - 4q < 0$
To the right of M dy/dx is always negative.

3. Determine any turning points of the curve:

$$y = 11x^2 + 14x + 3$$

Then: $$dy/dx = 22x + 14 = 2(11x + 7)$$

When $dy/dx = 0$, $2(11x + 7) = 0$, i.e. $11x = -7$, $x = -7/11$
When $x = -7/11$:

$$y = 11\left(\frac{49}{121}\right) - 14 \times \frac{7}{11} + 3$$

$$= \frac{49}{11} - \frac{98}{11} + \frac{33}{11}$$

$$= -\frac{16}{11} = -1\frac{5}{11}$$

The turning point is $\left(-\frac{7}{11}, -1\frac{5}{11}\right)$.
When $x < -\frac{7}{11}$, say $-\frac{8}{11}$ (i.e. to the left of the turning point), $dy/dx = 2(-8 + 7) = -2 < 0$
When $x > -\frac{7}{11}$, say $-\frac{6}{11}$ (to the right of the turning point), $dy/dx = 2(-6 + 7) = +2 > 0$

4. Determine any turning points of:

$$y = ax^2 + bx + c$$

Then: $$dy/dx = 2ax + b = 2a(x + b/2a)$$

When $dy/dx = 0$, $2ax = -b$, $x = -b/2a$

At the turning point:

$$y = a\left(\frac{b^2}{4a^2}\right) + b\left(-\frac{b}{2a}\right) + c$$

$$= \frac{b^2}{4a} - \frac{b^2}{2a} + c = -\frac{b^2}{4a} + c = -\frac{(b^2 - 4ac)}{4a}$$

Suppose $a > 0$

When $x < -b/2a$, $dy/dx = 2a(x + b/2a) = 2a \times \text{negative} < 0$
When $x > -b/2a$, $dy/dx = 2a(x + b/2a) = 2a \times \text{positive} > 0$ } Note Fig. 4.3

Suppose $a < 0$

When $x < -b/2a$, $dy/dx = 2a(x + b/2a) = 2a \times \text{negative} < 0$
When $x > -b/2a$, $dy/dx = 2a(x + b/2a) = 2a \times \text{positive} > 0$ } Note Fig. 4.4

4.2 The nature of a turning point

Although all points L, M, N and R in Figs 4.1, 4.2, 4.3 and 4.4 are turning points, L and N are *minimum points* and M and R are *maximum points.*

To distinguish between the two kinds, both from the graph and from the analysis (see the Examples above), we shall need to determine the gradient of the curve on each side of the point. Suppose there is a stationary point on the curve $y = f(x)$ at $x = a$. To determine the nature of that stationary point a procedure is followed such as that represented in Table 4.1.

Table 4.1 *The nature of a turning point*

Gradient of curve (dy/dx)		Nature of the point (at $x = a$)
To left ($x < a$)	To right ($x > a$)	
negative ($dy/dx < 0$)	positive ($dy/dx > 0$)	minimum } *4.1*
positive ($dy/dx > 0$)	negative ($dy/dx < 0$)	maximum } *4.1*

Exercise 4.1

Determine the co-ordinates and nature of the turning points of the following curves:

1. $y = x^2 - 3x + 2$
2. $y = x^2 - 5x + 6$

3. $y = x^2 - 12x + 35$
4. $y = x^2 - 6x + 8$
5. $y = x^2 + 4x + 3$
6. $y = x^2 + 8x + 12$
7. $y = 3 - 2x - x^2$
8. $y = 35 + 2x - x^2$
9. $y = 4x^2 + 7x - 9$
10. $y = \frac{1}{2}x^2 + \frac{3}{4}x - \frac{5}{8}$
11. $y = 18 - 11x - 3x^2$
12. $y = x^2 - 12x$
13. $y = 4x^2 + 5$
14. $y = -8 - 2x^2$
15. $y = -13 - 7x - 4x^2$

4.3 The turning points of cubic functions

Figs 4.8, 4.9, 4.10 and 4.11 represent sketches of different kinds of cubic functions. In Fig. 4.8 A and B are maximum and minimum points respectively. In Fig. 4.9 C and D are minimum and maximum points respectively. In Fig. 4.10 E is a stationary point (neither maximum nor minimum, but a point of inflexion). In Fig. 4.11 F is neither a stationary point nor a turning point. (It is a point of inflexion although $\mathrm{d}y/\mathrm{d}x \neq 0$.) In this chapter we are concerned solely with turning values.

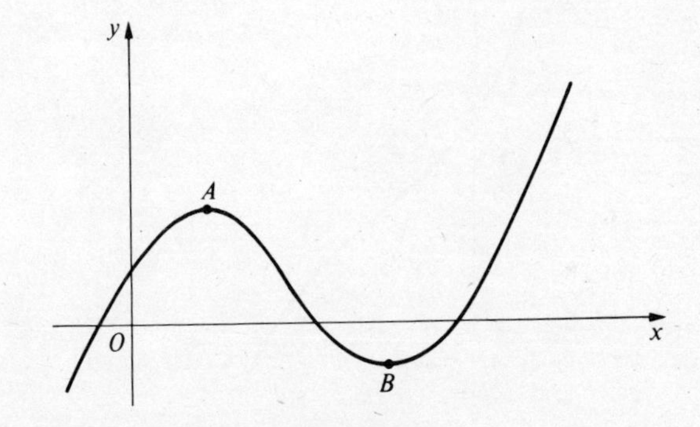

Figure 4.8

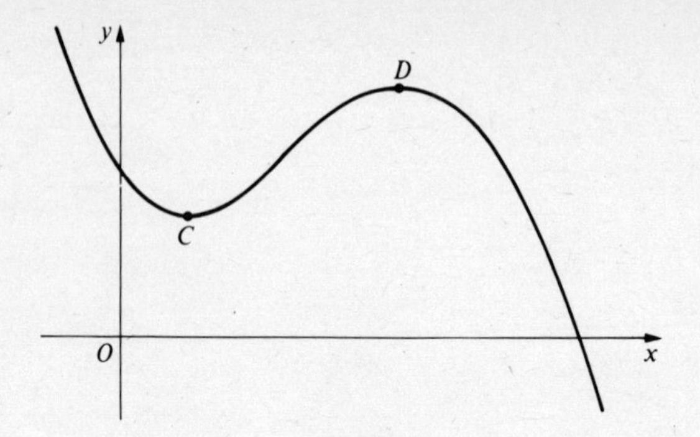

Figure 4.9

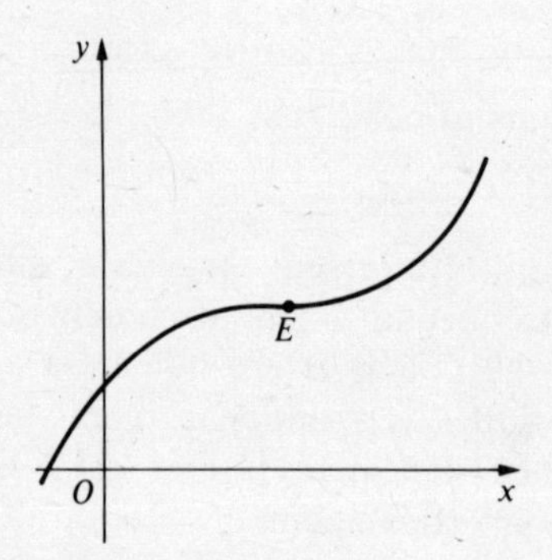

Figure 4.10

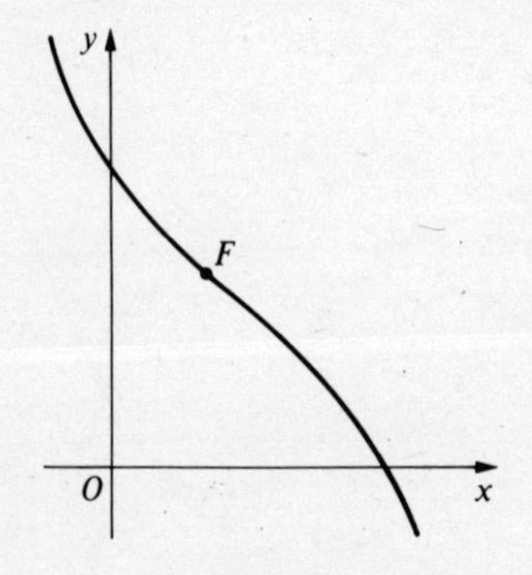

Figure 4.11

Examples

1. Determine the co-ordinates and nature of the turning points of the curve:

$$y = 2x^3 - 9x^2 + 12x - 7$$

Step 1 To determine the positions of the turning points:

$$dy/dx = 6x^2 - 18x + 12 = 6(x^2 - 3x + 2)$$
$$= 0 \text{ for stationary values}$$
$$\text{i.e.} \quad 6(x-1)(x-2) = 0$$

Then either $x = 1$ or $x = 2$.
When $x = 1$, $y = 2 - 9 + 12 - 7 = -2$. A turning point is $(1, -2)$.
When $x = 2$, $y = 16 - 36 + 24 - 7 = -3$. A turning point is $(2, -3)$.
Step 2 To determine the nature of the point $(1, -2)$:
When $x < 1$, then $x - 1 < 0$ and $x - 2 < 0$,
so $dy/dx = 6 \times - \times - = +, > 0$
When $x > 1$, then $x - 1 > 0$, $x - 2 < 0$,
so $dy/dx = 6 \times + \times - = -, < 0$
This gives a maximum point.
Step 3 To determine the nature of the point $(2, -3)$:
When $x < 2$, then $x - 1 > 0$, $x - 2 < 0$,
so $dy/dx = 6 \times + \times - = -, < 0$
When $x > 2$, then $x - 1 > 0$, $x - 2 > 0$,
so $dy/dx = 6 \times + \times + = +, > 0$
This gives a minimum point.
A convenient check of the relative positions of maximum and minimum points is the following. The graph of $y = ax^3 + bx^2 + cx + d$ will be:
similar to Fig. 4.8 if $a > 0$, because as $x \to +\infty$, $y \to +\infty$
similar to Fig. 4.9 if $a < 0$, because as $x \to +\infty$, $y \to -\infty$
(For large x the most important term in y is ax^3.)
In the case of $y = 2x^3 - 9x^2 + 12x - 7$, $a = 2$. So the curve is of type Fig. 4.8, with the maximum on the left and the minimum on the right.

2. Determine the positions of the turning points of the curve:

$$y = 8 + 12x + 5x^2 - 4x^3$$

Step 1 To determine the positions of the turning points:

$$dy/dx = 12 + 10x - 12x^2 = 2(6 + 5x - 6x^2)$$
$$= 2(3 - 2x)(2 + 3x) = 0 \text{ for stationary values}$$

Then either $x = 3/2$ or $x = -2/3$.

When $x = 3/2$:

$$\begin{aligned} y &= 8 + 12 \times \tfrac{3}{2} + 5 \times \tfrac{9}{4} - 4 \times \tfrac{27}{8} \\ &= 8 + 18 + 45/4 - 27/2 \\ &= 26 + 11\tfrac{1}{4} - 13\tfrac{1}{2} = 23\tfrac{3}{4} \end{aligned}$$

When $x = -2/3$:

$$\begin{aligned} y &= 8 - 12 \times \tfrac{2}{3} + 5 \times \tfrac{4}{9} + 4 \times \tfrac{8}{27} \\ &= 8 - 8 + 20/9 + 32/27 \\ &= 2\tfrac{2}{9} + 1\tfrac{5}{27} = 3\tfrac{11}{27} \end{aligned}$$

The turning points are $(1\tfrac{1}{2}, 23\tfrac{3}{4})$ and $(-\tfrac{2}{3}, 3\tfrac{11}{27})$

Step 2 To determine the nature of $(-\tfrac{2}{3}, 3\tfrac{11}{27})$:
When $x < -2/3$, $3 - 2x > 0$, $2 + 3x < 0$,
so $dy/dx = 2 \times + \times - = -, < 0$
When $x > -2/3$, $3 - 2x > 0$, $2 + 3x > 0$,
so $dy/dx = 2 \times + \times + = +, > 0$
This gives a minimum point.
Step 3 To determine the nature of $(1\tfrac{1}{2}, 23\tfrac{3}{4})$:
When $x < 1\tfrac{1}{2}$, $3 - 2x > 0$, $2 + 3x < 0$,
so $dy/dx = 2 \times + \times + = +, > 0$
When $x > 1\tfrac{1}{2}$, $3 - 2x < 0$, $2 + 3x > 0$,
so $dy/dx = 2 \times - \times + = -, < 0$
This gives a maximum point.

Exercise 4.2

Determine the positions and nature of the turning points of the following curves:

1. $y = x^3 - 6x^2 + 9x + 1$
2. $y = x^3 + 6x^2 + 9x + 11$
3. $y = x^3 - 9x^2 + 24x$
4. $y = x^3 + 3x^2 - 24x - 3$
5. $y = 2x^3 + 9x^2 - 60x$
6. $y = 4x^3 + 9x^2 + 6x - 1$
7. $y = x^3 + 4x^2 - 3x + 2$
8. $y = 11 - 9x + 6x^2 - x^3$
9. $y = 7 + 12x - \tfrac{3}{2}x^2 - \tfrac{1}{2}x^3$
10. $y = 9x - 21x^2 - 5x^3$

4.4 Alternative diagnosis of the nature of turning values

In certain circumstances an alternative to the above approach may be used to determine the nature of a turning value. For some problems the alternative method fails, but in those with which we are concerned at present it is satisfactory. The following examples illustrate the method.

Examples

1. When $y = x^2 - 4x + 3$
   Then $\mathrm{d}y/\mathrm{d}x = 2x - 4$
   From that $\mathrm{d}^2y/\mathrm{d}x^2 = 2 > 0$
   The turning point was a minimum (using the gradient diagnosis).
2. When $y = 2 + 3x - 2x^2$
   Then $\mathrm{d}y/\mathrm{d}x = 3 - 4x$
   From that $\mathrm{d}^2y/\mathrm{d}x^2 = -4 < 0$
   The turning point was a maximum (using the gradient diagnosis).
3. When $y = 2x^3 - 9x^2 + 12x - 7$
   Then $\mathrm{d}y/\mathrm{d}x = 6x^2 - 18x + 12$
   From that $\mathrm{d}^2y/\mathrm{d}x^2 = 12x - 18$
   When $x = 1$, $\mathrm{d}^2y/\mathrm{d}x^2 = 12 - 18 = -6 < 0$
   This turning point was a maximum.
   When $x = 2$, $\mathrm{d}^2y/\mathrm{d}x^2 = 24 - 18 = +6 > 0$
   This turning point was a minimum.
4. When $y = 8 + 12x + 5x^2 - 4x^3$
   Then $\mathrm{d}y/\mathrm{d}x = 12 + 10x - 12x^2$
   From that $\mathrm{d}^2y/\mathrm{d}x^2 = 10 - 24x$
   When $x = -2/3$, $\mathrm{d}^2y/\mathrm{d}x^2 = 10 + 16 = 26 > 0$
   This turning point was a minimum.
   When $x = 3/2$, $\mathrm{d}^2y/\mathrm{d}x^2 = 10 - 36 = -26 < 0$
   This turning point was a maximum.

If the evidence above is conclusive it would appear that at a turning point of a curve:

$$\left.\begin{array}{l}\text{when } \mathrm{d}^2y/\mathrm{d}x^2 > 0 \text{ that point is a minimum}\\ \text{when } \mathrm{d}^2y/\mathrm{d}x^2 < 0 \text{ that point is a maximum}\end{array}\right\} \qquad 4.2$$

Exercise 4.3

Assuming the truth of principle *4.2*, use that method to diagnose the nature of the turning points of the curves in Exercises 4.1 and 4.2. Check that the two methods agree.

Figs 4.12, 4.13, 4.14 and 4.15, grouped into sections I and II, illustrate graphically the equivalent of principle *4.2.* Figs 4.12 and 4.14 represent the graphs of $f(x)$ plotted against x, Fig. 4.12 with a minimum point, Fig. 4.14 with a maximum point. Figs 4.13 and 4.15 extract from the previous two figures some values for $f'(x)$ and plot them against x.

Figure 4.12 **Figure 4.14**

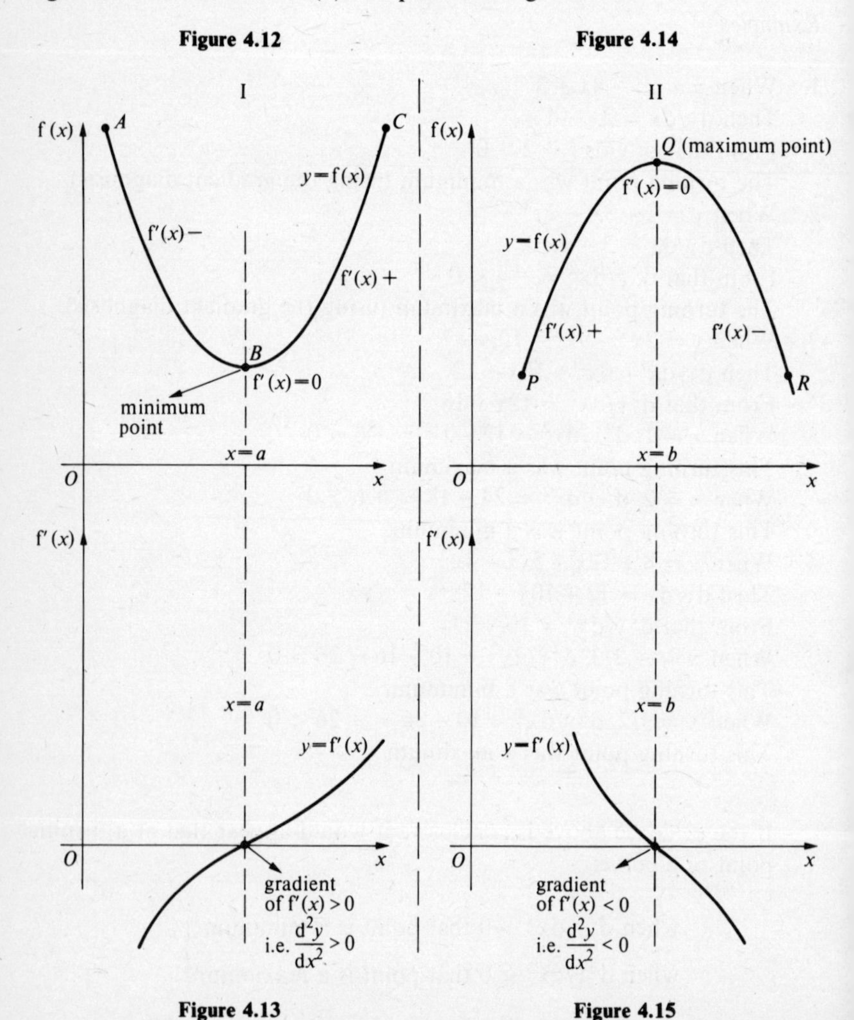

Figure 4.13 **Figure 4.15**

4.5 Turning values of other functions

Examples

Determine the turning values of the following functions:

1. $y = e^x + e^{-x}$
   $dy/dx = e^x - e^{-x} = 0$ for stationary values, i.e.

$$e^x = e^{-x}$$
$$e^{2x} = 1$$
$$x = 0$$

   When $x = 0$:

$$y = e^0 + e^0 = 2$$
$$d^2y/dx^2 = e^x + e^{-x}$$

   When $x = 0$, $d^2y/dx^2 = e^0 + e^0 = 2 > 0$
   The function has a minimum value at (0, 2).
2. $y = 6 \cos 2x$
   $dy/dx = -12 \sin 2x = 0$ for stationary values, i.e.

$$\sin 2x = 0$$
$$2x = 0°, 180°, 360° \text{ etc.}$$
$$x = 0°, 90°, 180° \text{ etc.}$$

   When $x = 0°$ $y = 6$
   When $x = 90°$ $y = -6$
   When $x = 180°$ $y = 6$
   When $x = 270°$ $y = -6$ and so on
   $d^2y/dx^2 = -24 \cos 2x$
   When $x = 0°$ $d^2y/dx^2 = -24$
   When $x = 90°$ $d^2y/dx^2 = +24$
   When $x = 180°$ $d^2y/dx^2 = -24$
   When $x = 270°$ $d^2y/dx^2 = +24$
   The function has maximum values at the points (0°, 6), (180°, 6), (360°, 6) etc.
   The function has minimum values at the points (90°, −6), (270°, −6), (450°, −6) etc.
3. $y = e^{-x} \sin 3x$
   $dy/dx = -e^{-x} \sin 3x + 3e^{-x} \cos 3x = 0$ for stationary values, i.e.

$$e^{-x}(3 \cos 3x - \sin 3x) = 0$$
$$\text{Since} \quad e^{-x} > 0, \sin 3x = 3 \cos 3x$$
$$\text{i.e.} \quad \tan 3x = 3$$

In radians, $3x = 1.2490458, \pi + 1.2490458, 2\pi + 1.2490458$ etc.
$x = 0.4163486, \pi/3 + 0.4163486, 2\pi/3 + 0.4163486$ etc.

When $x = 0.4163486$ $y = 0.6256095$
When $x = \pi/3 + 0.4163486$ $y = -0.2195388$
When $x = 2\pi/3 + 0.4163486$ $y = 0.0770405$

$$\begin{aligned} d^2y/dx^2 &= e^{-x}\sin 3x - 3e^{-x}\cos 3x - 3e^{-x}\cos 3x - 9e^{-x}\sin 3x \\ &= -6e^{-x}\cos 3x - 8e^{-x}\sin 3x \\ &= -2e^{-x}(3\cos 3x + 4\sin 3x) \end{aligned}$$

When $x = 0.4163486$ $d^2y/dx^2 = -6.256095 < 0$
When $x = \pi/3 + 0.4163486$ $d^2y/dx^2 = 2.1953877 > 0$
The function has a maximum value at $(0.4163486, 0.6256095)$.
The function has a minimum value at $(\pi/3 + 0.4163486, -0.2195388)$.

Exercise 4.4

Determine the turning points of the following functions:

1. $e^x + 2e^{-x}$
2. $2e^x + e^{-x}$
3. $5\sin 2x$
4. $4\cos 3x$
5. $-4\sin 4x$
6. $\sin x + \cos x$
7. $3\sin 2x - 4\cos 2x$
8. $\dfrac{1}{e^x + e^{-x}}$
9. $e^{-x}\sin x$
10. $e^{-x}\cos x$
11. $e^{-2x}\sin x$
12. $e^{-2x}\cos 4x$

4.6 Applications of maxima and minima

It is important to appreciate that the principles discussed and the conclusions reached apply only to curves of the form $y = f(x)$, i.e. to *functions of a single independent variable.* As yet we have not discovered any machinery for dealing with *functions of two or more independent variables*; that will come later. Therefore, where we are dealing with, for example:

$$A = xy \text{ (the area of a rectangle)} \qquad 4.3$$

where x and y are the length and breadth of the rectangle and each varies independently of the other (see Fig. 4.16), we are not able to proceed to such concepts as maxima and minima.

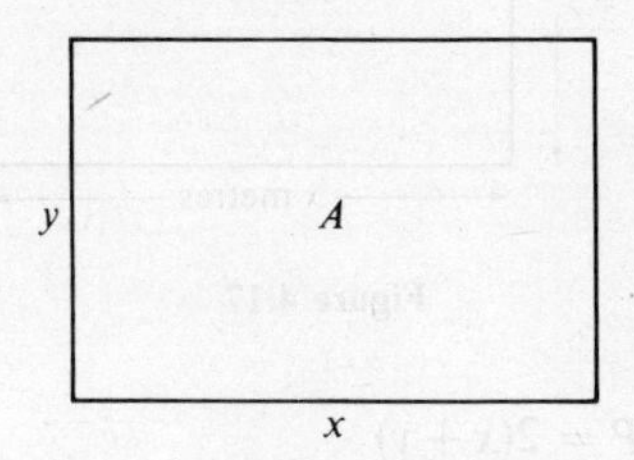

Figure 4.16

However, if there is a restriction on the way in which x and y may vary (this is called a *constraint*) then we have a different situation. For instance, if x and y are such that:

$$x + y = 50 \qquad 4.4$$

then: $$y = (50 - x)$$

Substitute for y in *4.3*:

$$A = x(50 - x)$$
$$= 50x - x^2 \qquad 4{\cdot}5$$

4.5 represents A as a function of a single variable, x. To determine any turning values of A:

$$\mathrm{d}A/\mathrm{d}x = 50 - 2x = 0 \text{ for stationary values}$$
$$\text{i.e.} \quad x = 25$$

When $x = 25$, $A = 25 \times 25 = 625$
When $x < 25$, $\mathrm{d}A/\mathrm{d}x = 2(25 - x) = 2 \times + = + > 0$
When $x > 25$, $\mathrm{d}A/\mathrm{d}x = 2(25 - x) = 2 \times - = - < 0$
The value 625 is a maximum value.
Check: $\mathrm{d}^2A/\mathrm{d}x^2 = -2 < 0$

Examples

1. Determine the dimension of a rectangular field enclosing an area of $16\,000\ \text{m}^2$ with the minimum of fencing surrounding it.
   Suppose the dimensions are as shown in Fig. 4.17.

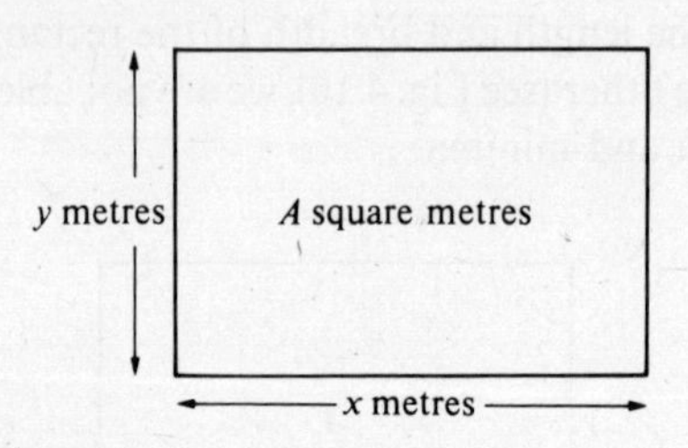

Figure 4.17

The perimeter, $P = 2(x + y)$ (i)

$A = xy = 16\,000$ (constraint) (ii)

Substituting from (ii) in (i), i.e. $y = 16\,000/x$:

$$P = 2\left(x + \frac{16\,000}{x}\right) \qquad \text{(iii)}$$

(iii) represents a function (P) of a single variable. Differentiate w.r.t. x:

$$\mathrm{d}P/\mathrm{d}x = 2 - \frac{32\,000}{x^2} = 0 \text{ for stationary values}$$

i.e. $x^2 = 16\,000$

$$x = \pm\sqrt{16\,000} = \pm 40\sqrt{10}$$

Reject the negative value, since it is not consistent with a real field. Then:

$$y = \frac{16\,000}{40\sqrt{10}} = \frac{400}{\sqrt{10}} = 40\sqrt{10} = 126.49111$$

Therefore $x \approx 126.6$ and $y \approx 126.6$. This means the field is square.

$$P = 2(x + y) = 2 \times 80\sqrt{10} = 160\sqrt{10} \approx 505.96443 \approx 506\,\text{m}$$

To prove the perimeter is a minimum:

$$\mathrm{d}^2P/\mathrm{d}x^2 = 0 + \frac{32\,000 \times 2}{x^3} = 64\,000/x^3 = \text{positive, since } x \text{ is positive.}$$

2. Fig. 4.18 represents a field the area of which is a rectangle together with a semicircle. If the total perimeter (P) is to be of length L determine the dimensions which will produce a maximum area.

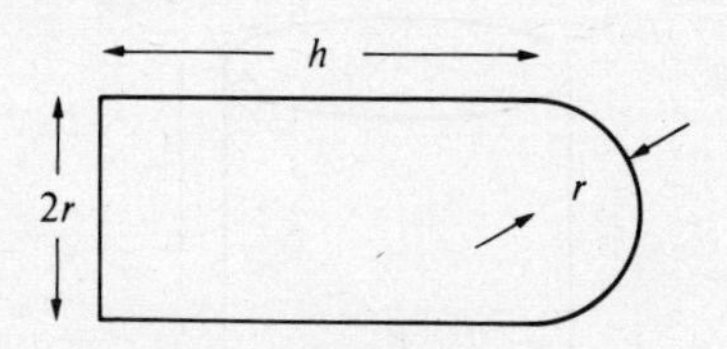

Figure 4.18

$$L = \pi r + 2r + 2h \text{ (constraint)} \qquad \text{(i)}$$

Then
$$2h = (L - \pi r - 2r) \qquad \text{(ii)}$$

$$\text{The area, } A = \tfrac{1}{2}\pi r^2 + 2rh = \tfrac{1}{2}\pi r^2 + (L - \pi r - 2r)r, \text{ from (ii)}$$
$$= \tfrac{1}{2}\pi r^2 + Lr - \pi r^2 - 2r^2$$
$$= Lr - \tfrac{1}{2}\pi r^2 - 2r^2 \qquad \text{(iii)}$$

Differentiate (iii) w.r.t. r:

$$\mathrm{d}A/\mathrm{d}r = L - \pi r - 4r = 0 \text{ for stationary values}$$

$$\text{i.e.} \quad r = \frac{L}{(\pi + 4)}$$

$$\mathrm{d}^2 A/\mathrm{d}r^2 = -\pi - 4 = -(\pi + 4)$$

$\mathrm{d}^2 A/\mathrm{d}r^2$ is negative. Then $r = L/(\pi + 4)$ gives a maximum area. From (ii):

$$2h = L - \pi r - 2r = L - (\pi + 2).\frac{L}{(\pi + 4)}$$

$$= \frac{L(\pi + 4) - L(\pi + 2)}{(\pi + 4)}$$

$$= \frac{2L}{(\pi + 4)}$$

For a maximum area $h = \dfrac{L}{(\pi + 4)} = r$.

3. A hollow cylindrical can is to be made from exactly 500 cm^2 of sheet

metal. Determine its dimensions, V, r and h, for it to enclose the maximum volume.

$$\text{Suppose } V = \pi r^2 h \text{ (Fig. 4.19)} \qquad \text{(i)}$$

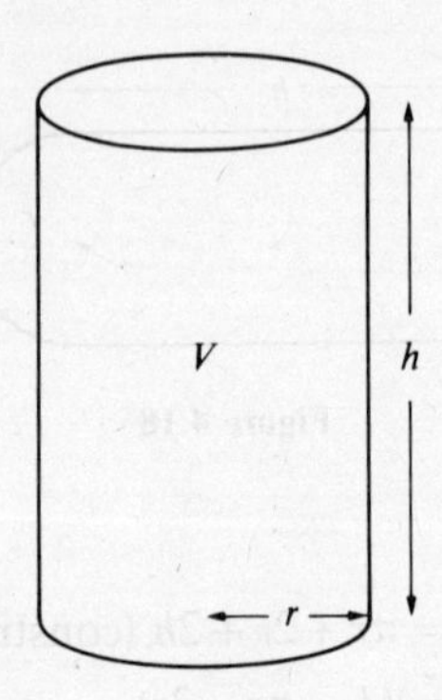

Figure 4.19

The total surface area, $\quad S = 2\pi r^2 + 2\pi rh \qquad$ (ii)

Substitute $S = 500$ in (ii):

$$2\pi rh + 2\pi r^2 = 500 \qquad \text{(iii)}$$

(iii) is the constraint equation, giving:

$$\pi rh + \pi r^2 = 250$$

$$\pi rh = 250 - \pi r^2 \qquad \text{(iv)}$$

Substitute for πrh from (iv) in (i):

$$V = r(250 - \pi r^2) = 250r - \pi r^3 \qquad \text{(v)}$$

(v) is now an equation for V, a function of a single independent variable, r, to which we may apply the principles of the chapter.
To find the turning values of V
Differentiate (v) w.r.t. r:

$$\mathrm{d}V/\mathrm{d}r = 250 - 3\pi r^2 = 0 \text{ for stationary values}$$

$$\text{i.e. } 3\pi r^2 = 250$$

$$r^2 = 250/3\pi$$

$$r = \pm\sqrt{\frac{250}{3\pi}} = \pm 5\sqrt{\frac{10}{3\pi}}$$

$$= 5.1503227 \approx 5.15\,\text{cm}$$

The negative value is ignored because it is not consistent with a real cylinder. When $r = 5\sqrt{\dfrac{10}{3\pi}}$, from (iv):

$$h = \frac{\overset{50}{\cancel{250}}}{\pi} \times \frac{1}{\cancel{5}_1}\sqrt{\frac{3\pi}{10}} - 5\sqrt{\frac{10}{3\pi}}$$

$$= 5\sqrt{\frac{30}{\pi}} - 5\sqrt{\frac{10}{3\pi}} = 5\sqrt{\frac{10}{3\pi}}\,.\,(3-1)$$

$$= 10\sqrt{\frac{10}{3\pi}} = 10.300645 \approx 10.3\,\text{cm}$$

$$V = \pi \times \frac{250}{3\pi} \times 10\sqrt{\frac{10}{3\pi}} = \frac{2500}{3}\sqrt{\frac{10}{3\pi}} = 858.38712 \approx 858.4\,\text{cm}^3$$

To prove this value is a maximum

$$\mathrm{d}^2V/\mathrm{d}r^2 = 0 - 6\pi r = -6\pi r$$

This must be negative because r is positive.

4. For a point moving in a straight line the displacement x, from a fixed point, at instant t is given by:

$$x = 3t^3 - 9t^2 - 27t + 1$$

Determine:

(a) the moments when it is stationary;
(b) the distances from 0 at those moments;
(c) the accelerations at those moments;
(d) when the acceleration is zero;
(e) the distance moved between $t = 1$ and $t = 3$;
(f) the distance moved in the 5th second.

$$x = 3t^3 - 9t^2 - 27t + 1$$

$$\mathrm{d}x/\mathrm{d}t = 9t^2 - 18t - 27 = 0 \text{ for stationary values}$$

$$\text{i.e.}\quad 9(t^2 - 2t - 3) = 0$$

$$9(t-3)(t+1) = 0$$

$$t = 3 \text{ or } -1$$

(a) It is stationary at $t = 3$ and $t = -1$.
(b) When $t = 3$, $x = 81 - 81 - 81 + 1 = -80$.
When $t = -1$, $x = -3 - 9 + 27 + 1 = 16$.
When $t = 3$, P is 80 units on the negative side of 0.
When $t = -1$, P is 16 units on the positive side of 0.

(c) When $t = 3$, $a = 18 \times 2 = 36$.
When $t = -1$, $a = 18(-2) = -36$.

(d) $a = d^2x/dt^2 = 18t - 18 = 18(t-1)$.
When $a = 0$, $18(t-1) = 0$, i.e. $t = 1$.

(e) When $t = 1$, $x = 3 - 9 - 27 + 1 = -32$.
When $t = 3$, $x = -80$.
The distance moved $= -80 - (-32) = -80 + 32 = -48$.
P moves 48 units further on the negative side of 0.

(f) The distance moved in the 5th second is the distance moved between $t = 4$ and $t = 5$.
When $t = 4$, $x = 192 - 144 - 108 + 1 = 193 - 252 = -59$.
When $t = 5$, $x = 375 - 225 - 135 + 1 = 376 - 360 = 16$.
The distance moved in the 5th second $= 16 - (-59) = 16 + 59 = 75$.

5. A source of e.m.f., E volts, with internal resistance r ohms is connected to a variable load, R ohms. Determine the power produced in the load and calculate the value of the load to maximize that power.
Suppose the current flowing through the circuit is i amperes. Then:

$$E = i(R + r)$$

$$i = \frac{E}{(R + r)}$$

The power in the external circuit is given by:

$$W = i^2 R$$

$$= \frac{E^2 . R}{(R + r)^2} \text{ watts}$$

To determine the value of R which produces maximum external power, i.e. produces the maximum power transfer, where W is a function of a single variable, R, we obtain:

$$\frac{dW}{dR} = E^2 \frac{(R + r)^2 . 1 - R . 2(R + r)}{(R + r)^4}$$

$$= E^2 \frac{(R + r) - 2R}{(R + r)^3}$$

$$= E^2 \frac{(r - R)}{(R + r)^3} = 0 \text{ for stationary values}$$

Therefore:

$$r - R = 0$$

$$r = R$$

To show that the value of R maximizes W

When $R < r, \dfrac{dW}{dR} > 0$

When $R > r, \dfrac{dW}{dR} < 0$

Therefore when $R = r$, W is a maximum.

Exercise 4.5

1. The displacement x of a moving point P from a fixed point along a line is given by $x = t^3 - 9t^2 + 15t - 11$. Determine: (a) the moments when P is stationary; (b) the corresponding distances from 0; (c) the accelerations at those moments; (d) the moment when the acceleration is zero; (e) its position when the acceleration is zero; (f) the distance moved between $t = 1$ and $t = 5$; (g) the distance moved in the 7th second.
2. The displacement x of a moving point P from a fixed point along a given line is given by $x = 3 + 11t + \frac{19}{2}t^2 - \frac{1}{2}t^3$. Determine: (a) the distance from 0 at $t = 0$; (b) the velocity and acceleration at $t = 0$; (c) the distance moved in the 4th second; (d) the distance moved between $t = 2$ and $t = 6$; (e) the moment when it is stationary (reject the negative value of t); (f) the moment when the acceleration is zero.
3. Determine the dimensions of a rectangular paved area of $50\,m^2$ with the minimum perimeter.
4. Determine the dimensions of a rectangular field to enclose the maximum area when the perimeter must be 400 m.
5. A rectangular field is enclosed between three fences and a straight river (Fig. 4.20). Determine the dimensions and minimum length of fencing to enclose an area of $80\,000\,m^2$.
6. A rectangular field similar to that in Fig. 4.20 has fencing of total length 800 m. Determine the dimensions of such a field to enclose the maximum area.
7. A hollow cylindrical vessel is made from exactly $1500\,cm^2$ of sheet metal. Determine the dimensions to enclose the maximum volume.
8. A solid cylinder of volume $1000\,cm^3$ is coated with a layer of melamine. Determine the dimensions of the cylinder when the surface area is a minimum.
9. A patio, similar in shape to Fig. 4.18, is surrounded by a low wall of total length (perimeter of area) 35 m. Determine the dimensions to produce the maximum area.

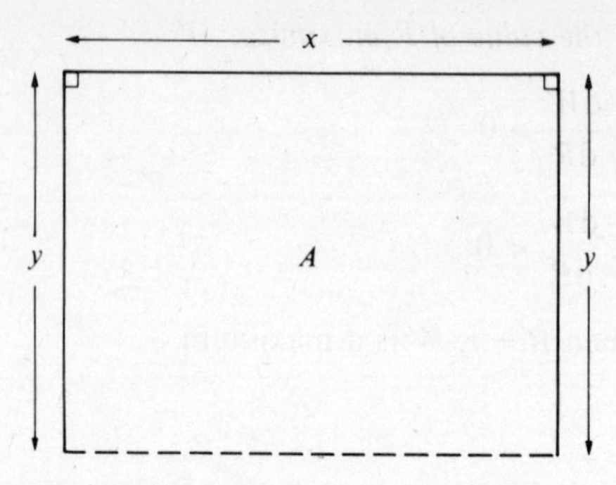

Figure 4.20

10. A patio, similar in shape to Fig. 4.18, of area $100\,\mathrm{m}^2$ is to be trimmed along the perimeter by edging stone. Determine the minimum length of edging stone required, allowing for $12\frac{1}{2}$ per cent wastage.
11. A heavy uniform beam of length 5 m is supported at both ends. The bending moment (M) at distance x m from one end is given by $M = \frac{1}{2}wx(5-x)$, where w is the weight per unit length. Determine the point of the beam where the bending moment is greatest.
12. A large, closed box-shaped container of volume $160\,\mathrm{m}^3$ is to be constructed so that the length of its base is twice its width. Calculate the dimensions of the box so that the total surface area is a minimum.
13. Paint is contained in a closed cylindrical drum. Determine the ratio of radius to height of the drum to produce the maximum volume of the container when the surface area is fixed.
14. An e.m.f. of 100 V and internal resistance r ohms supplies current to a circuit containing two resistors, R ohms and 50 Ω, in series. Determine the value of R to produce the maximum power transfer: (a) in the whole external circuit; (b) in the load, R.
15. An e.m.f. of E volts and internal resistance r ohms supplies current to a circuit containing two resistors, R ohms and A ohms, in parallel. Assuming that E, r and A are constant, determine the value of R to produce the maximum power transfer in the external circuit.
16. An alternating voltage of $300 \sin pt$ volts is applied to a constant resistance R ohms. Determine the values of t when the current is a maximum and when it is a minimum.
17. An alternating voltage of $(250 \sin 2\pi ft + 250 \cos 2\pi ft)$ volts is applied to a resistor of constant resistance 100 ohms. Determine the values of t when the current is a maximum and calculate that maximum value.
18. An amplifier is designed to give maximum power with a speaker of resistance 12.5 Ω. Determine the percentage loss in power when a speaker of resistance 6.5 Ω is used instead.

5

Integration of Trigonometric and Exponential Functions

5.1 The integral of cos x

Fig. 5.1 is a sketch of the method of drawing the graph of $\cos x$ using a rotating unit vector. The rotating vector starts at OA and rotates anticlockwise. OP is the position of the vector after rotating through angle x radians from OA. OQ is the position of the vector after rotating through angle $(x + \delta x)$ radians from OA.

P and Q project perpendicular to OA onto P' and Q' on the curve $y = \cos x$. $P'L'$ and $Q'M'$ are the ordinates of the curve $y = \cos x$. PL and QM are perpendicular to OB.

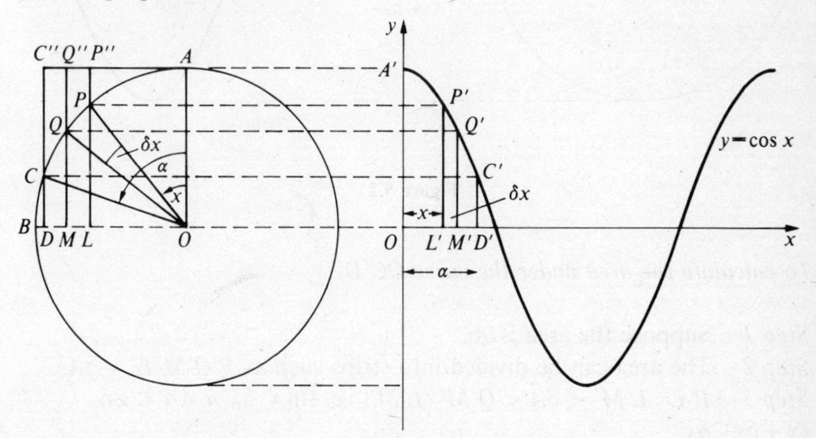

Figure 5.1

To calculate the area $OA'C'D'$

Step 1 Suppose the area is A.
Step 2 The area can be divided into strips such as $P'L'M'Q' = \delta A$.
Step 3 $Q'M'.L'M' < \delta A < P'L'.L'M'$,
i.e. $\cos(x+\delta x).\delta x < \delta A < \cos x.\delta x$.
Step 4 For small δx, $\delta A \approx \cos x.\delta x$.
Step 5 Area $P''Q''ML = ML.P''L = \delta x.\cos x \times 1 = \cos x.\delta x$.
Step 6 $\delta A = P''Q''ML$.
Step 7 By addition of the strips $A = OA'C'D' = OAC''D = OA.AC'' = 1 \times \sin\alpha$.
Step 8 Then $\int_0^\alpha \cos x.\mathrm{d}x = \sin\alpha$. This also means that $\int \cos x.\mathrm{d}x = \sin x$.

Fig. 5.2 illustrates how to determine $\int \sin x.\mathrm{d}x$. This time the rotating vector rotates from OA, in the circle on the left, in an anticlockwise direction.

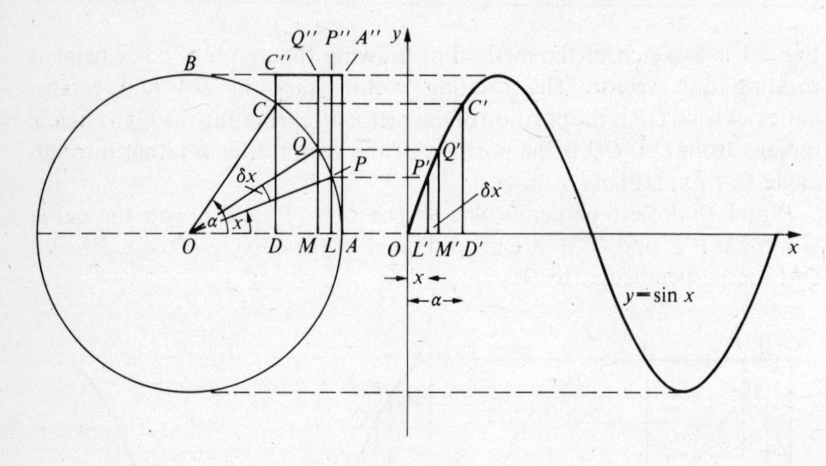

Figure 5.2

To calculate the area under the curve $OC'D'$

Step 1 Suppose the area is A.
Step 2 The area can be divided into strips such as $P'Q'M'L' = \delta A$.
Step 3 $P'L'.L'M' < \delta A < Q'M'.L'M'$, i.e. $\sin x.\delta x < \delta A < \sin (x+\delta x).\delta x$
Step 4 For small δx, $\delta A \approx \sin x.\delta x$
Step 5 Area $P''Q''ML = ML.P''L = x.\sin x \times 1 = \sin x.x$
Step 6 $\delta A = P''Q''ML$

Step 7 $A = OC'D' = AA''C''D = AA''.AD = 1 \times (1-\cos\alpha)$
Step 8 $\int_0^{\alpha} \sin x.\mathrm{d}x = 1-\cos\alpha$. This also means that $\int \sin x.\mathrm{d}x = -\cos x$.

5.2 Indefinite integrals of trigonometric and exponential functions

The list of general rules and standard forms for derivatives of special functions enables us to write down for integrals a set of rules which can be used to integrate a variety of functions. They are:

$$\int[\mathrm{f}(x)+\mathrm{g}(x)+\mathrm{h}(x)+\ldots].\mathrm{d}x = \int\mathrm{f}(x).\mathrm{d}x+\int\mathrm{g}(x).\mathrm{d}x+\int\mathrm{h}(x).\mathrm{d}x+\ldots \quad 5.1$$

$$\int a.\mathrm{f}(x).\mathrm{d}x = a\int\mathrm{f}(x).\mathrm{d}x, \text{ where } a \text{ is constant} \quad 5.2$$

$$\int x^n.\mathrm{d}x = \frac{x^{n+1}}{n+1}+c,\ n+1 \neq 0 \quad 5.3$$

$$\int\cos ax.\mathrm{d}x = \frac{1}{a}\sin ax+c \quad 5.4$$

$$\int\sin ax.\mathrm{d}x = -\frac{1}{a}\cos ax+c \quad 5.5$$

$$\int\mathrm{e}^{ax}.\mathrm{d}x = \frac{1}{a}\mathrm{e}^{ax}+c \quad 5.6$$

$$\int\frac{1}{x}.\mathrm{d}x = \ln x+c \quad 5.7$$

Examples

Determine the following indefinite integrals.

1. $$\int[5\sin 3x+6\cos\tfrac{1}{2}x+5x^2\sqrt{x}].\mathrm{d}x$$
$$= \int 5\sin 3x.\mathrm{d}x+\int 6\cos\tfrac{1}{2}x.\mathrm{d}x+\int 5x^{\frac{5}{2}}.\mathrm{d}x$$
$$= 5\int\sin 3x.\mathrm{d}x+6\int\cos\tfrac{1}{2}x.\mathrm{d}x+5\int x^{\frac{5}{2}}.\mathrm{d}x$$
$$= -\frac{5}{3}\cos 3x+\frac{6}{\frac{1}{2}}\sin\frac{1}{2}x+\frac{5}{\frac{7}{2}}x^{\frac{7}{2}}+c$$
$$= -\frac{5}{3}\cos 3x+12\sin\frac{1}{2}x+\frac{10}{7}x^3\sqrt{x}+c$$

Check: $\dfrac{d}{dx}\left[-\dfrac{5}{3}\cos 3x + 12\sin\dfrac{1}{2}x + \dfrac{10}{7}x^{\frac{7}{2}} + c\right]$

$$= -\frac{5}{3}(-3\sin 3x) + 12 \times \frac{1}{2}\cos\frac{1}{2}x + \frac{10}{7} \times \frac{7}{2}x^{\frac{5}{2}} + 0$$

$$= 5\sin 3x + 6\cos\tfrac{1}{2}x + 5x^2\sqrt{x}$$

2. $$\int \sin^2 ax\,.\,dx = \int \tfrac{1}{2}(1 - \cos 2ax)\,.\,dx$$

$$= \tfrac{1}{2}\int 1\,.\,dx - \tfrac{1}{2}\int \cos 2ax\,.\,dx$$

$$= \frac{1}{2}x - \frac{1}{2}\cdot\frac{1}{2a}\sin 2ax + c$$

$$= \frac{1}{2}x - \frac{1}{4a}\sin 2ax + c$$

Check: $\dfrac{d}{dx}\left[\dfrac{1}{2}x - \dfrac{1}{4a}\sin 2ax + c\right]$

$$= \frac{1}{2}\frac{d}{dx}(x) - \frac{1}{4a}\frac{d}{dx}(\sin 2ax) + 0$$

$$= \frac{1}{2}\,.\,1 - \frac{1}{4a}2a\,.\cos 2ax = \frac{1}{2} - \frac{1}{2}\cos 2ax$$

In this example use is made of a trigonometric identity to solve the problem. There are many cases where an integral can be solved only after a trigonometric identity is used to rearrange the integral. The following list is a sample.

$$\sin^2 ax = \tfrac{1}{2}(1 - \cos 2ax) \qquad 5.8$$

$$\cos^2 ax = \tfrac{1}{2}(1 + \cos 2ax) \qquad 5.9$$

$$\sin ax\,.\cos bx = \tfrac{1}{2}[\sin(a+b)x + \sin(a-b)x] \qquad 5.10$$

$$\sin ax\,.\sin bx = \tfrac{1}{2}[\cos(a-b)x - \cos(a+b)x] \qquad 5.11$$

$$\cos ax\,.\cos bx = \tfrac{1}{2}[\cos(a-b)x + \cos(a+b)x] \qquad 5.12$$

3. $$\int \cos^2 px\,.\,dx = \int \tfrac{1}{2}(1 + \cos 2px)\,.\,dx$$

$$= \tfrac{1}{2}\int 1\,.\,dx + \tfrac{1}{2}\int \cos 2px\,.\,dx$$

$$= \tfrac{1}{2}x + \frac{1}{4p}\sin 2px + c$$

4. $\int \sin ax \,.\, \cos bx \,.\, dx = \frac{1}{2}\int [\sin(a+b)x + \sin(a-b)x] \,.\, dx$

$$= \tfrac{1}{2}\int \sin(a+b)x \,.\, dx + \tfrac{1}{2}\int \sin(a-b)x \,.\, dx$$

$$= \frac{1}{2}\left[-\frac{1}{a+b}\cos(a+b)x\right] + \frac{1}{2}\left[-\frac{1}{a-b}\cos(a-b)x\right] + c$$

$$= -\frac{1}{2(a+b)}\cos(a+b)x - \frac{1}{2(a-b)}\cos(a-b)x + c$$

5. $\int \cos ax \,.\, \cos bx \,.\, dx = \int \frac{1}{2}[\cos(a+b)x + \cos(a-b)x] \,.\, dx$

$$= \tfrac{1}{2}\int \cos(a+b)x \,.\, dx + \tfrac{1}{2}\int \cos(a-b)x \,.\, dx$$

$$= \frac{1}{2}\frac{1}{(a+b)}\sin(a+b)x + \frac{1}{2}\frac{1}{(a-b)}\sin(a-b)x + c$$

$$= \frac{1}{2(a+b)}\sin(a+b)x + \frac{1}{2(a-b)}\sin(a-b)x + c$$

6. $\int (5e^{-x} - 6e^{-x/2}) \,.\, dx$

$$= \int 5e^{-x} \,.\, dx - \int 6e^{-x/2} \,.\, dx$$

$$= 5\int e^{-x} \,.\, dx - 6\int e^{-x/2} \,.\, dx$$

$$= 5 \,.\, \frac{e^{-x}}{-1} - 6 \,.\, \frac{e^{-x/2}}{-\frac{1}{2}} + c$$

$$= -5e^{-x} + 12e^{-x/2} + c$$

Exercise 5.1

Determine the following indefinite integrals.

1. $\int 4\cos 2x \,.\, dx$
2. $\int 6\cos \frac{1}{2}x \,.\, dx$
3. $\int \frac{4}{5}\cos \frac{2}{3}x \,.\, dx$
4. $\int 5 \,.\, \sin 2x \,.\, dx$
5. $\int 8\sin \frac{1}{2}x \,.\, dx$
6. $\int \frac{3}{4}\sin \frac{2}{3}x \,.\, dx$
7. $\int (\sin 2x + \cos 2x) \,.\, dx$
8. $\int (10\cos \frac{1}{4}x - 12\sin \frac{2}{3}x) \,.\, dx$
9. $\int p\cos ax \,.\, dx$
10. $\int q\sin ax \,.\, dx$

11. $\int(A\sin ax + B\cos bx).dx$
12. $\int(x+\cos x).dx$
13. $\int(1-\sin x).dx$
14. $\int(1-\cos 2x).dx$
15. $\int\frac{1}{2}(1+\cos 4x).dx$
16. $\int(2e^{x}+e^{2x}).dx$
17. $\int(2+3e^{-x}-4e^{-2x}).dx$
18. $\int\cos^2 x.dx$
19. $\int\sin^2 2x.dx$
20. $\int\frac{5}{4}\cos^2(\frac{2}{3}x).dx$
21. $\int\cos^2 3x.dx$
22. $\int 2.\sin 2x.\cos 4x.dx$
23. $\int 2.\cos x.\cos 2x.dx$
24. $\int\sin 4x.\cos 2x.dx$
25. $\int\sin 3x.\sin 5x.dx$
26. $\int 2e^{x}(3e^{2x}+e^{-x}).dx$
27. $\int\frac{(4e^{3x}-3e^{-2x})}{5e^{x}}dx$

5.3 Definite integrals of trigonometric and exponential functions

Examples

Evaluate the following definite integrals.

1.
$$I=\int_0^{\pi/4}(6\cos 4x-2\sin 2x).dx$$

$$I=6\int_0^{\pi/4}\cos 4x.dx-2\int_0^{\pi/4}\sin 2x.dx$$

$$=6\left[\frac{1}{4}\sin 4x\right]_0^{\pi/4}-2\left[-\frac{1}{2}\cos 2x\right]_0^{\pi/4}$$

$$=6[\tfrac{1}{4}\sin\pi-\tfrac{1}{4}\sin 0]+[\cos\pi/2-\cos 0]$$

$$=6\times(0-0)+(0-1)=-1$$

2.
$$I = \int_{-1}^{1} 4e^{3x}(3e^{2x} - 2e^{-4x}).dx$$

$$I = 12\int_{-1}^{1} e^{5x}.dx - 8\int_{-1}^{1} e^{-x}.dx$$

$$= 12\left[\frac{1}{5}e^{5x}\right]_{-1}^{1} - 8\left[-e^{-x}\right]_{-1}^{1}$$

$$= 12\left[\frac{1}{5}e^{5} - \frac{1}{5}e^{-5}\right] - 8\left[(-e^{-1}) - (-e^{1})\right]$$

$$= \frac{12}{5}(e^{5} - e^{-5}) + 8(e^{-1} - e)$$

For many purposes the answer may be left in this form. Alternatively:

$$I = \tfrac{12}{5}(148.41316 - 0.0067379) + 8(0.3678794 - 2.7182818)$$
$$= \tfrac{12}{5}(148.40642) + 8(-2.3504024)$$
$$= 356.17541 - 18.803219 = 337.37219$$

3.
$$I = \int_{0}^{\pi/10} \cos 2x . \cos 3x . dx$$

$$I = \tfrac{1}{2}\int_{0}^{\pi/10} (\cos 5x + \cos x).dx$$

$$= \tfrac{1}{2}\left[\frac{1}{5}\sin 5x + \sin x\right]_{0}^{\pi/10}$$

$$= \tfrac{1}{2}\left[\left(\frac{1}{5}\sin \pi/2 + \sin \pi/10\right) - \left(\frac{1}{5}\sin 0 + \sin 0\right)\right]$$

$$= \tfrac{1}{2}\left[\frac{1}{5} + \sin \pi/10\right] - 0$$

$$= \frac{1}{10} + \tfrac{1}{2}\sin \pi/10$$

$$\approx 0.1 + 0.5 \times 0.309017 = 0.1 + 0.1545085$$

$$= 0.2545085 \approx 0.255$$

Exercise 5.2

Evaluate the following definite integrals.

1. $\int_0^{\pi/2} \cos x \, . \, dx$
2. $\int_0^{\pi/4} \cos x \, . \, dx$
3. $\int_0^{\pi/3} \cos x \, . \, dx$
4. $\int_0^{\pi/4} \sin 2x \, . \, dx$
5. $\int_{-\pi/3}^{\pi/3} \cos x \, . \, dx$
6. $\int_0^{\pi/6} (4 \sin x + 6 \sin 2x) \, . \, dx$
7. $\int_{\pi/6}^{\pi/4} (2 \cos x - 3 \cos 2x) \, . \, dx$
8. $\int_0^{1} e^{2x} \, . \, dx$
9. $\int_0^{1} 4 \, . \, e^{-x} \, . \, dx$
10. $\int_{-1}^{1} \frac{5}{e^{2x}} dx$
11. $\int_0^{1} \frac{3e^{-x}}{4e^{x}} dx$
12. $\int_1^{2} 2e^{x}(3e^{x} + 4e^{-x}) \, . \, dx$
13. $\int_0^{1} \frac{1}{4}(x^2 + e^{-x}) \, . \, dx$
14. $\int_0^{1} (2 + 3e^{x})(4 - 5e^{x}) \, . \, dx$

15. $\int_0^{\pi/2} 2\cos^2 x\,.\,dx$

16. $\int_0^{\pi/4} \sin^2 x\,.\,dx$

17. $\int_0^{\pi/4} \frac{1}{2}\cos^2 2x\,.\,dx$

18. $\int_0^{\pi/4} \frac{2}{3}\sin^2 2x\,.\,dx$

19. $\int_0^{\pi/6} 4\sin 2x\,.\,dx$

20. $\int_0^{\pi/6} 4\sin 2x\,.\cos x\,.\,dx$

21. $\int_0^{\pi/4} (\cos^2 x+\cos x)\,.\,dx$

5.4 Mean and root mean square values

Fig. 5.3 represents the graph of $y=\sin x$ over the interval $x=0$ to $x=\pi/2$

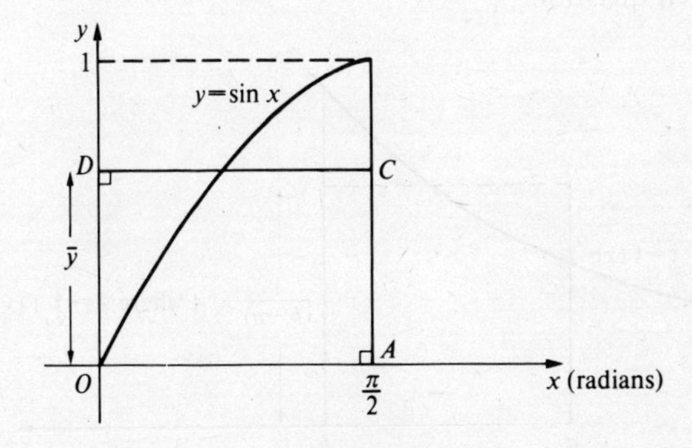

Figure 5.3

radians. The area under the graph is:

$$\int_0^{\pi/2} \sin x \, . \, dx$$

$$= \left[-\cos x \right]_0^{\pi/2}$$

$$= (-\cos \pi/2) - (-\cos 0)$$

$$= 0 - (-1) = 1$$

Suppose rectangle $OACD$ has an area exactly equal to that under the curve $y = \sin x$, where $OD = \bar{y}$. Then:

$$OD \times OA = 1$$

$$\text{i.e.} \quad \bar{y} \times \frac{\pi}{2} = 1$$

$$\bar{y} = \frac{2}{\pi}$$

$\bar{y}$ is the mean value of the curve $y = \sin x$ over the range 0 to $\pi/2$.

Definition

The mean value of $f(x)$ between $x = a$ and $x = b$, where $b > a$, is:

$$\frac{1}{b-a} \int_a^b f(x) \, . \, dx \qquad 5.13$$

Formula *5.13* represents the width of a rectangle of length $(b - a)$ whose area is equal to the area under the curve $y = f(x)$ between the limits $x = a$ and $x = b$. Fig. 5.4 illustrates the relationship between the mean of $f(x)$ and the area in question.

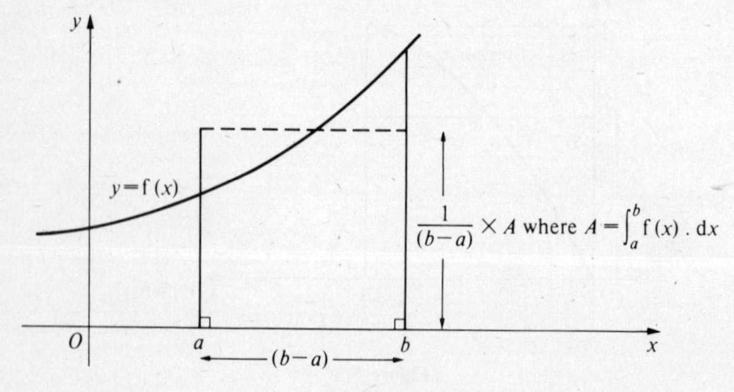

Figure 5.4

Examples

1. Calculate the mean value of the function $3x^2+2$ between $x=1$ and $x=4$.
   Mean value is:

$$\frac{1}{(4-1)}\int_1^4 (3x^2+2).\,dx$$

$$=\frac{1}{3}\Big[x^3+2x\Big]_1^4$$

$$=\frac{1}{3}\Big[(4^3+8)-(1^3+2)\Big]$$

$$=\frac{1}{3}\Big[72-3\Big]=\frac{1}{3}\times 69=23$$

2. Determine the mean value of $\tfrac{1}{4}e^{-2x}+\tfrac{2}{3}e^{-x}$ over the range $x=0$ to $x=3$.
   Mean value is:

$$\frac{1}{(3-0)}\int_0^3\left(\frac{1}{4}e^{-2x}+\frac{2}{3}e^{-x}\right).\,dx$$

$$=\frac{1}{3}\left[-\frac{1}{8}e^{-2x}-\frac{2}{3}e^{-x}\right]_0^3$$

$$=\frac{1}{3}\left[\left(-\frac{1}{8}e^{-6}-\frac{2}{3}e^{-3}\right)-\left(-\frac{1}{8}-\frac{2}{3}\right)\right]$$

$$=\frac{1}{3}\left[\frac{1}{8}+\frac{2}{3}-\frac{1}{8}e^{-6}-\frac{2}{3}e^{-3}\right]$$

$$=\frac{1}{3}\left[\frac{1}{8}(1-e^{-6})+\frac{2}{3}(1-e^{-3})\right]$$

$$=\frac{1}{24}(1-e^{-6})+\frac{2}{9}(1-e^{-3})$$

$$=0.0415634+0.2111584=0.2527218\approx 0.2527$$

3. Determine the mean value of $3\sin 2x+5\cos 4x$ over the range $x=0$ to $x=\pi/4$.

Mean value is:

$$\frac{1}{(\pi/4-0)}\int_0^{\pi/4}(3\sin 2x+5\cos 4x).\mathrm{d}x$$

$$=\frac{4}{\pi}\left[-\frac{3}{2}\cos 2x+\frac{5}{4}\sin 4x\right]_0^{\pi/4}$$

$$=\frac{4}{\pi}\left[\left(-\frac{3}{2}\cos\pi/2+\frac{5}{4}\sin\pi\right)-\left(-\frac{3}{2}\cos 0+\frac{5}{4}\sin 0\right)\right]$$

$$=\frac{4}{\pi}[0-(-3/2)]$$

$$=\frac{4}{\pi}\times\frac{3}{2}$$

$$=1.9098593\approx 1.910$$

Exercise 5.3

Determine the mean values of the following functions over the ranges indicated.

1. $\cos x$: $x=0$ to $x=\pi/2$
2. $(\cos x+\sin x)$: $x=0$ to $x=\pi/2$
3. $(4\cos x+3\sin x)$: $x=0$ to $x=\pi/2$
4. $(a\cos x+b\sin x)$: $x=0$ to $x=\pi/2$
5. $(a\cos x+b\sin x)$: $x=\pi/6$ to $x=\pi/3$
6. x^2+x: $x=0$ to $x=5$
7. $(x-2)^2$: $x=0$ to $x=4$
8. $1/x$: $x=1$ to $x=10$
9. $3/x$: $x=1$ to $x=2$
10. $4/5x$: $x=4$ to $x=8$
11. e^x: $x=0$ to $x=2$
12. e^{-x}: $x=0$ to $x=2$
13. $\dfrac{\mathrm{e}^x+\mathrm{e}^{-x}}{2}$: $x=0$ to $x=1$
14. $\sin 2x$: $x=0$ to $x=\pi/4$
15. $a\cos bx$: $x=0$ to $x=\pi/4b$
16. $\cos 2x$: $x=0$ to $x=\pi/2$
17. $p\sin qx$: $x=\pi/6q$ to $x=\pi/3q$

Root mean square values

The use of alternating currents and voltages is fundamental to many aspects of electrical technology. A concept called the root mean square value arises naturally as a result of certain aspects of these variable quantities.

Definition: The root mean square current of an alternating current is defined as follows. During one complete cycle of an alternating current which flows through a given resistor, energy is consumed. The direct current flowing through the same resistor which dissipates the same amount of energy in the same time is defined as the root mean square value of the current. In a similar way the root mean square voltage may be defined.

The concept of the root mean square value applies particularly to periodic functions, i.e. those which go through the same cycle of values over successive periods of the independent variable. Suppose that $f(x)$ is a periodic function of period from $x = a$ to $x = b$. The root mean square value of $f(x)$ is:

$$\sqrt{\frac{1}{(b-a)}\int_a^b [\,f(x)]^2 \, . \, dx} \qquad 5.14$$

Before we obtain root mean square values by integration methods it is instructive to look at a purely graphical approach. In Fig. 5.5 the continuous curve is $i = I_m \sin \omega t$, where i is the instantaneous value of the alternating current, I_m is the peak value of that current and ω is related to the period of the current. The dotted curve represents the curve obtained by squaring the corresponding values of i. This dotted curve is a sine curve whose minimum values lie along the t axis. The peak values have a value I_m^2 along the i axis. By the symmetry of the curves the shaded area is one half of the area of the

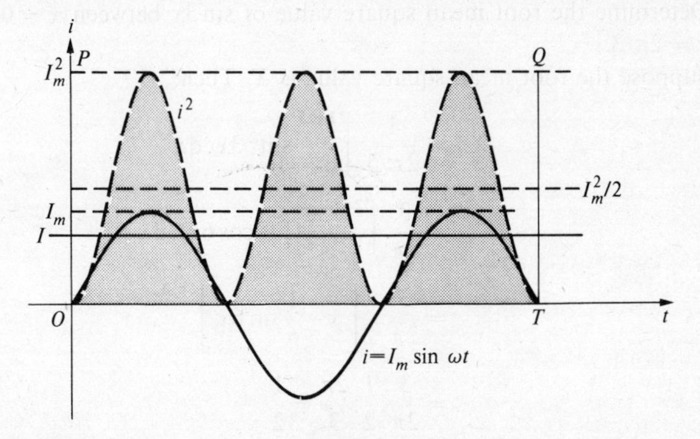

Figure 5.5

rectangle $OPQT$. That is, the shaded area $= (I_m^2/2) \times T$. Suppose the root mean square current is I. Then:

$$I^2 T = (I_m^2/2) \times T$$

$$\text{i.e.} \quad I = I_m/\sqrt{2}$$

Examples

1. Determine the root mean square value of $I_m \sin \omega t$ over the period $t = 0$ to $t = 2\pi/\omega$.
   Then, if I is the root mean square value:

$$I^2 = \frac{1}{2\pi/\omega} I_m^2 \int_0^{2\pi/\omega} \sin^2 \omega t \,.\, \mathrm{d}t$$

$$= \frac{\omega}{2\pi} I_m^2 \int_0^{2\pi/\omega} \tfrac{1}{2}(1 - \cos 2\omega t) \,.\, \mathrm{d}t$$

$$= \frac{\omega}{2\pi} I_m^2 \left[\tfrac{1}{2}t - \tfrac{1}{4\omega} \sin 2\omega t\right]_0^{2\pi/\omega}$$

$$= I_m^2 \frac{\omega}{2\pi} \times \frac{1}{2} \times \left[\frac{2\pi}{\omega} - 0\right]$$

$$I^2 = I_m^2 \frac{1}{2} \frac{\omega}{2\pi} \times \frac{2\pi}{\omega}$$

$$I = \frac{1}{\sqrt{2}} I_m$$

2. Determine the root mean square value of $\sin 3x$ between $x = 0$ and $x = 2\pi/3$.
   Suppose the root mean square value is X. Then:

$$X^2 = \frac{1}{2\pi/3} \int_0^{2\pi/3} \sin^2 3x \,.\, \mathrm{d}x$$

$$= \frac{3}{2\pi} \int_0^{2\pi/3} \tfrac{1}{2}(1 - \cos 6x) \,.\, \mathrm{d}x$$

$$= \frac{3}{2\pi} \cdot \frac{1}{2} \left[x - \frac{1}{6} \sin 6x\right]_0^{2\pi/3}$$

$$= \frac{3}{2\pi} \cdot \frac{1}{2} \cdot \frac{2\pi}{3} = \frac{1}{2}$$

$$X = 1/\sqrt{2}$$

3. Determine the root mean square value of $a \cos 4x$ between $x = 0$ and $x = \pi/4$.
Suppose the root mean square value of the function is A. Then:

$$A^2 = \frac{1}{\pi/4}\int_0^{\pi/4} a^2 \cos^2 4x \, . \, dx$$

$$= \frac{4}{\pi} . \frac{a^2}{2}\int_0^{\pi/4} (1 + \cos 8x) . dx$$

$$= \frac{4}{\pi} . \frac{a^2}{2}\left[x + \frac{1}{8}\sin 8x\right]_0^{\pi/4}$$

$$= \frac{4}{\pi} . \frac{a^2}{2} . \frac{\pi}{4}$$

$$= a^2/2$$

$$A = a/\sqrt{2}$$

Exercise 5.4

Determine the root mean square values of the following functions over the stated intervals.

1. $\sin x$: $x = 0$ to $x = \pi$
2. $\cos x$: $x = 0$ to $x = 2\pi$
3. $3 \sin 2x$: $x = 0$ to $x = \pi$
4. $5 \sin \frac{1}{2}x$: $x = 0$ to $x = 2\pi$
5. $6 \sin \frac{3}{4}x$: $x = 0$ to $x = 8\pi/3$
6. $2 \cos 3x$: $x = 0$ to $x = \pi/3$
7. $I_m \sin \omega t$: $t = 0$ to $t = \pi/\omega$
8. $I_m \cos \omega t$: $t = 0$ to $t = 2\pi/\omega$
9. $I_m \sin 2\omega t$: $t = 0$ to $t = 2\pi/\omega$
10. $V_m \sin 2\pi ft$: $t = 0$ to $t = 1/f$

6

The Use of Numerical Integration Formulae for the Evaluation of Definite Integrals

Numerical integration methods were introduced in Chapter 2 of Mensuration 2. The first of these is the trapezoidal rule or the trapezium rule.

6.1 The trapezium rule

Fig. 6.1 represents a curve $y = f(x)$ and the area $PQRS$ between that curve, the x axis and the ordinates $x = a$ and $x = b$. That area is divided into five strips of equal width, h, where $h = \frac{1}{5}(b - a)$. The ordinates dividing the area are FF', GG', HH' and KK'. The straight lines PF, FG, GH, HK and KQ are

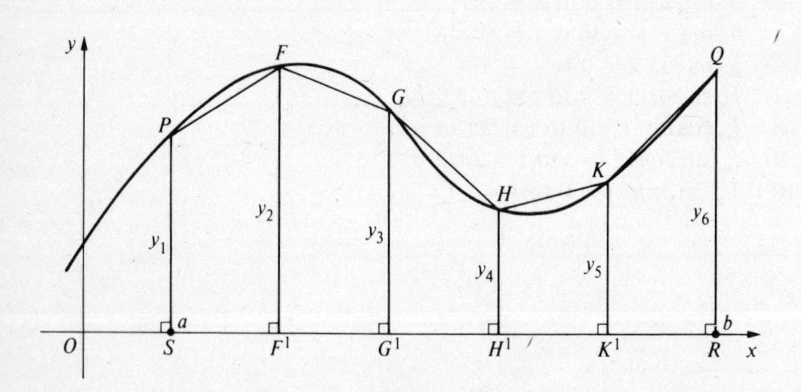

Figure 6.1

drawn. The area $PQRS$ is approximately equal to the sum of the areas of the trapezia $PFF'S$, $FGG'F'$, etc., i.e.

$$\int_a^b f(x).dx \approx \tfrac{1}{2}(y_1+y_2)h+\tfrac{1}{2}(y_2+y_3)h+\tfrac{1}{2}(y_3+y_4)h$$

$$+\tfrac{1}{2}(y_4+y_5)h+\tfrac{1}{2}(y_5+y_6)h$$

$$=\tfrac{1}{2}h[(y_1+y_6)+2(y_2+y_3+y_4+y_5)]$$

In general the area under the curve $y = f(x)$ between $x = a$ and $x = b$ is given by:

$$\int_a^b f(x).dx \approx \tfrac{1}{2}h[(y_1+y_{n+1})+2(y_2+y_3+y_4+\ldots+y_{n-1}+y_n)] \qquad 6.1$$

when the number of strips is n. In *6.1*:

$$h=(b-a)/n$$

This is the trapezium rule.

6.2 The mid ordinate rule

In Fig. 6.2 the area between $y = f(x)$, Ox and the ordinates $x = a$ and $x = b$ is divided into four strips of equal width, h. The mid ordinates of those strips have lengths Y_1, Y_2, Y_3 and Y_4. The area under the curve is approximately equal to the sum of four rectangular strips of lengths Y_1, Y_2, Y_3 and Y_4 and of width h,

$$\text{i.e.} \quad \int_a^b f(x).dx \approx h(Y_1+Y_2+Y_3+Y_4)$$

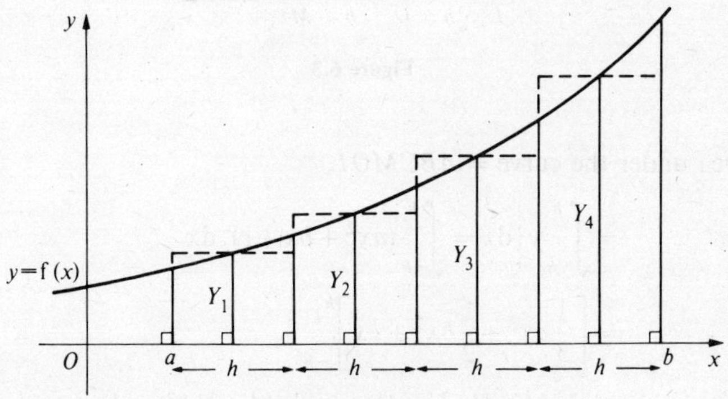

Figure 6.2

where $h = (b-a)/4$. When the number of strips is n and the mid ordinates are $Y_1, Y_2, Y_3, \ldots, Y_{n-1}, Y_n$, then:

$$\int_a^b f(x).dx \approx h(Y_1 + Y_2 + Y_3 + \ldots + Y_{n-1} + Y_n) \qquad 6.2$$

where $h = (b-a)/n$. This is the mid ordinate rule.

6.3 Simpson's rule

Simpson's rule is derived from Fig. 6.3. A, B and C represent three points of a curve $y = f(x)$ such that their co-ordinates are $(-h, y_1)$, $(0, y_2)$, (h, y_3). The three points A, B and C are assumed to lie on a parabola whose equation is of the form:

$$y = ax^2 + bx + c \qquad \text{(i)}$$

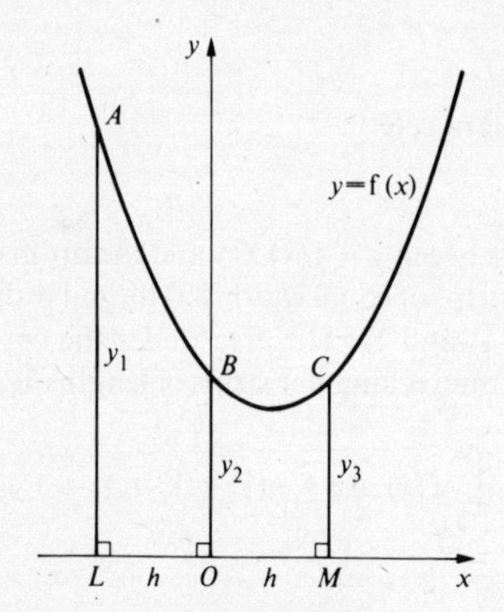

Figure 6.3

The area under the curve $= ABCMOL$

$$= \int_{-h}^{h} y.dx = \int_{-h}^{h} (ax^2 + bx + c).dx$$

$$= \left[\frac{1}{3}ax^3 + \frac{1}{2}bx^2 + cx\right]_{-h}^{h}$$

$$= (\tfrac{1}{3}ah^3 + \tfrac{1}{2}bh^2 + ch) - (-\tfrac{1}{3}ah^3 + \tfrac{1}{2}bh^2 - ch)$$

$$= \tfrac{2}{3}ah^3 + 2ch \qquad \text{(ii)}$$

The values of y_1, y_2 and y_3 are obtained by substituting in (i) the values $-h$, 0 and h respectively for x. Therefore:

$$y_1 = ah^2 - bh + c \qquad \text{(iii)}$$

$$y_2 = c \qquad \text{(iv)}$$

$$y_3 = ah^2 + bh + c \qquad \text{(v)}$$

(iii) + (v) gives:

$$y_1 + y_3 = 2(ah^2 + c)$$

From (ii):

$$\begin{aligned} \text{Area} &= \tfrac{2}{3}ah^3 + \tfrac{2}{3}ch + \tfrac{4}{3}ch \\ &= \tfrac{1}{3}(y_1 + y_3)h + \tfrac{4}{3}y_2 h \\ &= \tfrac{1}{3}h[y_1 + 4y_2 + y_3] \qquad 6.3 \end{aligned}$$

Formula *6.3* is the basic form of Simpson's rule for an area which is subdivided into two strips of equal width h.

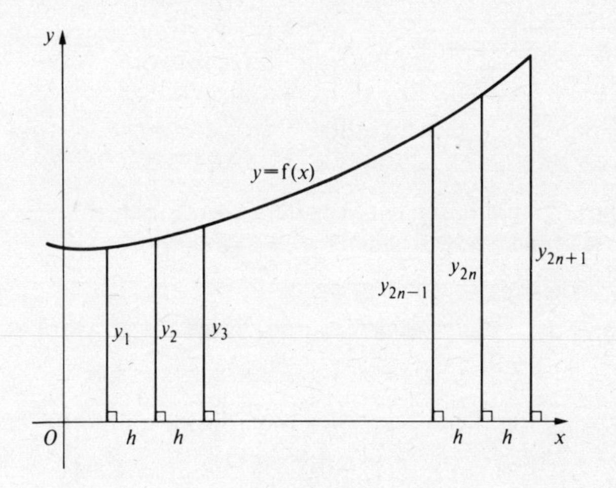

Figure 6.4

Extension of Simpson's rule to areas divided into 2n strips of equal width

From Fig. 6.4:

$$\begin{aligned} \text{Area} &\approx \tfrac{1}{3}h[(y_1 + 4y_2 + y_3) + (y_3 + 4y_4 + y_5) + \ldots \\ &\quad + (y_{2n-3} + 4y_{2n-2} + y_{2n-1}) + (y_{2n-1} + 4y_{2n} + y_{2n+1})] \\ &= \tfrac{1}{3}h[y_1 + 4y_2 + 2y_3 + 4y_4 + 2y_5 + \ldots + 4y_{2n-2} \\ &\quad + 2y_{2n-1} + 4y_{2n} + y_{2n+1}] \end{aligned}$$

$$= \tfrac{1}{3}h[(y_1 + y_{2n+1}) + 4(y_2 + y_4 + \ldots + y_{2n}) + 2(y_3 + y_5 + \ldots + y_{2n-1})] \qquad 6.4a$$

$$= \tfrac{1}{3}h[(\text{first} + \text{last}) + 4 \times \text{sum of even ordinates} + 2 \times \text{sum of intermediate odd ordinates}] \qquad 6.4b$$

$$\text{or Area} = \tfrac{1}{3}h[X + 4E + 2O] \qquad 6.4c$$

where X = first + last ordinates, E = sum of even ordinates and O = sum of intermediate odd ordinates.

Examples

1. Evaluate the area under the curve $y = 10e^{-x}$ between $x = 0$ and $x = 6$ by: (a) the mid ordinate rule; (b) the trapezium rule; and (c) Simpson's rule. Take six strips each of width 1 unit. Then:

$$y_1 = 10e^{-0} = 10$$
$$y_2 = 10e^{-1} = 3.6787944$$
$$y_3 = 10e^{-2} = 1.3533528$$
$$y_4 = 10e^{-3} = 0.4978707$$
$$y_5 = 10e^{-4} = 0.1831564$$
$$y_6 = 10e^{-5} = 0.0673795$$
$$y_7 = 10e^{-6} = 0.0247875$$

(a) For the mid ordinate rule assume there are three strips of width $h = 2$ divided by ordinates y_1, y_3, y_5 and y_7.

$$\text{The area} \approx 2[y_2 + y_4 + y_6]$$
$$= 2[3.6787944 + 0.4978707 + 0.0673795]$$
$$= 2 \times 4.2440446 = 8.4880892 \approx 8.49$$

(b) By the trapezium rule, using six strips of width $h = 1$:

$$\text{Area} \approx \tfrac{1}{2} \times 1[(10 + 0.0247875) + 2(3.6787944 + \ldots + 0.0673795)]$$
$$= \tfrac{1}{2}[10.024788 + 2 \times 5.7794538]$$
$$= \tfrac{1}{2} \times 21.583696 = 10.791848$$

(c) By Simpson's rule, using six strips of width $h = 1$:

$$\text{Area} \approx \tfrac{1}{3} \times 1[(10 + 0.0247875) + 4(3.6787944 + 0.4978707 + 0.0673795) + 2(1.3533528 + 0.1831564)]$$
$$= \tfrac{1}{3}[10.024788 + 16.976178 + 3.0730184]$$
$$= 10.024661$$

By direct integration:

$$\text{Area} = \int_0^6 10e^{-x}.dx = -\left[10e^{-x}\right]_0^6$$

$$= (-10e^{-6}) - (-10)$$

$$= 10 - 10e^{-6} = 10(1 - e^{-6})$$

$$= 10(1 - 0.0024788) = 9.9752125$$

The percentage error in Simpson's rule is:

$$\frac{10.024661 - 9.9752125}{9.9752125} \times 100$$

$$= 0.495714 \approx 0.5$$

2. Evaluate the area of example 1 by Simpson's rule using 12 strips of equal width $= \frac{1}{2}$.
The ordinates are to be calculated at $x = 0, \frac{1}{2}, 1, 1\frac{1}{2}, \ldots 6$.

$y_1 =$	10		
$y_2 =$		6.0653066	
$y_3 =$			3.6787944
$y_4 =$		2.2313016	
$y_5 =$			1.3533528
$y_6 =$		0.82085	
$y_7 =$			0.4978707
$y_8 =$		0.3019738	
$y_9 =$			0.1831564
$y_{10} =$		0.11109	
$y_{11} =$			0.0673795
$y_{12} =$		0.0408677	
$y_{13} =$	0.0247875		
	10.024788	9.5713897	5.7805538

$$\text{Area} \approx \tfrac{1}{3} \times \tfrac{1}{2}[10.024788 + 4 \times 9.5713897 + 2 \times 5.7805538]$$

$$= \tfrac{1}{6}[10.024788 + 38.285559 + 11.561108]$$

$$= \tfrac{1}{6} \times 59.871454 = 9.9785757$$

Percentage error $= 0.0337156 \approx 0.03$.

3. Using Simpson's rule determine an approximate value of $\int_0^{\pi/2} 5 \sin x\,.dx$: (a) using four strips of equal width; and (b) using eight strips of equal width. In each case calculate the percentage error.

(a)

$y_1 = 5\sin 0$	$= 0$		
$y_2 = 5\sin \pi/8$	$=$	1.9134172	
$y_3 = 5\sin \pi/4$	$=$		3.5355339
$y_4 = 5\sin 3\pi/8$	$=$	4.6193977	
$y_5 = 5\sin \pi/2$	$= 5$		
	5	6.5328149	3.5355339

$$\text{Area} \approx \frac{1}{3} \times \frac{\pi}{8}[5 + 4 \times 6.5328149 + 2 \times 3.5355339]$$

$$= \frac{1}{3} \times \frac{\pi}{8}[5 + 26.13126 + 7.0710678]$$

$$= \frac{1}{3} \times \frac{\pi}{8} \times 38.202328$$

$$= 5.000673$$

By integration:

$$\text{Area} = \Big[-5\cos x\Big]_0^{\pi/2} = (-5 \times 0) - (-5 \times 1)$$

$$= 5$$

Percentage error = 0.0134585

(b)

$y_1 = 5\sin 0$	$= 0$		
$y_2 = 5\sin \pi/16$	$=$	0.9754516	
$y_3 = 5\sin \pi/8$	$=$		1.9134172
$y_4 = 5\sin 3\pi/16$	$=$	2.7778512	
$y_5 = 5\sin \pi/4$	$=$		3.5355339
$y_6 = 5\sin 5\pi/16$	$=$	4.1573481	
$y_7 = 5\sin 3\pi/8$	$=$		4.6193977
$y_8 = 5\sin 7\pi/16$	$=$	4.9039264	
$y_9 = 5\sin \pi/2$	$= 5$		
	5	12.814577	10.068349

$$\text{Area} \approx \frac{1}{3} \times \frac{\pi}{16}[5 + 4 \times 12.814577 + 2 \times 10.068349]$$

$$= \frac{1}{3} \times \frac{\pi}{16}[5 + 51.258309 + 20.136697]$$

$$= \frac{1}{3} \times \frac{\pi}{16} \times 76.395006$$

$$= 5.0000415$$

Percentage error = 0.00083

4. Using Simpson's rule determine an approximate value of $\int_2^{10} (x^2 + x + 1/x)\,.\,dx$: (a) using two strips of width 4; (b) using four strips of width 2; and (c) using eight strips of width 1.

(a)

when $x = 2$,	$y_1 = 2^2 + 2 + 1/2$	$= 6.5$	
when $x = 6$,	$y_2 = 6^2 + 6 + 1/6$	$=$	42.166667
when $x = 10$,	$y_3 = 10^2 + 10 + 1/10$	$= 110.1$	
		116.6	42.166667

$$\text{Area} \approx \tfrac{1}{3} \times 4[116.6 + 4 \times 42.166667]$$
$$= \tfrac{4}{3}[116.6 + 168.66667]$$
$$= \tfrac{4}{3} \times 285.26667 = 380.35556$$

(b)

when $x = 2$, $y_1 =$	6.5		
when $x = 4$, $y_2 =$		20.25	
when $x = 6$, $y_3 =$			42.166667
when $x = 8$, $y_4 =$		72.125	
when $x = 10$, $y_5 =$	110.1		
	116.6	92.375	42.166667

$$\text{Area} \approx \tfrac{1}{3} \times 2[116.6 + 4 \times 92.375 + 2 \times 42.166667]$$
$$= \tfrac{2}{3}[116.6 + 369.5 + 84.333334]$$
$$= \tfrac{2}{3} \times 570.43333 = 380.2889$$

(c)

when $x = 2$, $y_1 =$	6.5		
when $x = 3$, $y_2 =$		12.333333	
when $x = 4$, $y_3 =$			20.25
when $x = 5$, $y_4 =$		30.2	
when $x = 6$, $y_5 =$			42.166667
when $x = 7$, $y_6 =$		56.142857	
when $x = 8$, $y_7 =$			72.125
when $x = 9$, $y_8 =$		90.111111	
when $x = 10$, $y_9 =$	110.1		
	116.6	188.787301	134.541667

$$\text{Area} \approx \tfrac{1}{3} \times 1[116.6 + 4 \times 188.787301 + 2 \times 134.541667]$$
$$= \tfrac{1}{3} \times [116.6 + 755.1492 + 269.083324]$$
$$= \tfrac{1}{3} \times 1140.8325 = 380.27751$$

The integral is

$$\int_2^{10} (x^2 + x + 1/x).\mathrm{d}x$$

$$= \left[\frac{1}{3}x^3 + \frac{1}{2}x^2 + \ln x\right]_2^{10}$$

$$= \left(\frac{1}{3} \times 1000 + \frac{1}{2} \times 100 + \ln 10\right) - \left(\frac{1}{3} \times 8 + \frac{1}{2} \times 4 + \ln 2\right)$$

$$= \frac{1}{3}(1000 - 8) + \frac{1}{2}(100 - 4) + \ln 10 - \ln 2$$

$$= \frac{1}{3} \times 992 + \frac{1}{2} \times 96 + \ln 5$$

$$= 330\frac{2}{3} + 48 + \ln 5$$

$$= 378\frac{2}{3} + \ln 5 = 380.2761$$

The percentage error in (a) = 0.0208941 ≈ 0.021.
The percentage error in (b) = 0.0033633 ≈ 0.0034.
The percentage error in (c) = 0.0003708 ≈ 0.00037.

5. By Simpson's rule evaluate $\int_0^1 \frac{1}{1+x^2}.\mathrm{d}x$: (a) using two strips; (b) using four strips; and (c) using six strips. In each case determine the percentage error.

(a)

when $x = 0$, $y_1 = 1/1 = 1$		
when $x = 1/2$, $y_2 =$		0.8
when $x = 1$, $y_3 =$	0.5	
	1.5	0.8

$$\text{Area} \approx \tfrac{1}{3} \times \tfrac{1}{2}[1.5 + 4 \times 0.8]$$

$$= \tfrac{1}{6} \times [1.5 + 3.2] = \tfrac{1}{6} \times 4.7 = 0.7833333$$

(b)

when $x = 0$, $y_1 = 1$			
when $x = \frac{1}{4}$, $y_2 =$		0.9411765	
when $x = \frac{1}{2}$, $y_3 =$			0.8
when $x = \frac{3}{4}$, $y_4 =$		0.64	
when $x = 1$, $y_5 = 0.5$			
	1.5	1.5811765	0.8

$$\text{Area} \approx \tfrac{1}{3} \times \tfrac{1}{4}[1.5 + 4 \times 1.5811765 + 2 \times 0.8]$$
$$= \tfrac{1}{12} \times [1.5 + 6.324706 + 1.6]$$
$$= \tfrac{1}{12} \times 9.424706 = 0.7853922$$

(c)

when $x = 0$,	$y_1 = 1$		
when $x = \frac{1}{6}$,	$y_2 =$	0.972973	
when $x = \frac{1}{3}$,	$y_3 =$		0.9
when $x = \frac{1}{2}$,	$y_4 =$	0.8	
when $x = \frac{2}{3}$,	$y_5 =$		0.6923077
when $x = \frac{5}{6}$,	$y_6 =$	0.5901633	
when $x = 1$,	$y_7 = 0.5$		
	1.5	2.3631369	1.5923077

$$\text{Area} \approx \tfrac{1}{3} \times \tfrac{1}{6}[1.5 + 4 \times 2.3631369 + 2 \times 1.5923077]$$
$$= \tfrac{1}{18}[1.5 + 9.4525476 + 3.1846154]$$
$$= \tfrac{1}{18} \times 14.137163 = 0.7853979$$

The integral is:

$$\int_0^1 \frac{1}{1+x^2}\,dx = \Big[\text{arc tan}\,x\Big]_0^1$$
$$= \text{arc tan}\,(1) - \text{arc tan}\,(0) = \pi/4$$
$$= 0.7853982.$$

Note: this integral does not come within the scope of those included in this syllabus. Simpson's rule provides one method of obtaining an approximate value of such an integral.
The percentage error in (a) = 0.2629112.
The percentage error in (b) = 0.0007639.
The percentage error in (c) = 0.0000382.

6.4 Errors in results obtained by Simpson's rule

By reference to the results obtained in the examples above it is obvious that there is no uniform pattern of error for all functions. In fact with some functions Simpson's rule produces an answer which is exactly the integral. It is clear that the greater the number of strips taken in the application of the rule the higher the degree of accuracy. It is also clear that as the number of strips increases the more complicated some of the calculations become.

A compromise has to be reached to relate accuracy to the difficulty and length of the calculation process. For a strip of width one unit it is unusual for the error to be greater than 1 per cent, and it is often much less than that.

When the number of strips is doubled the percentage error is usually reduced to at least one tenth of its previous value. Consequently, the following is a very rough guide:

1. To provide an answer with approximately 1 per cent error take a strip width of one unit.
2. To reduce a given error to about 1/10 of its value double the number of strips.

Exercise 6.1

Using Simpson's rule, determine approximate values of the following integrals. In each case calculate the percentage error in the result.

1. $\int_0^2 \frac{1}{10}e^x.dx$, using two strips
2. $\int_0^{\pi/2} 10\cos x.dx$, (a) using two strips, (b) using four strips
3. $\int_0^{\pi} 10\sin x.dx$, (a) using four strips, (b) using six strips
4. $\int_2^{10} (x^2 - x + 1/x).dx$, (a) using two strips, (b) using four strips, (c) using eight strips
5. $\int_{-2}^{2} \frac{1}{2}(e^x + e^{-x}).dx$, (a) using two strips, (b) using four strips
6. $\int_0^{\pi/2} x.\sin x.dx$, (a) using two strips, (b) using four strips. Assume the real value is 1.
7. $\int_0^{\pi} x.\sin x.dx$, using six strips. Assume the real value is π.
8. $\int_0^2 x.e^x.dx$, (a) using two strips, (b) using four strips. Assume the real value is $e^2 + 1$.
9. $\int_0^{\pi/2} (3\sin x + 4\cos x).dx$, (a) using two strips, (b) using four strips
10. $\int_2^{14} \ln x.dx$, (a) using four strips, (b) using six strips, (c) using 12 strips. Assume the real value = 23.560508.
11. Evaluate $\int_0^8 10.e^{-2x}.dx$ to an approximate degree of accuracy of 0.1 per cent error
12. Evaluate $\int_0^{\pi/2} 10\cos x.dx$ to an approximate degree of accuracy of 0.01 per cent error

13. Evaluate $\int_{2}^{10} (2x^2 - 3x + 5/x) \, .dx$ to an approximate degree of accuracy of 0.001 per cent error

14 Evaluate $\int_{0}^{2} \frac{1}{1+x^2} \, .dx$ to an approximate degree of accuracy of 0.0001 per cent error

7

The Solution of Differential Equations of the Type $\dfrac{dy}{dx} = f(x)$

At Level 2 and in previous chapters of the present book we have dealt at some length with the operation of differentiation, that is, finding the derived function of a given function, or, given $f(x)$, to find $f'(x)$. The operation which is the inverse process, i.e. integration, has also been introduced.

When given $$dy/dx = 2x \qquad 7.1$$

we determine $$y = x^2 + c$$

Because $$y = \int 2x.dx = 2\int x.dx = 2 \times \tfrac{1}{2}x^2 + c = x^2 + c \qquad 7.2$$

7.1 is a differential equation. It describes a relation between the derivative of a variable, y, w.r.t. an independent variable x, and x itself.

7.2 is the result of carrying out an integration on *7.1* and produces an equation which relates y to x. What does *7.2* represent, and how is it connected with *7.1*? For different values of c, *7.2* represents a family of curves, all parallel to $y = x^2$. Some of these curves are sketched in Fig. 7.1. The family is the general solution of *7.1*. Note that for a given value of x the gradients of all the curves are the same (the tangents are parallel, the derivatives have the same values).

In an earlier chapter integration was described as indefinite integration when it occurs in the form of changing *7.1* into *7.2*. It specifies a whole family of curves without picking out any particular one.

Through any point in the plane there cannot be more than one curve of the family. If there were two, for instance, they would have to cross each other at that point, and so would have different gradients there, and we have seen that, for a given value of x, all curves must be parallel. This means, then, that for any point in the plane there is just one curve of the family which

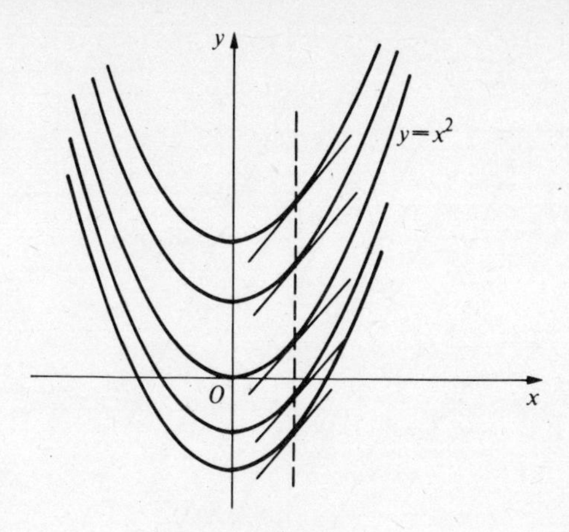

Figure 7.1

passes through it. Consequently, when we wish to pick out a particular curve of the family we merely have to select one of the points through which it passes. For instance, suppose we wish to select the member of the family $y = x^2 + c$ which passes through the point $(2, -5)$. In other words, the machine $y = x^2 + c$, which determines the values of y when we are given values of x, tells us this value when $x = 2$.

$$y = 2^2 + c = 4 + c, \text{ and this value must be } -5$$

Therefore:

$$4 + c = -5$$
$$c = -9$$

The particular curve of the family which passes through $(2, -5)$ is $y = x^2 - 9$. It is a particular solution of *7.1*.

Examples

1. Sketch a few members of the family of curves denoted by $dy/dx = -2$ and determine the equation of the particular curve which passes through $(3, -7)$.
   Step 1 The differential equation is $dy/dx = -2$.
   Step 2 Integrate w.r.t. x: $y = -2x + c$.
   Step 3 The family of curves represented by $y = -2x + c$ is a set of parallel lines of gradient -2.

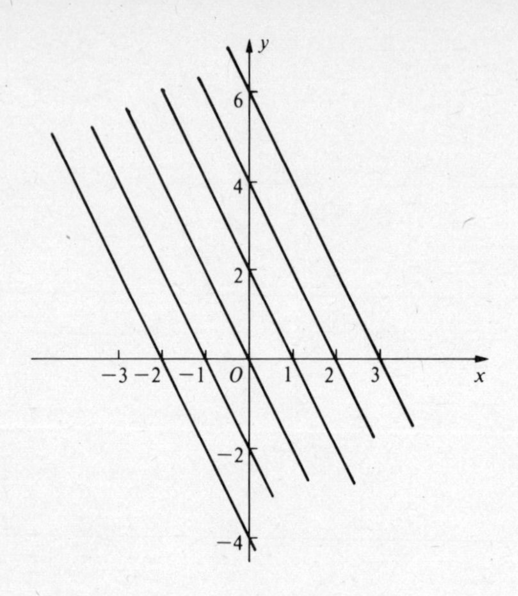

Figure 7.2

Step 4 Fig. 7.2 is a sketch of the family.
Step 5 The equation of the line which passes through $(3, -7)$ is satisfied by those co-ordinates, i.e.

$$-7 = -2 \times 3 + c$$
$$-7 = -6 + c$$
$$-1 = c$$

The member of the family required is $y = -2x - 1$.

2. Determine the family of curves represented by $dy/dx = 3x^2 - 12x + 5$. Sketch a few members of the family and determine the equation of the particular curve which passes through $(4, 17)$.
Step 1 The differential equation is $dy/dx = 3x^2 - 12x + 5$.
Step 2 Integrate w.r.t. x: $y = x^3 - 6x^2 + 5x + c$. (i)
Step 3 The family represented by (i) is a set of cubic curves parallel to one another.
Step 4 Fig. 7.3 is a sketch of the family.
Step 5 One member $(y = x^3 - 6x^2 + 5x)$ intersects Ox at $x = 0, 1, 5$.
Step 6 The equation of the curve which passes through $(4, 17)$ is

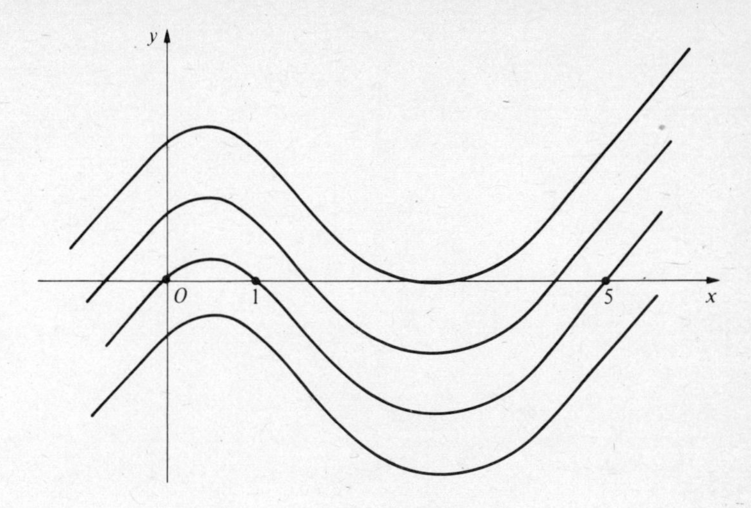

Figure 7.3

satisfied by those co-ordinates, i.e.

$$17 = 64 - 96 + 20 + c$$
$$17 = c - 12$$
$$29 = c$$

The equation of the required curve is $y = x^3 - 6x^2 + 5x + 29$.

3. Integrate the differential equation $\dfrac{dy}{dx} = 3e^{-x} + 4e^{-2x}$.

Then:

$$y = \int (3e^{-x} + 4e^{-2x}).dx$$
$$= 3\int e^{-x}.dx + 4\int e^{-2x}.dx$$
$$= -3e^{-x} - \tfrac{4}{2}e^{-2x} + c$$
$$y = -3e^{-x} - 2e^{-2x} + c$$

To determine the particular member of the family which passes through the point $(-1, 4)$:

$$4 = -3e^1 - 2e^2 + c$$
$$c = 4 + 3e + 2e^2$$
$$= 4 + 8.1548455 + 14.778112$$
$$= 26.932958 \approx 26.93$$

The particular curve is $y = -3e^{-x} - 2e^{-2x} + 26.93$

4. Integrate $\dfrac{dy}{dx} = -4/x$.

$$y = \int -\frac{4}{x}.dx$$
$$= -4\int \frac{1}{x}.dx$$
$$= -4 \ln x + c$$

Write $c = 4 \ln a$, $y = 4 \ln a - 4 \ln x$

$$= 4 \ln\left(\tfrac{a}{x}\right)$$

To determine a given that when $x = 5$, $y = 2$:

$$2 = 4 \ln\left(\tfrac{a}{5}\right)$$
$$\tfrac{2}{4} = \ln\left(\tfrac{a}{5}\right)$$
$$\tfrac{1}{2} = \ln\left(\tfrac{a}{5}\right)$$
$$\tfrac{a}{5} = e^{\frac{1}{2}}$$
$$a = 5e^{\frac{1}{2}} = 8.2436064 \approx 8.244$$
$$y = 4 \ln\left(\frac{8.244}{x}\right)$$

5. Integrate $\dfrac{dy}{dx} = 5 \sin 2x - 3 \cos x$.

$$y = \int (5 \sin 2x - 3 \cos x).dx$$
$$= 5\int \sin 2x.dx - 3\int \cos x.dx$$
$$= -\tfrac{5}{2} \cos 2x - 3 \sin x + c$$

To determine the member of the family which passes through $(\pi/2, 6)$:

$$6 = -\tfrac{5}{2}\cos \pi - 3 \sin\frac{\pi}{2} + c$$
$$= \tfrac{5}{2} - 3 + c$$
$$c = 6 + 3 - 5/2 = 6\tfrac{1}{2}$$

The particular curve is $y = 6\frac{1}{2} - \frac{5}{2} \cos 2x - 3 \sin x$.

6. Integrate $dy/dx = \sec^2 5x$.

$$y = \int \sec^2 5x.dx$$
$$= \tfrac{1}{5}\int 5 \sec^2 5x.dx$$
$$= \tfrac{1}{5} \tan 5x + c$$

Suppose that $y = 8$ when $x = \pi/4$:

$$\begin{aligned} 8 &= \tfrac{1}{5}\tan\tfrac{5\pi}{4} + c \\ &= \tfrac{1}{5} + c \\ c &= 7\tfrac{4}{5} \end{aligned}$$

The particular solution is $y = \frac{1}{5}\tan 5x + 7\frac{4}{5}$.

Exercise 7.1

Obtain the equations of the families of curves defined by the following differential equations. Determine the members of the families which pass through the given points.

1. $dy/dx = 2x + 3$: $(2, 0)$. Sketch the family.
2. $dy/dx = 5 - 4x$: $(-1, 0)$. Sketch the family.
3. $dy/dx = 3x^2 + 4x - 3$: $(4, 5)$. Sketch the family.
4. $dy/dx = 1/x$: $(5, 2)$. Sketch the family.
5. $dy/dx = 3\cos 2x$: $(\pi/8, -3/2)$. Sketch the family.
6. $dy/dx = 6e^{-x}$: $(3, -10)$
7. $dy/dx = \frac{2}{3}e^{2x} - \frac{4}{5}e^{-3x}$: $(1, 5)$
8. $dy/dx = 4x - 8/x^2$: $(10, 7)$
9. $6 \cdot \dfrac{dv}{dt} = 3\sqrt{t} - \dfrac{2}{\sqrt{t}}$: $(t = 4, v = 3)$
10. $\dfrac{di}{dt} - \dfrac{3}{5}t - 8 = 0$: $(t = 6, i = 8)$
11. $\dfrac{dA}{dx} = 5x + 2\sec^2 x$: $(\pi/4, -7)$
12. $\dfrac{dy}{dx} + 3e^{-x} = 5e^{-2x}$: $(1, 2)$
13. $t \cdot \dfrac{di}{dt} = 5t + 3$: $(t = 2, i = 1)$
14. $10\dfrac{di}{dt} = 3\cos\left(\dfrac{5}{2}t\right)$: $(t = 2, i = 8)$
15. $dy/dx = 7\cos 2x - 3\sin 3x$: $(\pi/4, 5)$

Boundary conditions

When we integrate a differential equation of the type $dy/dx = f(x)$ into the form $y = F(x) + c$, where

$$F(x) = \int f(x).dx \qquad 7.3$$

we do not always wish to sketch or draw accurately members of the general family. Nor do we wish to sketch the one curve of that family which passes through a given point. We merely wish to obtain $y = F(x) + c$ in the form $y = F(x) + a$, where a is a particular value which c takes to conform with the restriction that our curve must pass through a given point, say (x_1, y_1). In that case we substitute $x = x_1$ in *7.3*, so obtaining the value of y_1 for y. That is:

$$y_1 = F(x_1) + c$$
$$\text{or } c = y_1 - F(x_1)$$

And this must be the constant we called a. Our particular curve could then be described as:

$$y = F(x) + y_1 - F(x_1) \qquad 7.4$$

This is the equation which complies with the restriction that it passes through (x_1, y_1). *Such a restriction is called a boundary condition.*

Examples

1. Solve the equation $dy/dx = 5x^3$ when the boundary condition is that when $x = 1$, $y = 4$.
   The indefinite integral of $dy/dx = 5x^3$ is $y = 5.\frac{1}{4}x^4 + c = \frac{5}{4}x^4 + c$
   When $x = 1$, $y = \frac{5}{4}.1^4 + c = \frac{5}{4} + c$.
   This must produce the result $y = 4$.
   Therefore $\frac{5}{4} + c = 4$, i.e., $c = 4 - 1\frac{1}{4} = 2\frac{3}{4} = \frac{11}{4}$.
   The particular integral to satisfy both the differential equation and the boundary condition is $y = \frac{5}{4}x^4 + \frac{11}{4}$.
2. Determine the general solution of the equation:

$$x^2 \frac{dy}{dx} = 4x\sqrt{x} - 2x \qquad \text{(i)}$$

   and obtain the particular solution which satisfies the boundary condition that when $x = 4$, $y = 8$.
   Rewrite (i):

$$\frac{dy}{dx} = \frac{4x\sqrt{x}}{x^2} - \frac{2x}{x^2}$$

$$= \frac{4}{\sqrt{x}} - \frac{2}{x}$$
$$= 4x^{-\frac{1}{2}} - 2/x$$
$$y = \int (4x^{-\frac{1}{2}} - 2/x) . dx$$
$$= 4\int x^{-\frac{1}{2}} . dx - 2\int \tfrac{1}{x} dx$$
$$= 4 . 2x^{\frac{1}{2}} - 2 \ln x + c$$
$$= 8\sqrt{x} - 2 \ln x + c$$

This is the general solution.
To determine the particular solution.
When $x = 4$, $y = 8$:

$$8 = 8 \times 4^{\frac{1}{2}} - 2 \ln 4 + c$$
$$= 16 - 2 \ln 4 + c$$
$$c = 8 - 16 + 2 \ln 4$$
$$= 2 \ln 4 - 8$$

The particular solution is $y = 8(\sqrt{x} - 1) + 2 \ln (4/x)$.

3. A body oscillating about an equilibrium position and moving in a line through that point has a displacement, s, given by:

$$\frac{ds}{dt} = \frac{3}{4} \sin 2t + \frac{1}{5} \cos t$$

Determine the general solution of the equation and obtain the particular solution satisfying the boundary condition that when $t = \pi/6$, $s = 1$.
To determine the general solution:

$$s = \int \left(\frac{3}{4} \sin 2t + \frac{1}{5} \cos t \right) . dt$$
$$= \frac{3}{4} \int \sin 2t . dt + \frac{1}{5} \int \cos t . dt$$
$$= -\frac{3}{4} . \frac{1}{2} \cos 2t + \frac{1}{5} \sin t + c$$
$$= -\frac{3}{8} \cos 2t + \frac{1}{5} \sin t + c$$

To determine the particular solution:

When $t = \pi/6$, $s = 1$:

$$1 = -\frac{3}{8}\cos\frac{\pi}{3} + \frac{1}{5}\sin\frac{\pi}{6} + c$$

$$= -\frac{3}{8}\cdot\frac{1}{2} + \frac{1}{5}\cdot\frac{1}{2} + c$$

$$c = 1 + \frac{3}{16} - \frac{1}{10}$$

$$= 1 + \frac{15}{80} - \frac{8}{80} = 1\frac{7}{80}$$

The particular solution is $s = -\frac{3}{8}\cos 2t + \frac{1}{5}\sin t + 1\frac{7}{80}$.

4. A particle is constrained to move along a straight line so that its velocity, v, is given by:

$$\frac{dv}{dt} = ae^{-pt} + be^{-qt}$$

where a, b, p and q are constants.
Determine the general solution and the particular solution given that $v = A$ when $t = T$.

$$v = \int (ae^{-pt} + be^{-qt}).dt$$

$$= -\frac{a}{p}e^{-pt} - \frac{b}{q}e^{-qt} + c$$

This is the general solution.
When $t = T$, $v = A$:

$$A = -\frac{a}{p}e^{-pT} - \frac{b}{q}e^{-qT} + c$$

$$c = A + \frac{a}{p}e^{-pT} + \frac{b}{q}e^{-qT}$$

The particular solution is:

$$v = -\frac{a}{p}e^{-pt} - \frac{b}{q}e^{-qt} + A + \frac{a}{p}e^{-pT} + \frac{b}{q}e^{-qT}$$

$$= A + \frac{a}{p}(e^{-pT} - e^{-pt}) + \frac{b}{q}(e^{-qT} - e^{-qt})$$

5. The bending moment of a beam, M, at a point whose position is distance x from one end, is expressed by the relationship:

$$dM/dx = w(L - x)$$

where L is the total length of the beam and w is the weight per unit

length. Determine the general solution and the particular solution given that when $x = \frac{1}{4}L$, $M = -9wL^2/32$.
Integrating:

$$\begin{aligned} M &= wLx - \tfrac{1}{2}wx^2 + c \\ &= -\tfrac{1}{2}wL^2 + wLx - \tfrac{1}{2}wx^2 + \tfrac{1}{2}wL^2 + c \\ &= -\tfrac{1}{2}w(L^2 - 2Lx + x^2) + k, \text{ where } k = \tfrac{1}{2}wL^2 + c \\ &= -\tfrac{1}{2}w(L-x)^2 + k \end{aligned}$$

This is the general solution.
When $x = \frac{1}{4}L$, $M = -9wL^2/32$:

$$\begin{aligned} -\frac{9wL^2}{32} &= -\tfrac{1}{2}w\left(\frac{3L}{4}\right)^2 + k \\ &= -\frac{9wL^2}{32} + k \\ k &= 0 \end{aligned}$$

The particular solution is $M = -\frac{1}{2}w(L-x)^2$

Exercise 7.2

Determine the general solutions of the following equations and the particular solutions which satisfy the given boundary conditions.

1. $dy/dx = 4x + 2$: when $x = 1$, $y = 2$
2. $dy/dx = 6x^2 - 8x + 3$: when $x = 2$, $y = -5$
3. $\dfrac{2}{x^2} \cdot \dfrac{dy}{dx} = \dfrac{4}{x} + 3 - 5x$: when $x = -1$, $y = -4$
4. $e^t \dfrac{ds}{dt} = 5 + 6e^t$: when $t = 0$, $s = 0$
5. $4 + \dfrac{ds}{dt} = 3e^{-2t} - 4e^{-t}$: when $t = 1$, $s = 0$
6. $\sec x \dfrac{dy}{dx} = 2 + 3\tan x$: when $x = \pi/3$, $y = -4$
7. $\cos x . \dfrac{dy}{dx} = 3\sec x$: when $x = \pi/4$, $y = 2$
8. $(1 + x) . \dfrac{dy}{dx} = 1 + 2x + x^2$: when $x = 5$, $y = 3$
9. $(2 + 3x) . \dfrac{dy}{dx} = (2 + 3x)^2 - 5(2 + 3x)$: when $x = \frac{1}{4}$, $y = -2/3$
10. $(1 - 2x) . \dfrac{dy}{dx} = -7 - 20x + 12x^2$: when $x = 10$, $y = -23$

8

First Order Differential Equations of the Form $dQ/dt = kQ$

A first order differential equation is one in which the highest order derivative is a first order derivative, e.g. dy/dx, ds/dt, di/dt and so on.

Fig. 8.1 represents a capacitor, of capacitance C and initial potential difference V_0, which is connected in series with a resistor, R. Assume that t seconds after the switch is closed the quantity of charge on the plates is given by:

$$Q = A.e^{-kt}$$

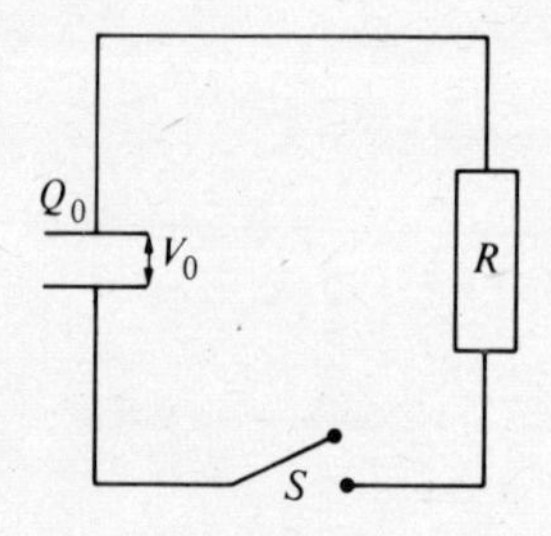

Figure 8.1

At that moment the p.d. between the plates is V. When $t = 0$, $Q = A.e^0 = A = CV_0 = Q_0$, the charge on the plates before the switch is closed. Then:

$$Q = Q_0.e^{-kt}$$

When, more generally,

$$Q = A \,.\, e^{kt}, \text{ where } A \text{ and } k \text{ are constants,} \qquad 8.1$$

$$\begin{aligned} dQ/dt &= \frac{d}{dt}(A \,.\, e^{kt}) \\ &= A \,.\, \frac{d}{dt}(e^{kt}) \\ &= A \,.\, k \,.\, e^{kt} \\ &= k \,.\, A \,.\, e^{kt} \\ &= k \,.\, Q \end{aligned} \qquad 8.2$$

And this is a differential equation.

In Chapter 7 it was shown that the indefinite integral represented a family of curves. The equation of that family contained an arbitrary constant, and each value of the constant determines a particular member of the family. From the above, equation *8.1* gives rise to a differential equation, *8.2*. Formula *8.1* contains one arbitrary constant, A. Then *8.1* must represent a family of curves. Fig. 8.2 represents a few members of that family. Note that: (a) no two curves intersect; and (b) their intercepts on OQ = the values of A.

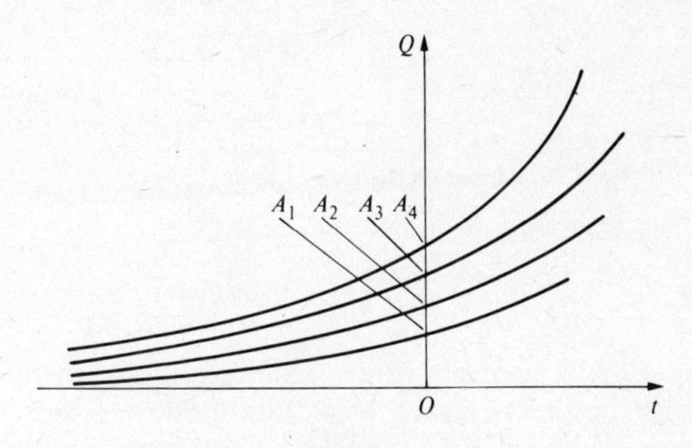

Figure 8.2

The question to be decided is whether it is true that *8.1* is an indefinite integral of *8.2*. As a preliminary step, substitute for Q from $Q = A \,.\, e^{kt}$ in *8.2* and see if it fits. When $Q = A \,.\, e^{kt}$ then $dQ/dt = k \,.\, A \,.\, e^{kt}$. Substitute for Q and dQ/dt in *8.2*: the LHS $= k \,.\, A \,.\, e^{kt}$ and the RHS $= k \,.\, A \,.\, e^{kt}$. This verifies that $Q = A \,.\, e^{kt}$ does satisfy *8.2*. For further evidence rewrite *8.2* as:

$$\frac{1}{Q}\frac{dQ}{dt} = k \qquad 8.3$$

Suppose the LHS of *8.3* were dy/dt where y is some function $f(Q)$, which is to be determined. Q must be a function of t because dQ/dt occurs in the differential equation. So y is a function of Q and Q is a function of t. Therefore y is a function of a function of t, i.e.

$$\frac{dy}{dt} = \frac{dy}{dQ} \cdot \frac{dQ}{dt}$$

Then: $$\frac{dy}{dQ} \cdot \frac{dQ}{dt} = \frac{1}{Q} \cdot \frac{dQ}{dt}, \text{ by substituting in } 8.3$$

Therefore: $$\frac{dy}{dQ} = \frac{1}{Q}$$

Integrating: $$y = \int \frac{1}{Q} dQ = \ln Q$$

8.3 is equivalent to: $$\frac{dy}{dt} = k$$

Then: $$y = kt + c$$

i.e. $$\ln Q = kt + c$$

Therefore:

$$\begin{aligned} Q &= e^{kt + c} \\ &= e^{kt} . e^{c} \\ &= A . e^{kt}, \text{ where } A = e^{c} \end{aligned}$$

Applications

There are many problems arising in nature and technology where a variable which is changing with time is such that its rate of change at any moment is proportional to the value of the variable, e.g. where:

$$dx/dt = k . x$$

where x is the variable and k is a constant. All such equations are satisfied by a solution of the type:

$$x = A . e^{kt}$$

For such a solution:

$$dx/dt = A . k . e^{kt} = k . Ae^{kt} = k . x$$

Examples

1. Verify that $y = a\,.\,\mathrm{e}^{px}$ is a solution of $\mathrm{d}y/\mathrm{d}x = py$, where a is a constant. Differentiate y w.r.t. x.

$$\frac{\mathrm{d}y}{\mathrm{d}x} = \frac{\mathrm{d}}{\mathrm{d}x}(a\,.\,\mathrm{e}^{px}) = a\,.\frac{\mathrm{d}}{\mathrm{d}x}(\mathrm{e}^{px}) = a\,.\,p\,.\,\mathrm{e}^{px} = p\,.\,a\mathrm{e}^{px} = p\,.\,y$$

2. Show that $T = T_0\,.\,\mathrm{e}^{-k\theta}$ is a solution of $\dfrac{\mathrm{d}T}{\mathrm{d}\theta} = -k\,.\,T.$

Differentiate $T = T_0\mathrm{e}^{-k\theta}$ w.r.t. θ.

$$\frac{\mathrm{d}T}{\mathrm{d}\theta} = \frac{\mathrm{d}}{\mathrm{d}\theta}(T_0\,.\,\mathrm{e}^{-k\theta}) = T_0\,.\frac{\mathrm{d}}{\mathrm{d}\theta}(\mathrm{e}^{-k\theta})$$

$$= T_0\,.\,-k\mathrm{e}^{-k\theta} = -k(T_0\,.\,\mathrm{e}^{-k\theta}) = -k\,.\,T$$

3. Show that $P = P_0\,.\,\mathrm{e}^{\frac{1}{a}t}$ is a solution of $\dfrac{\mathrm{d}P}{\mathrm{d}t} = \dfrac{1}{a}P.$

Differentiate $P = P_0\,.\,\mathrm{e}^{\frac{1}{a}t}$ w.r.t. t.

$$\frac{\mathrm{d}P}{\mathrm{d}t} = \frac{\mathrm{d}}{\mathrm{d}t}(P_0\,.\,\mathrm{e}^{\frac{1}{a}t}) = P_0\,.\frac{\mathrm{d}}{\mathrm{d}t}(\mathrm{e}^{\frac{1}{a}t}) = P_0\,.\frac{1}{a}\mathrm{e}^{\frac{1}{a}t}$$

$$= \frac{1}{a}\,.\,P$$

In example 1 the solution $y = a\,.\,\mathrm{e}^{px}$ contains an arbitrary constant which may take any value. Therefore the equation represents a family of curves. Consequently it is a general solution.

Exercise 8.1

Show that the following functions are solutions of the associated differential equations.

1. $y = b\,.\,\mathrm{e}^{2t}$; $\mathrm{d}y/\mathrm{d}t = 2y$
2. $y = c\,.\,\mathrm{e}^{-3t}$; $\mathrm{d}y/\mathrm{d}t = -3y$
3. $y = p\,.\,\mathrm{e}^{\frac{2}{3}t}$; $\mathrm{d}y/\mathrm{d}t = 2y/3$
4. $T = A\,.\,\mathrm{e}^{-\theta}$; $\mathrm{d}T/\mathrm{d}\theta = -T$
5. $T = B\,.\,\mathrm{e}^{-\theta/4}$; $\mathrm{d}T/\mathrm{d}\theta + \frac{1}{4}T = 0$
6. $i = I_0\,.\,\mathrm{e}^{-\frac{R}{L}t}$; $L.\dfrac{\mathrm{d}i}{\mathrm{d}t} + Ri = 0$

7. $i = I.e^{-5t}$; $di/dt = -5i$

8. $T = C.e^{a\theta}$; $\frac{1}{a}\frac{dT}{d\theta} - T = 0$

9. $\theta = a.e^{-\frac{1}{10}t}$; $d\theta/dt = -\frac{1}{10}\theta$

10. $\theta = b.e^{-pt}$; $\frac{1}{p}\frac{d\theta}{dt} + \theta = 0$

11. $M = M_0.e^{-kt}$; $dM/dt = -kM$

12. $y = B.e^{-x/c}$; $c.\frac{dy}{dx} + y = 0$

The following examples illustrate some of the instances where differential equations of this type arise in various technologies.

Examples

1. A radioactive substance decays (i.e. loses its mass through radiation) at a rate which is proportional to the mass of the substance at the moment in question. Suppose the mass at time t is M kilograms. Then:

$$dM/dt = -kM \qquad \text{(i)}$$

where t is measured in a suitable unit of time. For those substances which decay rapidly this might be a second or even a fraction of a second. In other cases the unit might be a minute, an hour, a day, or even a year. Suppose, in this case, it is a year. The reason for the negative sign in (i) is that the mass is decreasing, and the normal meaning of a derivative is that it is positive for an increasing function.
The solution of (i) is:

$$M = M_0 e^{-kt} \qquad \text{(ii)}$$

When $t = 0$, $M = M_0$, i.e. M_0 is the mass at $t = 0$. k is called the decay constant.
Suppose $k = 3.08 \times 10^{-2}$ per year.
Then $M = M_0.e^{-3.08 \times 10^{-2}.t}$
Calculate the time for the mass to be reduced to $\frac{1}{2}M_0$.
Suppose the time is T.

$$\tfrac{1}{2}M_0 = M_0.e^{-3.08 \times 10^{-2}.t}$$

Take natural logs of both sides:

$$\ln \tfrac{1}{2} = -3.08 \times 10^{-2}.T$$

$$T = -\frac{10^2}{3.08} \times \ln\left(\tfrac{1}{2}\right)$$

$$= 22.504779, \text{ by calculator}$$

The half life is approximately 22.5 years.

2. During a certain period the number of bacteria in a given population increases at a rate proportional to the number of bacteria at the time. Suppose the number of bacteria at time t hour is N. Then:

$$\frac{\mathrm{d}N}{\mathrm{d}t} = k.N$$

where k is positive and a constant, because the population is increasing. The general solution is $N = N_0.\mathrm{e}^{kt}$. N_0 represents the population at $t = 0$.

If $N_0 = 6.85 \times 10^{10}$ and $N = 8.45 \times 10^{15}$ at $t = 2$, determine the value of k and calculate the length of time for the population to double in size.

Substitute for N, N_0 and t in the general solution:

$$8.45 \times 10^{15} = 6.85 \times 10^{10}.\mathrm{e}^{2k}$$

$$\mathrm{e}^{2k} = \frac{8.45}{6.85} \times 10^5$$

Take logs to base e of both sides:

$$2k = \ln\left(\frac{8.45}{6.85} \times 10^5\right)$$

$$k = \tfrac{1}{2}\ln\left(\frac{8.45}{6.85} \times 10^5\right)$$

$$= 5.8614216 \approx 5.86$$

When $N = 2N_0$ suppose $t = t_1$. Then:

$$2N_0 = N_0\mathrm{e}^{kt_1}$$

Take natural logs of both sides:

$$kt_1 = \ln(2)$$

$$t_1 = \frac{1}{k}\ln(2)$$

$$= \frac{1}{5.8614216}\cdot\ln(2)$$

$$= 0.1182588 \approx 0.12 \text{ hour}$$

3. Fig. 8.3 represents a stationary rope wrapped partly round a rough cylindrical pole. The tension, T, tends to pull the rope in a clockwise direction.

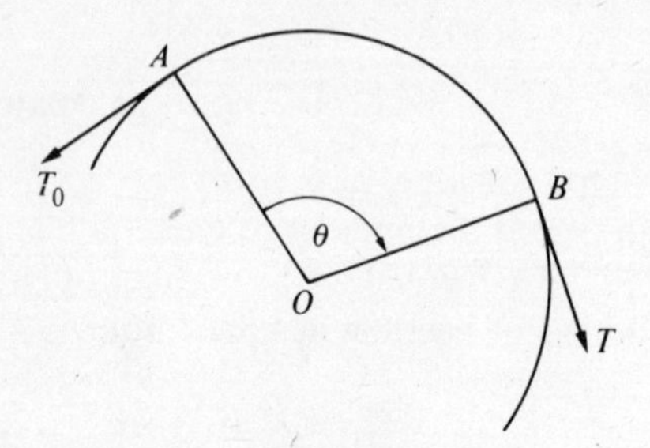

Figure 8.3

Then $T = T_0 . e^{\mu\theta}$, where μ is the coefficient of friction between the rope and the pole.
Consequently, $dT/d\theta = \mu T$, i.e. the rate of increase of T with angle (measured in radians) is proportional to T. The coefficient of friction is the constant of proportionality.
When $\mu = 0.425$ and T_0 (i.e. the value of T when $\theta = 0$) = 1000 N, determine the value of T when $\theta = 3\pi/4$.

$$T = 1000 . e^{0.425 \times \frac{3\pi}{4}} = 2722.0429 \approx 2720 \text{ N}$$

That is, a tension of 1000 N will just hold a pull or load of 2720 N.

4. A body at a temperature of θ °C above that of its environment cools at a rate proportional to that excess temperature, i.e.

$$d\theta/dt = -k\theta$$

A body at temperature 120 °C cools to 90 °C in 10 minutes when the environmental temperature is a constant 20 °C. Determine the temperature after 20 minutes and find out how long it takes for the temperature to fall to 60 °C.

$$\theta = \theta_0 . e^{-kt} \qquad \text{(i)}$$

where k is constant and θ_0 is the excess temperature at $t = 0$. Substituting in (i):

$$(90 - 20) = (120 - 20)e^{-k.10}$$
$$70 = 100 . e^{-10k}$$
$$e^{-10k} = 70/100 = 0.7$$
$$-10k = \ln (0.7) = -0.3566749$$

$$k = \frac{0.3566749}{10} = 0.0356675$$

When $t = 20$:

$$\begin{aligned}\theta &= 100\,.\,\mathrm{e}^{-20k}\\ &= 48.999995 \approx 49\,°\end{aligned}$$

The temperature of the body is $(49\,° + 20\,°)\mathrm{C} = 69\,°\mathrm{C}$.
When $\theta = (60 - 20) = 40\,°\mathrm{C}$, the time, t_1, is given by:

$$\begin{aligned}40 &= 100\,.\,\mathrm{e}^{-kt_1}\\ \mathrm{e}^{-kt_1} &= \frac{40}{100} = 0.4\\ kt_1 &= -\ln(0.4)\\ t_1 &= 25.68974 \approx 25.7 \text{ minutes}\end{aligned}$$

5. An electrical circuit contains an inductor, L henry, and a resistor, R ohm, in series. The current, i amps, at time t seconds, is given by:

$$L\,.\frac{\mathrm{d}i}{\mathrm{d}t} + Ri = 0 \qquad \text{(i)}$$

When $R = 5$ and $L = 0.65\,\mathrm{H}$, the current after $5.2 \times 10^{-1}\,\mathrm{s}$ is $0.25\,\mathrm{A}$. Determine the value of the current after $8.3 \times 10^{-1}\,\mathrm{s}$, and how long it takes for the current to reach $0.15\,\mathrm{A}$.

$$i = i_0\,.\,\mathrm{e}^{-\frac{R}{L}t} \qquad \text{(ii)}$$

Substitute in (ii):

$$\begin{aligned}0.25 &= i_0\,\mathrm{e}^{-\frac{5}{0.65} \times 5.2 \times 10^{-1}}\\ 0.25 &= i_0 \times 0.0183156\\ i_0 &= 13.649566\\ &\approx 13.65\,\mathrm{A}\end{aligned}$$

When $t = 8.3 \times 10^{-1}$:

$$\begin{aligned}i &= 13.649566 \times \mathrm{e}^{-\frac{5}{0.65} \times 8.3 \times 10^{-1}}\\ &= 0.0230311\\ &\approx 0.023\,\mathrm{A}\end{aligned}$$

When $i = 0.15\,\mathrm{A}$:

$$0.15 = 13.649566 \times \mathrm{e}^{-\frac{5}{0.65}t}$$

$$\mathrm{e}^{-\frac{5}{0.65}t} = \frac{0.15}{13.649566}$$

$$\frac{5}{0.65}t = -\ln\left(\frac{0.15}{13.649566}\right)$$

$$= 4.5108277$$

$$t = 0.5864076$$

$$\approx 5.86 \times 10^{-1} \text{ seconds}$$

Exercise 8.2

Determine the general solutions of the following differential equations and obtain the particular solutions which satisfy the given boundary conditions.

1. $dy/dx = y$: given $y = 2$ when $x = 0$
2. $dy/dx = 2y$: given $y = -5$ when $x = 0$
3. $dy/dx = -3y$: given $y = 20$ when $x = 1$
4. $dy/dx = -3y/4$: given $y = 10$ when $x = -1$
5. $8\dfrac{dy}{dx} + 5y = 0$: given $y = 3$ when $x = 2$
6. $a\dfrac{dy}{dx} + 2y = 0$: given $y = 8$ when $x = a$
7. $p\dfrac{dx}{dt} + qx = 0$: given $x = 5a$ when $t = -2p/q$
8. $dT/d\theta + \frac{1}{4}T = 0$: given $T = 450$ N when $\theta = 0$ rad
9. $2\dfrac{di}{dt} + 3i = 0$: given $i = 0.25$ A when $t = 0$
10. $L\dfrac{di}{dt} + Ri = 0$: given $L = 4.5$ H, $R = 2.5\,\Omega$ and $i = 0.35$ A when $t = 0$.
11. $RC\dfrac{di}{dt} + i = 0$: given $R = 2.5 \times 10^3\,\Omega$, $C = 1.6 \times 10^{-6}$ F, and $i = 0.85$ A when $t = 0$.
12. $\dfrac{1}{a}\cdot\dfrac{dM}{dt} + M = 0$: given $a = 1.56 \times 10^{-2}$ per year and $M = 485$ kg when $t = 0$. Calculate the value of M when $t = 30.5$ years and calculate the time when the mass is 220 kg.
13. A body at temperature 140 °C cools to 100 °C in 15 minutes when the surrounding temperature is a steady 18 °C. Determine the temperature of the body after 25 minutes and the time taken for the temperature to fall to 82 °C.
14. A culture of bacteria increases from 2.09×10^8 in number to 1.85×10^9 one hour later. Determine the length of time for the population

to reach a size of 3.29×10^{10} and calculate the magnitude of the population 3.5 hours after the initial observation.

15. A rope wrapped once round a rough circular pole supports a load of 2000 N at one end. Determine the tension at the other end of the rope if the coefficient of friction between the rope and the pole is 0.52. At what point of the rope will the tension be 1000 N?
16. An electrical circuit contains an inductor, 6.2 H, and a resistor, 3.8 Ω, in series. A battery is allowed to charge the inductor until the current ceases. The switch is thrown, thus allowing the inductor to discharge through the resistor. At time $t = 0$ seconds the discharging current is 0.55 A. Determine the current after 2×10^{-1} seconds and calculate the moment when the current drops to 0.35 A.

9

The Evaluation of Expressions Involving Exponentials and Natural Logarithms

There are many constants which arise in mathematics, science, statistics and many branches of technology. One such is that designated by the symbol e. It is incommensurate (cannot be measured exactly) and has an approximate value of 2.7182818285. It is not recurring. In addition to common logarithms, i.e. logarithms to base 10, we need to be able to use logarithms to base e. These logarithms are variously called natural logarithms, Napierian logarithms, and hyperbolic logarithms. In some branches of our work it is essential to know how to deal with them.

9.1 The range of values of e^x

A few calculations will suffice to show us the range of values of e^x as x varies from $-\infty$ to $+\infty$.

Examples

By calculator:

1. $e^0 = 1$. First key in 0; then key in e^x.
2. $e^{0.25} = 1.284025$. First key in 0.25; then key in e^x.
3. $e^{0.5} = 1.6487213$. First key in 0.5; then key in e^x.
4. $e^1 = 2.7182818$.
5. $e^{10} = 22026.466$.

Exercise 9.1

By calculator determine:

1. $e^{0.2}$
2. $e^{0.75}$
3. $e^{1.5}$
4. e^{5}
5. e^{50}
6. e^{100}
7. e^{150}
8. e^{200}
9. e^{220}
10. e^{230}
11. e^{231}

Note that, as x increases from 0, e^x increases from 1. Our calculator will determine readily e^{230}: 7.7220185×10^{99}, but cannot cope with e^{231}. In other words, for $0 < x < \infty$, $1 < e^x < \infty$. As $x \to \infty$ $e^x \to \infty$.

Examples

By calculator determine:

1. $e^0 = 1$
2. $e^{-0.25} = 0.7788008$
3. $e^{-0.5} = 0.6065307$
4. $e^{-1} = 0.3678794$
5. $e^{-5} = 0.00637379$
6. $e^{-10} = 0.0000454$

Exercise 9.2

Calculate:

1. $e^{-0.75}$
2. e^{-2}
3. e^{-20}
4. e^{-50}
5. e^{-100}
6. e^{-200}
7. e^{-220}

8. e^{-227}
9. e^{-228}

Note that as x decreases from 0, e^x decreases from 1. Our calculator will readily determine e^{-227}: $2.6010734 \times 10^{-99}$, but cannot cope with e^{-228} or a calculation of e^x for any value of x less than -228. In all cases where $0 > x > -\infty$, then e^x is positive and tends to 0 as $x \to -\infty$. Consequently, for all x, $e^x > 0$. In particular:

$$e^x > 1 \text{ for positive } x$$
$$1 > e^x > 0 \text{ for negative } x$$
$$e^x = 1 \text{ for } x = 0$$

9.2 Evaluation of expressions involving e

In the solution of many problems arising in engineering, mechanics, electrical engineering, physics, chemistry, biology, statistics and so on we need to be able to calculate expressions involving e.

Examples

1. Evaluate ae^x when $a = 11.75$ and $x = 2.09$.

$$\begin{array}{ll} A = x & A = 2.09 \\ B = e^A & B = 8.0849152 \\ C = a \times B & C = 94.997753 \approx 95.00 \end{array}$$

2. Evaluate pe^{ax} when $p = 7.47$, $a = -3.26$ and $x = 0.54$.

$$\begin{array}{ll} A = x & A = 0.54 \\ B = a \times A & B = -1.7604 \\ C = e^B & C = 0.1719761 \\ D = p \times C & D = 1.2846612 \approx 1.28 \end{array}$$

3. Evaluate $a(1 - e^{-pt})$ when: (a) $a = 240.5$, $p = 51.6$ and $t = 0.75$; (b) $a = 240.5$, $p = 51.6$ and $t = 0.1$; (c) $a = 240.5$, $p = 51.6$ and $t = 0.001$.

	(a)	(b)	(c)
$A = t$	$A = 0.75$	$A = 0.1$	$A = 0.001$
$B = p \times A$	$B = 38.7$	$B = 5.16$	$B = 0.0516$

$C = e^{-B}$	$C = 1.5588472 \times 10^{-17}$	$C = 0.0057417$	$C = 0.9497087$
$D = 1 - C$	$D = 1$	$D = 0.9942583$	$D = 0.0502913$
$E = a \times D$	$E = 240.5$	$E = 239.11912$	$E = 12.095064$
		≈ 239.1	≈ 12.10

4. Evaluate $V = V_0 e^{-at}$ given $V_0 = 250$ and $a = 0.014$, when $t =$ (a) 0.5; (b) 1.0; (c) 2.5; and (d) 10.0.

	(a)	(b)
$A = t$	$A = 0.5$	$A = 1.0$
$B = a \times A$	$B = 0.007$	$B = 0.014$
$C = e^{-B}$	$C = 0.9930244$	$C = 0.9860975$
$D = V_0 \times C$	$D = 248.25611$	$D = 246.52439$
	≈ 248.3	≈ 246.5

(c)	(d)
$A = 2.5$	$A = 10.0$
$B = 0.035$	$B = 0.14$
$C = 0.9656054$	$C = 0.8693582$
$D = 241.40135$	$D = 217.33956$
≈ 241.4	≈ 217.3

Note that as t increases V decreases. This corresponds to the decaying voltage across the terminals of a capacitor which is slowly discharging through a circuit.

Exercise 9.3

Evaluate by calculator the following expressions, giving answers correct to the degree indicated.

1. e^{ax} to three decimal places, when $x = 0.5$, 1.0, 5.0 and 10.0, given $a = 2.5$. How does e^{ax} vary as x increases?
2. e^{ax} to three decimal places, given $a = 1.3$, when $x = 0.5$, 1.5, 4.0 and 12.0. How does e^{ax} vary with an increase in x?
3. ae^t to two decimal places, given $a = 120.5$, when $t = 0.12, 0.56, 3.48$ and 10.29. State how ae^t varies as t increases.
4. be^{-t} to two decimal places, given $b = 325.6$, when $t = 0.23$, 0.86, 2.67 and 8.38. How does be^{-t} vary as t increases?
5. $I_0 e^{-pt}$ to three decimal places, given $I_0 = 3.25$ and $p = 0.012$, when $t = 0.15$, 0.46, 4.82 and 10.77. Note the variation as t increases.
6. ae^{bx} to three decimal places, given $a = 10.79$ and $b = 0.74$ when $x = 0.15$, 0.5, 5, 20 and 50. State the variation as x increases.

7. $a(1-e^{-pt})$ to three decimal places, given $a = 150.6$ and $p = 0.74$, when $t = 0.001$, 0.02 and 0.5. Determine the variation as t increases.

9.3 Natural or Napierian logarithms

In *Mathematics Level 1*, when we defined logarithms to base 10 we wrote down two different statements which postulated the same relationship, that between a number and its logarithm to base 10. They were:

$$\text{(a) } N = 10^x \quad \text{and} \quad \text{(b) } x = \log_{10} N \qquad 9.1$$

The first made the number the subject of the relationship; the second made the logarithm the subject. Remember that a logarithm is an index. We must be as much at ease with the one as with the other. Further, we must realize that each of the functions expressed in *9.1* is the inverse of the other. For example, by substituting for x from (b) in (a) we obtain:

$$N = 10^{\log_{10} N} \qquad 9.2$$

The operations carried out on N on the right of *9.2* produce N on the left. *The operations have left N unchanged.* Again, by substituting for N from (a) in (b) we obtain:

$$x = \log_{10}(10^x) \qquad 9.3$$

The operations carried out on x on the right of *9.3* produces x on the left. *These operations have left x unchanged.*

Napierian logarithms or natural logarithms involve e as a base instead of 10. The relationships corresponding to *9.1* are:

$$\text{(a) } N = e^x \text{ and (b) } x = \log_e N \text{ or } \ln N \qquad 9.4$$

The counterpart of *9.2* is:

$$N = e^{\ln N} \qquad 9.5$$

The counterpart of *9.3* is:

$$x = \ln(e^x) \qquad 9.6$$

From section 9.1 it is important to note that:

$$N = e^x > 0 \text{ for all values of } x$$

As $0 < N < \infty$ then $-\infty < x < \infty$. In particular, for $0 < N < 1$, then $-\infty < x < 0$, and for $1 < N < \infty$, then $0 < x < \infty$.

There is one important difference between tables of natural logarithms and those of common logarithms. Natural logarithm tables give both the integral and the decimal parts of the logarithms at one and the same time. When we wish to determine the number for a given logarithm we have to either look up the same tables in reverse or refer to tables for e^x and e^{-x}, according to whether the logarithm is positive or negative.

9.4 The use of tables and calculator to determine natural logarithms

Some sets of tables (four-figure) give more detail on natural logarithms than others. The more comprehensive ones are set out like tables of common logarithms giving logarithms to four figures from 1.000 to 9.999. The left-hand column gives digits from 1.0 to 9.9. The third digits are given in further columns headed from 0 to 9 inclusive. The fourth digit is given on the extreme right of the page in columns headed 1 to 9 inclusive. The numbers in the last column are to be added to those on the main part of the page. A few examples will show how the method works.

Examples

1. $\ln 1.8 = 0.5878$; $\ln 1.83 = 0.6043$; $\ln 1.834 = 0.6043 + 0.0022 = 0.6065$
2. $\ln 4.6 = 1.5261$; $\ln 4.68 = 1.5433$; $\ln 4.687 = 1.5433 + 0.0015 = 1.5448$
3. $\ln 2.7 = 0.9933$; $\ln 2.74 = 1.0080$ (note the change in integral value between 2.7 and 2.74, usually noted in tables by a change in the typeface); $\ln 2.747 = 1.0080 + 0.0025 = 1.0105$

Now check the above results by keying into your calculator first x and then $\ln x$.

Exercise 9.4

Use four-figure tables to determine the natural logarithms of the following numbers. In each case check the result by calculator.

1. 1.4
2. 2.8
3. 3.6

4. 4.8
5. 6.2
6. 8.7
7. 1.49
8. 2.89
9. 3.65
10. 4.81
11. 6.22
12. 8.76
13 1.431
14. 2.894
15. 3.652
16. 4.817
17. 6.226
18. 8.769
19. 7.396
20. 9.928

For values of N outside the range 1.0 to 9.999, for example 365.2, we proceed as follows:

$$\begin{aligned} N &= 365.2 = 3.652 \times 10^2 \\ \ln N &= \ln (3.652 \times 10^2) = \ln 3.652 + \ln 10^2 \\ &= \ln 3.652 + 2 \ln 10 \end{aligned}$$

Taking ln 10 to be 2.3026, we have:

$$\begin{aligned} \ln 365.2 &= \ln 3.652 + 2 \times 2.3026 \\ &= 1.2952 + 4.6052 = 5.9004 \end{aligned}$$

The values of $\ln 10^x$ for $x = 1, 2, 3, 4, 5, 6$, etc. may be taken from the calculator and corrected to four significant figures or from the following table:

x	1	2	3	4	5	6
$\ln x$	2.3026	4.6052	6.9078	9.2103	11.5129	13.8155

Where we would wish to determine the natural logarithms for numbers between 0 and 1, e.g. 0.07523, we would write $N = 7.523 \times 10^{-2}$. Then:

$$\begin{aligned} \ln 0.07523 &= \ln 7.523 + \ln 10^{-2} \\ &= 2.0180 - 2 \times 2.3026 \\ &= 2.0180 - 4.6052 = -2.5872 \end{aligned}$$

Again, values of $\ln 10^{-x}$ for $x = 1, 2, 3, 4, 5, 6$, etc. may be taken from the calculator and corrected to four significant figures or from the above tables by multiplying by -1. For instance, $\ln 10^{-5} = -\ln 10^5$. Some books of tables give $\ln 10^{-x}$ in the following form:

x	1	2	3	4	5	6
$\ln 10^{-x}$	$\overline{3}.6974$	$\overline{5}.3948$	$\overline{7}.0922$	$\overline{10}.7897$	$\overline{12}.4871$	$\overline{14}.1845$

This system reverts to the method for common logarithms of keeping the decimal part of the logarithm positive and making the integral part negative. Yet a calculator will always give us a complete logarithm negative.

Exercise 9.5

Use four-figure tables to determine the natural logarithms of the following numbers. In each case check the result by calculator.

1. 14.31
2. 28.94
3. 36.52
4. 48.17
5. 62.26
6. 87.69
7. 73.96
8. 99.28
9. 103.4
10. 0.2894
11. 0.3652
12. 0.4817
13. 0.8769
14. 0.7396
15. 0.9928
16. 317.5
17. 7256
18. 0.002549
19. 7 152 900
20. 0.00006423

To determine the number whose natural logarithm is known

Here we must use *9.4*(a), i.e. $N = e^x$.
For this we must refer to tables of e^x.
Where $N > 1$, then $x > 0$, i.e. x is positive.
Where $0 < N < 1$, then $x < 0$, i.e. x is negative.

We will deal with x positive first.
The better kind of tables will tabulate x from 0.00 to 4.00 in steps of 0.01 (i.e. three digits). The left-hand column gives the first two digits of x starting at 0.0 and ending at 4.0. The third digit is given in columns headed .00, .01, .02, . . . , .09.

Examples

Determine the numbers whose natural logarithms are the following (use e^x tables and check by calculator):

1. 0.7: $N = 2.0138$; calculator 2.0137527
2. 2.9: $N = 18.174$; calculator 18.174145
3. 3.6: $N = 36.598$; calculator 36.598234
4. 4.0: $N = 54.598$; calculator 54.59815
5. 0.3: $N = 1.3499$; calculator 1.3498588
6. 0.36: $N = 1.4333$; calculator 1.4333294
7. 1.74: $N = 5.6973$; calculator 5.6973434
8. 2.89: $N = 17.993$; calculator 17.99331
9. 3.43: $N = 30.877$; calculator 30.876643
10. 3.97: $N = 52.985$; calculator 52.984531

When x is negative we may use tables for e^{-x} in a way similar to the above or, if there are no tables for e^{-x}, we can use tables for e^x and then use reciprocal tables. For example, suppose we wish to determine $e^{-2.65}$. We write:

$$e^{-2.65} = \frac{1}{e^{2.65}} = \frac{1}{14.154} = 0.07068 \text{ (by four-figure tables)}$$

The calculator gives 0.0706514. Alternatively, we may use ln tables in reverse.

Examples

By tables, determine the numbers whose natural logarithms are:

1. 6.0809; this is greater than 4.6052, i.e. $\ln 10^2$. We write:

$$\begin{aligned} \ln N &= 6.0809 \\ \ln 10^2 &= 4.6052 \\ \text{Subtract:} \quad & \ 1.4757 \end{aligned}$$

From ln tables in reverse $1.4757 = \ln 4.374$.

Then $N = 4.374 \times 10^2 = 437.4$. The calculator gives 437.4227.

2. -2.1842. Then:

$$\begin{aligned} \ln N &= -2.1842 \\ \ln 10^{-1} &= -2.3026 \\ \text{Subtract:} \quad & \quad 0.1184 \end{aligned}$$

Now $0.1184 = \ln 1.126$. Therefore $N = 1.126 \times 10^{-1} = 0.1126$. By calculator, 0.1125678.

3. -10.4523.

$$\begin{aligned} \ln N &= -10.4523 \\ \ln 10^{-5} &= -11.5129 \\ \text{Subtract:} \quad & \quad 1.0606 \end{aligned}$$

Now $1.0606 = \ln 2.888$. Therefore $N = 2.888 \times 10^{-5}$
$= 0.00002888$. By calculator, 0.0000289.

Exercise 9.6

By tables determine the numbers whose natural logarithms are the following. Check the results by calculator.

1. 0.052
2. 0.894
3. −0.894
4. −0.052
5. −1.43
6. −2.89
7. −3.49
8. 0.0517
9. 0.8948
10. 3.9120
11. 4.4602
12. 6.2461
13. 8.0021
14. 16.0537
15. −2.1163
16. −1.6441
17. −2.0130
18. −2.0753
19. −1.2730
20. −0.2784
21. −2.5810

22. -7.518
23. -9.2896
24. -12.7945

9.5 The relationship between common and natural logarithms

Where we have no source of natural logarithms at our disposal, yet we do have access to common logarithms, we are still able to calculate natural logarithms. Suppose that a given number, N, is written in the following forms:

$$N = 10^x \qquad 9.7$$
$$N = e^y \qquad 9.8$$

9.7 means:

$$x = \log_{10} N = \log N \qquad 9.9$$

9.8 means:

$$y = \log_e N = \ln N \qquad 9.10$$

From *9.7* and *9.8*:

$$10^x = e^y \qquad 9.11$$

Take logs to base 10 of *9.11*:

$$x = \log e^y$$
$$= y \log e \qquad 9.12$$

Now substitute for x and y from *9.9* and *9.10* in *9.12*:

$$\log_{10} N = \log_e N \,.\, \log_{10} e \text{ or } \log N = \ln N \,.\, \log e$$

giving $$\ln N = \frac{\log N}{\log e} \qquad 9.13$$

Now $\log e = 0.4343$ (by calculator 0.4342945). *9.13* becomes:

$$\ln N = \frac{1}{0.4343} \times \log N \qquad 9.14$$

$$\text{or } \ln N = 2.3026 \times \log N \qquad 9.15$$

9.15 is rather easier to handle than *9.14*.
If we wish to be more accurate, $\ln N = 2.3025851 \times \log N$.

Examples

Determine the natural logarithms of the following numbers using *9.15*. Check by calculator.

1. 1.425; $\ln 1.425 = 2.3026 \times \log 1.425 = 2.3026 \times 0.1538 = 0.3541$
   By calculator, 0.3541718.
2. 0.1425; $\ln (0.1425) = 2.3026 \times (\bar{1}.1538) = 2.3026 \times (-0.8462) = -1.9484$
   By calculator, -1.9484133.

Exercise 9.7

Determine the natural logarithms of the following numbers using *9.15*. Check by calculator.

1. 1.378
2. 2.463
3. 37.85
4. 60.29
5. 234.7
6. 400.9
7. 80 200
8. 0.6253
9. 0.5217
10. 0.04267
11. 0.01074
12. 0.002305

9.6 Applications to other technological units

Examples

1. Fig. 9.1 represents a rope in contact with a rough surface whose coefficient of friction is μ, where the tensions at two points, A and B, of the rope are T_0 and T respectively and where the normals to the surface, OA and OB, make an angle θ (in radians) with one another. Then $T = T_0 e^{\mu\theta}$, assuming that the tension T is attempting to pull the rope slowly from A to B. Calculate T given $T_0 = 503.5$ N, $\mu = 0.432$ and θ

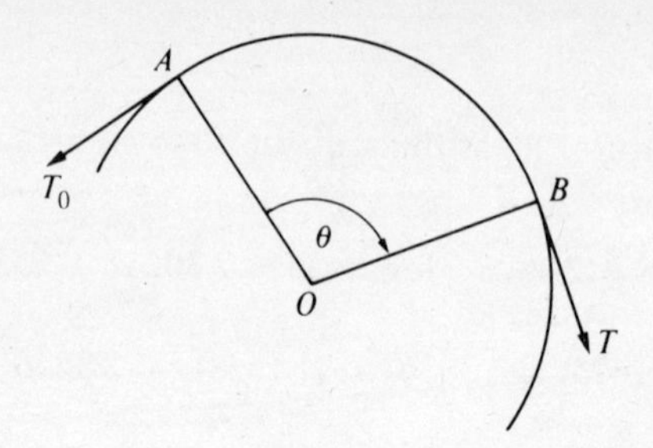

Figure 9.1

takes the values: (a) 0.5; (b) 1.0; (c) 1.5; (d) 2.0; (e) π; (f) 2π; (g) 3π rad.

	(a)	(b)
$A = \theta$	$A = 0.5$	$A = 1.0$
$B = \mu \times A$	$B = 0.216$	$B = 0.432$
$C = e^B$	$C = 1.2411024$	$C = 1.5403351$
$D = T_0 \times C$	$D = 624.89505$	$D = 775.55873$
	≈ 624.9	≈ 775.6

(c)	(d)	(e)
$A = 1.5$	$A = 2.0$	$A = \pi$
$B = 0.648$	$B = 0.864$	$B = 1.357168$
$C = 1.9117136$	$C = 2.3726323$	$C = 3.885175$
$D = 962.54779$	$D = 1194.6203$	$D = 1956.1856$
≈ 962.5	≈ 1195	≈ 1956

(f)	(g)
$A = 2\pi$	$A = 3\pi$
$B = 2.7143361$	$B = 4.0715041$
$C = 15.094585$	$C = 58.645103$
$D = 7600.1234$	$D = 29\,527.809$
≈ 7600	$\approx 29\,530$

Note how rapidly the value of T increases the more the rope is wrapped round the rough surface. If a rope is wrapped just once round such a rough circular pole then a given tension, T_0, is able to resist a pull of 15 times its magnitude at the other end.

2. Fig. 9.2 represents a certain kind of electrical circuit in which the quantity of charge on the capacitor at time t after the switch is closed is given by the formula:

$$Q = CE\,(1 - e^{-t/CR})$$

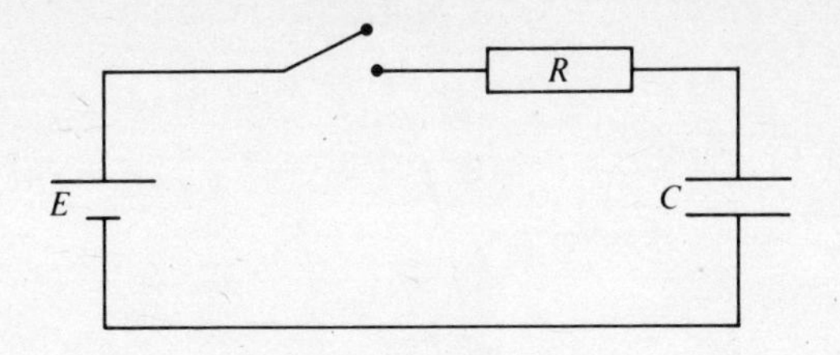

Figure 9.2

where E is measured in volts, C in farads, R in ohms, t in seconds and Q in coulombs. Determine the values of Q when t = (a) 0.001; (b) 0.01; (c) 0.1 seconds given that $E = 250$, $C = 1.04 \times 10^{-5}$ and $R = 10^3$.

	(a)	(b)
$A = t$	$A = 0.001$	$A = 0.01$
$B = C$	$B = 0.0000104$	$B = 0.0000104$
$D = B \times R$	$D = 0.0104$	$D = 0.0104$
$F = -A/D$	$F = -0.0961538$	$F = -0.9615385$
$G = e^F$	$G = 0.9083243$	$G = 0.3823043$
$H = 1 - G$	$H = 0.0916757$	$H = 0.6176957$
$J = E \times H$	$J = 22.918934$	$J = 154.42393$
$K = B \times J$	$K = 0.0002384$	$K = 0.001606$

(c)

$A = 0.1$
$B = 0.0000104$
$D = 0.0104$
$F = -9.6153846$
$G = 0.0000667$
$H = 0.9999333$
$J = 249.98333$
$K = 0.0025998$

The entries under J represent the voltages across the terminals of the capacitor. Note how they increase with time, as do the entries under K, the charge on the plates of the capacitor.

3. Fig. 9.3 represents a steel rod, in a vertical position with its base fixed, which is set vibrating by pulling the top laterally away from the equilibrium position. The displacement, x, of the free end, after time t, is given by:

$$x = ae^{-pt} \cos \omega t$$

Figure 9.3

Given $a = 2.52\,\text{cm}$, $p = 0.2$ and $\omega = 314.2$, determine the values of x when $t =$ (a) 0.1; (b) 0.2; (c) 1.0; (d) 10.0; (e) 50.0; (f) 100.0.

	(a)	(b)
$A = t$	$A = 0.1$	$A = 0.2$
$B = \omega \times A$	$B = 31.42$	$B = 62.84$
$C = \cos \text{B}$	$C = 0.9999917$	$C = 0.9999668$
$D = p \times A$	$D = 0.02$	$D = 0.04$
$E = \text{e}^{-D}$	$E = 0.9801987$	$E = 0.9607894$
$F = E \times C$	$F = 0.9801905$	$F = 0.9607576$
$G = a \times F$	$G = 2.4700802$	$G = 2.421109$

(c)	(d)	(e)
$A = 1.0$	$A = 10.0$	$A = 50.0$
$B = 314.2$	$B = 3142$	$B = 15\,710$
$C = 0.9991705$	$C = 0.9181753$	$C = -0.4492594$
$D = 0.2$	$D = 2.0$	$D = 10$
$E = 0.8187308$	$E = 0.1353353$	$E = 0.0000454$
$F = 0.8180516$	$F = 0.1242615$	$F = -0.0000204$
$G = 2.06149$	$G = 0.313139$	$G = -0.0000514$

(f)

$A = 100.0$
$B = 31\,420$
$C = -0.596332$
$D = 20$
$E = 2.0611536 \times 10^{-9}$
$F = -1.2291319 \times 10^{-9}$
$G = -3.0974124 \times 10^{-9}$

Note the decrease in the amount of vibration as the time increases, due to the damping factor e^{-pt}.

4. From the formula $I = I_0\,e^{-Rt/L}$ determine t, given $L = 2.8$, $R = 0.61$ and $I_0 = 7.87$, when $I =$ (a) 7.52; (b) 5.43; (c) 2.07; (d) 0.38.

Rearrange $\quad I/I_0 = e^{-Rt/L}$

giving $\quad e^{Rt/L} = I_0/I$

Then $\quad Rt/L = \ln(I_0/I)$

Therefore $\quad t = \dfrac{L}{R}\ln(I_0/I)$

	(a)
$A = I_0$	$A = 7.87$
$B = A/I$	$B = 1.0465426$
$C = \ln B$	$C = 0.045919$
$D = L$	$D = 2.8$
$E = D/R$	$E = 4.5901639$
$F = E \times C$	$F = 0.2088154$
	≈ 0.209

(b)	(c)	(d)
$A = 7.87$	$A = 7.87$	$A = 7.87$
$B = 1.4493554$	$B = 3.8019324$	$B = 20.710526$
$C = 0.3711189$	$C = 1.3355095$	$C = 3.0306421$
$D = 2.8$	$D = 2.8$	$D = 2.8$
$E = 4.5901639$	$E = 4.5901639$	$E = 4.5901639$
$F = 1.7034967$	$F = 6.1302073$	$F = 13.911144$
≈ 1.703	≈ 6.130	≈ 13.911

Exercise 9.8

1. Evaluate $A = ae^t$, given $a = 10.9$, when $t = 0.1, 0.5, 1.5, 5.0, 10.0$.
2. Evaluate $M = M_0e^{-t}$ given $M_0 = 13.27$, when $t = 0.5, 2.5, 10, 50$.
3. Evaluate $x = ae^{-pt}$ given $a = 2.5$ and $p = 0.2$, when $t = 0.001, 0.01, 0.1$.
4. Calculate $y = ae^{bx}$ given $a = 102.7$ and $b = 0.36$, when $x = 0.01, 0.1, 1.0, 10$.
5. Determine the values of $y = a(1 - e^{-bx})$ for $a = 239.4$ and $b = 0.75$, when x takes the values 0.001, 0.01, 0.1, 1.0, 10.
6. From the formula $I = I_0\,e^{-Rt/L}$ evaluate I, given $I_0 = 4.23$, $L = 0.4$ and $R = 2.95$, when $t = 0.5, 1.0, 5.0, 10.0$.

7. Where $Q = Q_0\,e^{-t/CR}$ determine Q given $Q_0 = 0.05$, $C = 3.2 \times 10^{-7}$ and $R = 2 \times 10^4$, when $t = 0.002, 0.03, 0.4$.
8. Given $T = T_0 e^{\mu\theta}$ determine T, given $T_0 = 37.8$ N and $\mu = 0.417$, when $\theta = 0.32, 0.85, 2.96$ and 6.43 radians.
9. Using the formula $I = \dfrac{V}{R}(1 - e^{-Rt/L})$ evaluate I, given $V = 30.25$, $R = 525$ and $L = 8.6$, when $t = 0.01, 0.05, 0.1, 0.5$ seconds.
10. From the formula $A(e^{a\theta} - 1) = Pe^{a\theta}$ evaluate P, given $A = 107.2$ and $a = 0.0032$, when $\theta = 0.5, 2.5, 5.0, 10.0, 100$.
11. The displacement, x, in a damped oscillation, is given by $x = ae^{-bt}\sin(nt)$ where t is the time in seconds and nt is measured in radians. Determine x, given $a = 3.42$, $b = 0.35$ and $n = 20$, when t takes the values 0.1, 3.1, 5.1, 10.1 and 20.1 seconds.
12. For a certain period of a plant's growth its mass M is given by the formula $M = M_0\,e^{at}$, where M_0 is the mass at time $t = 0$. Determine M given that $M_0 = 0.031$ and $a = 0.52$, when $t = 1.2, 3.5, 6.2, 10.7$.

9.7 The solution of equations involving e^x and $\ln x$

Examples

1. The basic equation $e^x = a$ is solved by converting to the form $x = \ln a$, providing $a > 0$, using formula *9.6*.
2. A slightly more difficult equation is $ae^x + b = c$.
   Rearranging:

$$ae^x = (c - b)$$

giving:

$$e^x = \left(\frac{c-b}{a}\right)$$

Therefore:

$$x = \ln\left(\frac{c-b}{a}\right)$$

which gives real values of x only if $\dfrac{c-b}{a} > 0$

3. Equations frequently encountered are of the type $3e^x - 10 + 3e^{-x} = 0$. Multiply by e^x:

$$3e^{2x} - 10e^x + 3 = 0$$

Put $e^x = X$:

$$3X^2 - 10X + 3 = 0 \text{ (a quadratic equation)}$$

By factorization:

$$\begin{aligned} (3X-1)(X-3) &= 0 \\ 3X - 1 = 0 &\quad \text{or} \quad X - 3 = 0 \\ X = 1/3 &\quad \text{or} \quad X = 3 \end{aligned}$$

Substituting back:

$$\begin{aligned} e^x = 1/3 &\quad \text{or} \quad e^x = 3 \\ x = \ln(1/3) &\quad \text{or} \quad x = \ln 3 \\ x = -1.0986123 &\quad \text{or} \quad x = 1.0986123 \\ \approx -1.099 &\quad \text{or} \quad \approx 1.099 \end{aligned}$$

4. Solve $2e^x - 12e^{-x} = 5$.
Rearrange:

$$2e^x - 5 - 12e^{-x} = 0$$

Multiply by e^x:

$$2e^{2x} - 5e^x - 12 = 0$$

Put $e^x = X$:

$$2X^2 - 5X - 12 = 0$$

By factorization:

$$\begin{aligned} (2X+3)(X-4) &= 0 \\ 2X + 3 = 0 &\quad \text{or} \quad X - 4 = 0 \\ X = -3/2 &\quad \text{or} \quad X = 4 \\ e^x = -3/2 &\quad \text{or} \quad e^x = 4 \end{aligned}$$

But e^x cannot be negative for real values of x, therefore $e^x = -3/2$ has no real solution. However, $e^x = 4$ gives:

$$x = \ln 4 = 1.3862944 \approx 1.386$$

5. $\ln x = a$ is solved by conversion into the form $x = e^a$, using formula *9.5*.

6. $\ln(x-1)+\ln(x-2)=0.$

$$\begin{aligned}\ln(x-1)(x-2)&=0\\(x-1)(x-2)&=e^0=1\\x^2-3x+2&=1\\x^2-3x+1&=0\end{aligned}$$

This cannot be solved by factorization. By the formula:

$$x=\frac{3\pm\sqrt{9-4}}{2}=\frac{3\pm\sqrt{5}}{2}$$

$$=\frac{3+2.236068}{2}\text{ or }\frac{3-2.236068}{2}$$

$$=2.618034\text{ or }0.381966$$
$$\approx 2.618\text{ or }0.382$$

7. $\ln(2x-3)^2=-5.$

$$\begin{aligned}2\ln(2x-3)&=-5\\\ln(2x-3)&=-2.5\\2x-3&=e^{-2.5}\\2x&=3+e^{-2.5}\\x&=\tfrac{1}{2}(3+e^{-2.5})\\&=\tfrac{1}{2}(3+0.082085)\\&=\tfrac{1}{2}\times 3.082085\\&=1.5410425\approx 1.541\end{aligned}$$

8. $\ln(ax+b)^{p/q}=c.$

$$\begin{aligned}\frac{p}{q}\ln(ax+b)&=c\\\ln(ax+b)&=cq/p\\ax+b&=e^{cq/p}\\ax&=e^{cq/p}-b\\x&=\frac{1}{a}(e^{cq/p}-b)\end{aligned}$$

Exercise 9.9

Solve the following equations.

1. $e^x=5$
2. $2e^x=7$

3. $e^x = -2$
4. $3e^x + 2 = 0$
5. $2e^{2x} = 3$
6. $5e^{-x} = 4$
7. $3e^{-x} + 5 = 0$
8. $4e^{-2x} - 7 = 0$
9. $3e^{-2x/3} = 2$
10. $e^{x^2} = 4$
11. $e^x - 4 + 3e^{-x} = 0$
12. $e^x - 6 + 8e^{-x} = 0$
13. $e^{2x} - 3e^x = 0$
14. $3e^x + 5e^{-x} = 9$
15. $3e^{3x} + 5e^x = 8e^{2x}$
16. $4e^x - 11 = 3e^{-x}$
17. $e^x + 2 + e^{-x} = 0$
18. $e^x + 2e^{-x} = -3$
19. $(e^{x/2} + e^{-x/2})^2 = 0$
20. $\ln x = 3$
21. $5 \ln x = 7$
22. $\ln x = -5$
23. $4 \ln x = -11$
24. $2 \ln x + 5 = 3$
25. $(2 \ln x + 1)^2 = 9$
26. $(2 \ln x - 1)(3 \ln x + 2) = 0$
27. $\ln (x-2)^2 = 1$
28. $\ln (x-2) + \ln (x-3) = 0$
29. $\ln (2x-3) - \ln (3x-5) = 0$
30. $\ln (2x+1) = -\ln (3x-2)$
31. $\ln 2 + \ln (x+3) = \ln 3 + \ln (x-2)$
32. $\ln (3x+1)^2 + 4 = 0$
33. $\ln (x-2) = 5$
34. $\ln (x-3)^3 = -3/4$
35. $\ln (4x+1)^{2/3} = 7$
36. $\ln x^2 + \ln x = 3$
37. $\ln (4x+1)^2 + \ln (4x+1) = 3$
38. $\ln (px+q)^{a/b} = c$
39. $\ln (px+q)^a + \ln (px+q)^b = c$
40. $2 \ln x + \ln (1/x) = -2$

10

Partial Fractions

10.1 Identities

The determination of partial fractions makes frequent use of identities. The concept of identities was introduced in *Mathematics Level 1*. Where two algebraic expression are identical they are equal for all values of the independent variable. For instance, if

$$f(x) \equiv g(x) \qquad 10.1$$

then the two functions f(x) and g(x) are equal for all real values of x between $-\infty$ and $+\infty$. Further, if f(x) and g(x) can be expressed as polynomials so that *10.1* may be rewritten:

$$a_0x^n + a_1x^{n-1} + a_2x^{n-2} + \ldots + a_n$$
$$\equiv b_0x^n + b_1x^{n-1} + b_2x^{n-2} + \ldots + b_n \qquad 10.2$$

then this requires that:

$$a_0 = b_0$$
$$a_1 = b_1$$
$$a_2 = b_2$$
$$\cdot$$
$$\cdot$$
$$\cdot$$
$$a_r = b_r$$
$$\cdot$$
$$\cdot$$
$$\cdot$$
$$a_n = b_n$$

There are $n+1$ equations above. They are obtained by equating the coefficients of x on the two sides of identity *10.2*. Both principles will be used in the following examples. In these examples it is usually much simpler to use the first method where applicable, that is, to substitute a particular value of x in both $f(x)$ and $g(x)$ and equate the results. Where that is not possible then the principle of equating coefficients is applied.

10.2 Algebraic fractions

In arithmetic p/q is a proper fraction if and only if the numerator is numerically less than the denominator, e.g. 10/11 is a proper fraction, while 12/11 is an improper fraction (it is > 1).

In similar fashion, $p(x)/q(x)$, where $p(x)$ and $q(x)$ are polynomials, is a proper algebraic fraction if and only if *the degree of* $p(x)$ *is less than the degree of* $q(x)$.

Examples

1. The following are proper algebraic fractions:

$$\frac{3x+2}{4x^2-5x+2} \qquad \frac{ax+b}{Ax^2+Bx+C} \qquad \frac{p}{Px+Q} \qquad \frac{p}{ax^2+bx+c}$$

$$\frac{3x+2}{(x-4)(x+9)} \qquad \frac{2x-5}{(4x-1)(3x+2)(5x-7)}$$

2. The following are not proper fractions in algebra:

$$\frac{3x+2}{4x-5} \qquad \frac{2x^2+3x-5}{7x+11} \qquad \frac{7(x-2)(2x+1)(3x-5)}{x(x+2)(4x-3)}$$

$$\frac{3x^4}{x^2+2} \qquad \frac{7x^3}{(x-1)(x-2)(x+3)}$$

Where a function is not a proper algebraic fraction a rearrangement of it is essential before the following methods may be applied.

Examples

1. Convert $\dfrac{4x-2}{2x+3}$ into a form involving a proper algebraic fraction.

Method 1. By division technique:

$$\begin{array}{r} 2x+3)\overline{4x-2}(2 \\ \underline{4x+6} \\ -8 \end{array}$$

The remainder -8 is of degree 1 less than the divisor. Then:

$$\frac{4x-2}{2x+3} \equiv 2 \text{ (i.e. the quotient)} + \frac{-8}{2x+3}$$

Note the identity. The two sides are not just equal. They are equal in all respects. They are alternative forms of the same expression. Then:

$$\frac{4x-2}{2x+3} \equiv 2 - \frac{8}{2x+3}$$

On the right, $8/(2x+3)$ is a proper algebraic fraction.
Method 2. Since $4x-2$ and $2x+3$ are of the same degree, assume that:

$$\frac{4x-2}{2x+3} \equiv A + \frac{B}{2x+3}$$

where A and B are constants. Multiply both sides by $(2x+3)$. Then:

$$4x-2 \equiv A(2x+3)+B$$

To determine the values of A and B:
equate coefficients of x: $\quad 4 = 2A;\ A = 2 \quad$ (i)
equate constants: $\quad -2 = 3A + B \quad$ (ii)
Substitute in (ii) for A:

$$-2 = 6 + B;\ B = -8$$

Therefore:

$$\frac{4x-2}{2x+3} \equiv 2 - \frac{8}{2x+3}$$

2. Convert the fraction $\dfrac{3x^2-5x+7}{2x+7}$ into a form involving a proper algebraic fraction.
Method 1.

$$\begin{array}{r} 2x+7)\overline{3x^2-5x+7}(\frac{3}{2}x-\frac{31}{4} \\ \underline{3x^2+\frac{21}{2}x} \\ -\frac{31}{2}x+7 \\ \underline{-\frac{31}{2}x-\frac{217}{4}} \\ \frac{245}{4} \end{array}$$

Therefore $$\frac{3x^2-5x+7}{2x+7} \equiv \frac{3}{2}x - \frac{31}{4} + \frac{245}{4(2x+7)}$$

Method 2. Since $3x^2-5x+7$ is of degree 1 more than $2x+7$ assume that:

$$\frac{3x^2-5x+7}{2x+7} \equiv Ax+B+\frac{C}{2x+7}$$

$$3x^2-5x+7 \equiv (Ax+B)(2x+7)+C$$

Equate coefficients of x^2:

$$3 = 2A$$
$$A = 3/2$$

Equate coefficients of x: $\quad -5 = 7A+2B$
Substitute for A:

$$2B+21/2 = -5$$
$$2B = -31/2$$
$$B = -31/4$$

Equate constants: $\quad 7 = 7B+C$
Substitute for B:

$$C-217/4 = 7$$
$$C = 7+217/4$$
$$= 245/4$$

Then $$\frac{3x^2-5x+7}{2x+7} \equiv \frac{3}{2}x - \frac{31}{4} + \frac{245}{4(2x+7)}$$

The expression on the left is identical to a linear function plus a proper algebraic fraction.

3. Convert $\dfrac{4x^3+11x^2-9x-17}{(x-1)(x+2)}$ into a form involving a proper algebraic fraction.

Method 1. First write $(x-1)(x+2) \equiv x^2+x-2$

$$
\begin{array}{r}
x^2+x-2)\,4x^3+11x^2-9x-17\,(4x+7 \\
4x^3+4x^2-8x \\
\hline
7x^2-x-17 \\
7x^2+7x-14 \\
\hline
-8x-3
\end{array}
$$

The remainder, $-8x-3$, is of degree 1 less than the divisor. Therefore:

$$\frac{4x^3+11x^2-9x-17}{(x-1)(x+2)} \equiv 4x+7-\frac{(8x+3)}{(x-1)(x+2)}$$

Method 2. Assume that:

$$\frac{4x^3+11x^2-9x-17}{(x-1)(x+2)} \equiv Px+Q+\frac{Rx+S}{(x-1)(x+2)}$$

$Rx+S$ is taken to be the numerator of the proper fraction because it is a polynomial of *highest degree*, which is, in degree, still 1 less than that of the denominator.

$$4x^3+11x^2-9x-17 \equiv (Px+Q)(x-1)(x+2)+(Rx+S)$$

Put $x=1$:

$$4+11-9-17=0+R+S$$
$$R+S=-11 \qquad \text{(i)}$$

Put $x=-2$:

$$-32+44+18-17=0-2R+S$$
$$2R-S=-13 \qquad \text{(ii)}$$

(i) + (ii) gives:

$$3R=-24$$
$$R=-8$$

Substitute for R in (i):

$$S=-11+8=-3$$

Equate coefficients of x^3: $\quad 4=P$
Equate coefficients of x^2: $\quad 11=P+Q$
Substitute for P: $\quad Q=11-4=7$
Then:

$$\frac{4x^3+11x^2-9x-17}{(x-1)(x+2)} \equiv 4x+7-\frac{(8x+3)}{(x-1)(x+2)}$$

Exercise 10.1

Express each of the following as a polynomial (that includes the case of a constant) plus a proper algebraic fraction.

1. $\dfrac{2x+1}{x-1}$

2. $\dfrac{2x+1}{x+1}$

3. $\dfrac{3x-2}{x-2}$

4. $\dfrac{4x+3}{x+3}$

5. $\dfrac{5x+6}{2x-3}$

6. $\dfrac{10x-7}{5x-2}$

7. $\dfrac{x^2+x+2}{x+1}$

8. $\dfrac{x^2-4x+7}{x-1}$

9. $\dfrac{x^2+3x+2}{x-1}$

10. $\dfrac{x^2+5x+5}{x-3}$

11. $\dfrac{2x^2+3x-4}{2x-1}$

12. $\dfrac{x^3+x^2+x+7}{x-1}$

13. $\dfrac{x^3+x^2+x+1}{x-2}$

14. $\dfrac{x^3-4x^2+5x-6}{x-3}$

15. $\dfrac{6x^3-7x^2+11x-4}{3x-2}$

16. $\dfrac{x^2+5x+6}{x^2-2x-3}$

17. $\dfrac{6x^2+11x+19}{2x^2-3x+5}$

18. $\dfrac{6x^3+17x^2-15x-9}{3x^2+12x-5}$

19. $\dfrac{6x^3+17x^2-15x-9}{(x-2)(x+3)}$

20. $\dfrac{6x^3+17x^2-15x-9}{(x+1)(x+2)(x-3)}$

It should always be possible now for us to convert an expression $\dfrac{P(x)}{Q(x)}$, where the degree of $P(x)$ is at least equal to the degree of $Q(x)$, into the form developed above: a polynomial plus a proper algebraic fraction $\dfrac{p(x)}{q(x)}$, where $p(x)$ is at most of degree 1 less than $q(x)$. In certain cases the fraction $\dfrac{p(x)}{q(x)}$ may be expressed as the sum of other fractions whose denominators are factors of $q(x)$. When these further fractions are themselves proper algebraic fractions then they are called partial fractions of $\dfrac{p(x)}{q(x)}$.

Examples

1. Express $\dfrac{p(x)}{(x-a)(x-b)}$ as the sum of two fractions whose denominators are $(x-a)$ and $(x-b)$ respectively, where $p(x)$ is at most of degree 1, and does not possess either $(x-a)$ or $(x-b)$ as a factor.
   We assume:

$$\frac{p(x)}{(x-a)(x-b)} \equiv \frac{A}{(x-a)} + \frac{B}{(x-b)} \qquad 10.3$$

   where A and B are constants. To evaluate A and B multiply *10.3* by $(x-a)(x-b)$. Then:

$$p(x) \equiv A(x-b)+B(x-a)$$

Put $x = a$:

$$\mathrm{p}(a) = A(a-b)+0$$

$$A = \frac{\mathrm{p}(a)}{(a-b)} \qquad \textit{10.4}$$

Put $x = b$:

$$\mathrm{p}(b) = 0+B(b-a)$$

$$B = \frac{\mathrm{p}(b)}{(b-a)} \qquad \textit{10.5}$$

2. Use example 1 to express $\dfrac{3x+5}{(x-1)(x-2)}$ in partial fractions.

Assume:

$$\frac{3x+5}{(x-1)(x-2)} \equiv \frac{A}{(x-1)}+\frac{B}{(x-2)} \qquad \text{(i)}$$

Multiply (i) by $(x-1)(x-2)$:

$$3x+5 \equiv A(x-2)+B(x-1) \qquad \text{(ii)}$$

In (ii) put $x = 1$:

$$8 = A(-1)+0$$
$$A = -8$$

In (ii) put $x = 2$:

$$11 = 0+B(1)$$
$$B = 11$$

Then:

$$\frac{3x+5}{(x-1)(x-2)} \equiv \frac{11}{(x-2)}-\frac{8}{(x-1)} \qquad \text{(iii)}$$

Check: put $x = 0$ in (iii):

$$\text{LHS} = \frac{5}{(-1)(-2)} = 5/2; \quad \text{RHS} = \frac{11}{-2}-\frac{8}{-1} = 5/2$$

Alternatively the values of A and B may be found by direct substitution in *10.4* and *10.5* above. These two formulae are also obtained by what is called the cover-up rule. To find the numerator of the partial fraction of

$\dfrac{p(x)}{(x-a)(x-b)}$ with denominator $(x-a)$ cover that factor up in the denominator of the original expression and in the remaining part of it substitute the value a for x. To find the numerator of the fraction with denominator $(x-b)$ cover up that factor in the original expression and in the remaining part of it substitute the value b for x. That procedure gives the two formulae *10.4* and *10.5*. The cover-up rule may also be extended to the determination of the partial fractions of an expression with any number of simple factors in the denominator.

3. Obtain the partial fractions of $\dfrac{13}{(x-1)(x+2)}$.

Assume:

$$\frac{13}{(x-1)(x+2)} \equiv \frac{A}{x-1} + \frac{B}{x+2}$$

By the cover-up rule, to find A, cover up the factor $(x-1)$ in the denominator of the original expression, thus leaving $\dfrac{13}{(x+2)}$. In that put $x = 1$, giving:

$$\frac{13}{1+2} = \frac{13}{3}$$

To find B:

$$B = \frac{13}{(x-1)} \text{ when } x = -2$$

$$= \frac{13}{-2-1} = -\frac{13}{3}$$

Then:

$$\frac{13}{(x-1)(x+2)} \equiv \frac{13}{3(x-1)} - \frac{13}{3(x+2)} \qquad \text{(i)}$$

Check: put $x = 0$ in (i):

$$\text{LHS} = \frac{13}{(-1)(2)} = -13/2; \quad \text{RHS} = \frac{13}{-3} - \frac{13}{6} = -13/2$$

Exercise 10.2

Express the following in partial fractions.

1. $\dfrac{2x+3}{(x-1)(x-2)}$

2. $\dfrac{3x+5}{(x-1)(x-2)}$

3. $\dfrac{4x-7}{(x-1)(x-2)}$

4. $\dfrac{7}{(x-1)(x+2)}$

5. $\dfrac{8}{(x+2)(x+1)}$

6. $\dfrac{3}{(x+2)(x+3)}$

7. $\dfrac{5x+4}{x(x+5)}$: Express as $\dfrac{A}{x}+\dfrac{B}{x+5}$

8. $\dfrac{11x-10}{(x-2)(x+5)}$

9. $\dfrac{3x}{(x-2)(x-6)}$

10. $\dfrac{11}{x(x+3)}$

Examples

1. Express $\dfrac{\mathrm{p}(x)}{(x-a)(x-b)(x-c)}$ in partial fractions, where $\mathrm{p}(x)$ is, at most, of degree 2.
   Assume:

$$\frac{\mathrm{p}(x)}{(x-a)(x-b)(x-c)} \equiv \frac{A}{(x-a)}+\frac{B}{(x-b)}+\frac{C}{(x-c)}$$

Then, multiplying throughout by $(x-a)(x-b)(x-c)$:

$$\mathrm{p}(x) \equiv A(x-b)(x-c)+B(x-c)(x-a)+C(x-a)(x-b)$$

Put $x = a$:

$$\mathrm{p}(a) = A(a-b)(a-c)+0+0$$

$$A = \frac{\mathrm{p}(a)}{(a-b)(a-c)}$$

Put $x = b$:

$$\mathrm{p}(b) = 0+B(b-c)(b-a)+0$$

$$B = \frac{\mathrm{p}(b)}{(b-c)(b-a)}$$

Put $x = c$:

$$\mathrm{p}(c) = 0+0+C(c-a)(c-b)$$

$$C = \frac{\mathrm{p}(c)}{(c-a)(c-b)}$$

Again the values of A, B and C may be found by using the cover-up rule. Note that in the above example it is assumed that $\mathrm{p}(x)$ does not possess a factor equal to $(x-a)$, $(x-b)$ or $(x-c)$.

2. Use example 1 to express $\dfrac{5}{(x-1)(x-2)(x+2)}$ in partial fractions.

Assume:

$$\frac{5}{(x-1)(x-2)(x+2)} \equiv \frac{A}{(x-1)}+\frac{B}{(x-2)}+\frac{C}{(x+2)}$$

Therefore:

$$5 \equiv A(x-2)(x+2)+B(x-1)(x+2)+C(x-1)(x-2)$$

Put $x = 1$:

$$5 = A(-1)(3)+0+0$$
$$A = -5/3$$

Put $x = 2$:

$$5 = 0+B(1)(4)+0$$
$$B = 5/4$$

Put $x = -2$:

$$5 = 0 + 0 + C(-3)(-4)$$
$$C = 5/12$$

Therefore:

$$\frac{5}{(x-1)(x-2)(x+2)} \equiv -\frac{5}{3(x-1)} + \frac{5}{4(x-2)} + \frac{5}{12(x+2)}$$

Check: put $x = 0$:

$$\text{LHS} = \frac{5}{(-1)(-2)(2)} = 5/4$$

$$\text{RHS} = \frac{-5}{-3} + \frac{5}{-8} + \frac{5}{24} = \frac{5}{3} - \frac{5}{8} + \frac{5}{24} = \frac{40 - 15 + 5}{24} = \frac{30}{24} = \frac{5}{4}$$

3. Use example 1 to express $\frac{3x+4}{x(x-2)(x+4)}$ in partial fractions.

Assume:

$$\frac{3x+4}{x(x-2)(x+4)} \equiv \frac{A}{x} + \frac{B}{x-2} + \frac{C}{x+4}$$

By the cover-up rule:

$$A = \frac{0+4}{(-2)(4)} = -\frac{1}{2}$$

$$B = \frac{6+4}{2(2+4)} = \frac{10}{12} = \frac{5}{6}$$

$$C = \frac{-12+4}{-4(-4-2)} = \frac{-8}{24} = -\frac{1}{3}$$

Then:

$$\frac{3x+4}{x(x-2)(x+4)} \equiv -\frac{1}{2x} + \frac{5}{6(x+2)} - \frac{1}{3(x+4)}$$

Check: put $x = 1$:

$$\text{LHS} = \frac{7}{1(-1)(5)} = -7/5$$

$$\text{RHS} = -\frac{1}{2} + \frac{5}{-6} - \frac{1}{15} = \frac{-15 - 25 - 2}{30} = -\frac{42}{30} = -\frac{7}{5}$$

4. Express $\dfrac{3x^2-4x+5}{(x-1)(x+3)(x-5)}$ in partial fractions.

Assume:

$$\frac{3x^2-4x+5}{(x-1)(x+3)(x-5)} \equiv \frac{A}{x-1}+\frac{B}{x+3}+\frac{C}{x-5}$$

$$\text{i.e. } 3x^2-4x+5 \equiv A(x+3)(x-5)+B(x-1)(x-5)+C(x-1)(x+3)$$

Put $x = 1$:

$$3-4+5 = A(4)(-4)+0+0$$
$$A = -1/4$$

Put $x = -3$:

$$27+12+5 = 0+B(-4)(-8)+0$$
$$B = 11/8$$

Put $x = 5$:

$$75-20+5 = 0+0+C(4)(8)$$
$$C = 15/8$$

Then:

$$\frac{3x^2-4x+5}{(x-1)(x+3)(x-5)} \equiv -\frac{1}{4(x-1)}+\frac{11}{8(x+3)}+\frac{15}{8(x-5)}$$

Check: put $x = 0$:

$$\text{LHS} = \frac{5}{(-1)(3)(-5)} = \frac{1}{3}$$

$$\text{RHS} = \frac{-1}{-4}+\frac{11}{24}+\frac{15}{-40} = \frac{30+55-45}{120} = \frac{40}{120} = \frac{1}{3}$$

Exercise 10.3

Express the following in partial fractions.

1. $\dfrac{3}{x(x-1)(x+1)}$

2. $\dfrac{4}{x(x-1)(x-2)}$

3. $\dfrac{17}{(x-1)(x-2)(x-3)}$

4. $\dfrac{11}{(x+1)(x+4)(x+6)}$

5. $\dfrac{x+2}{x(x-1)(x+1)}$

6. $\dfrac{3x+5}{(x-1)(x-4)(x-7)}$

7. $\dfrac{11-6x}{x(x-2)(x-8)}$

8. $\dfrac{25x-73}{(x+3)(x-9)(x+11)}$

9. $\dfrac{x^2+2}{x(x-1)(x+1)}$

10. $\dfrac{x^2-2x}{(x-1)(x+2)(x-3)}$

11. $\dfrac{x^2+x+1}{x(x-1)(x+1)}$

12. $\dfrac{x^2+8x+10}{(x-2)(x+6)(x-5)}$

13. $\dfrac{3x^2-5}{x(x+2)(x+5)}$

14. $\dfrac{4x^2+7}{(x-5)(x+5)(x-7)}$

15. $\dfrac{5x^2+11x-19}{x(x-4)(x+10)}$

16. $\dfrac{24-13x-7x^2}{(x+11)(x-5)(x+2)}$

10.3 When q(x) involves a repeated factor

Examples

1. Where $p(x)/q(x)$ is of the form $\dfrac{ax+b}{(x-c)^2}$ we assume it may be written:

$$\frac{A}{(x-c)}+\frac{B}{(x-c)^2}$$

This is equivalent to:

$$\frac{A(x-c)}{(x-c)^2}+\frac{B}{(x-c)^2}$$

$$\equiv\frac{A(x-c)+B}{(x-c)^2}\equiv\frac{Ax+(B-Ac)}{(x-c)^2}$$

which is a proper algebraic fraction: the numerator is of degree less than the denominator.

2. Using example 1 express $\dfrac{4x+5}{(x-2)^2}$ in partial fractions.

Assume:

$$\frac{4x+5}{(x-2)^2}\equiv\frac{A}{(x-2)}+\frac{B}{(x-2)^2}$$

Then:

$$4x+5\equiv A(x-2)+B$$

On this occasion the cover-up rule is not applicable.
Put $x=2$:

$$8+5=0+B$$
$$B=13$$

Equate coefficients of x: $4=A$. Therefore:

$$\frac{4x+5}{(x-2)^2}\equiv\frac{4}{(x-2)}+\frac{13}{(x-2)^2}$$

Check: put $x=0$:

$$\text{LHS}=\frac{5}{(-2)^2}=5/4$$

$$\text{RHS}=\frac{4}{-2}+\frac{13}{(-2)^2}=-2+\frac{13}{4}=-2+3\tfrac{1}{4}=1\tfrac{1}{4}$$

3. Express $\dfrac{3}{(x-4)^2}$ in partial fractions.

   This is already in the form required. No further work needs to be done on it.

4. Express $\dfrac{ax^2+bx+c}{(x-p)^2\,(x-q)}$ in partial fractions.

   Assume it may be written:

$$\frac{A}{(x-p)}+\frac{B}{(x-p)^2}+\frac{C}{(x-q)}$$

5. Express $\dfrac{7x-8}{(x+3)^2\,(x-7)}$ in partial fractions.

   Assume:

$$\frac{7x-8}{(x+3)^2\,(x-7)} \equiv \frac{A}{(x+3)}+\frac{B}{(x+3)^2}+\frac{C}{(x-7)}$$

   Then:

$$7x-8 \equiv A(x+3)(x-7)+B(x-7)+C(x+3)^2$$

   Put $x = 7$:

$$41 = 0+0+C(10)^2$$

$$C = \frac{41}{100}$$

   Note that the value of C could have been obtained by the cover-up rule: $C =$ the value of $\dfrac{7x-8}{(x+3)^2}$ when $x=7$.

   Put $x = -3$:

$$-29 = 0+B(-10)+0$$

$$B = \frac{29}{10}$$

   Equate coefficients of x^2: $0 = A+C$.
   Substitute for C:

$$A = -\frac{41}{100}$$

   Note that the values of B and A may not be found by the cover-up rule.
   Then:

$$\frac{7x-8}{(x+3)^2\,(x-7)} \equiv \frac{-41}{100(x+3)}+\frac{29}{10(x+3)^2}+\frac{41}{100(x-7)}$$

Check: put $x = 0$:

$$\text{LHS} = \frac{11}{9 \times 1} = 11/9$$

$$\text{RHS} = \frac{-22}{-147} + \frac{11}{63} + \frac{44}{49} = \frac{22}{147} + \frac{11}{63} + \frac{44}{49}$$

$$= \frac{66 + 77 + 396}{441} = \frac{539}{441} = 11/9$$

Exercise 10.4

Express the following in partial fractions.

1. $\dfrac{x+1}{(x-1)^2}$
2. $\dfrac{x-1}{(x+1)^2}$
3. $\dfrac{2x+3}{(x+1)^2}$
4. $\dfrac{5x+6}{(x-2)^2}$
5. $\dfrac{11x-12}{(x+3)^2}$
6. $\dfrac{2x-3}{(x-1)^2}$
7. $\dfrac{15x+44}{(x-6)^2}$
8. $\dfrac{3x+2}{(2x-1)^2}$
9. $\dfrac{4x-5}{(2x-3)^2}$
10. $\dfrac{2x+3}{(3x-2)^2}$
11. $\dfrac{10}{(x-1)^2\,(x-2)}$

12. $\dfrac{9}{(x-2)^2(x-1)}$

13. $\dfrac{16}{(x+1)^2(x+2)}$

14. $\dfrac{8}{x^2(x-1)}$: assume it to be $\dfrac{A}{x}+\dfrac{B}{x^2}+\dfrac{C}{x-1}$

15. $\dfrac{3x+5}{x(4x-3)^2}$

16. $\dfrac{5x-8}{(2x+3)(x-4)^2}$

17. $\dfrac{9-4x}{(5x+1)^2(3x-7)}$

18. $\dfrac{x^2+x+3}{(x-1)^2(x-2)}$

19. $\dfrac{7x^2-11x+4}{(x+2)^2(x-3)}$

20. $\dfrac{2-9x-13x^2}{(3x-5)^2(6x+1)}$

10.4 When q(x) contains a quadratic factor

Examples

1. When $\mathrm{p}(x)/\mathrm{q}(x)$ is of the form $\dfrac{\mathrm{p}(x)}{x^2+a^2}$, where a is a constant and $\mathrm{p}(x)$ is of degree, at most, 1.
   Then $\mathrm{p}(x)/(x^2+a^2)$ must be of the form $\dfrac{Ax+B}{x^2+a^2}$, which is a proper algebraic fraction. This is also true when $A=0$, i.e. $\mathrm{p}(x)$ is a constant. When that is so we need take the process no further.
2. Examples of 1 are:

$$\frac{3}{x^2+1} \qquad -\frac{9}{x^2+4} \qquad \frac{11}{4x^2+9} \qquad \frac{3x-5}{x^2+5} \qquad \frac{7-8x}{9x^2+16}$$

3. When $p(x)/q(x)$ is of the form $\dfrac{p(x)}{(x^2+a^2)(x-b)}$ and $p(x)$ is, at most, of degree 2, we may assume that it may be written:

$$\frac{Ax+B}{(x^2+a^2)}+\frac{C}{(x-b)}$$

4. Express $\dfrac{11}{(x^2+1)(x-2)}$ in partial fractions.

   By example 3 assume it may be written $\dfrac{Ax+B}{x^2+1}+\dfrac{C}{x-2}$

   Then $11 \equiv (Ax+B)(x-2)+C(x^2+1)$

   Put $x = 2$:

$$11 = 0+C(5)$$
$$C = \frac{11}{5}$$

   Note that the value of C may be found by the cover-up rule.
   Equate coefficients of x^2: $0 = A+C$.
   Substitute for C:

$$A = -\frac{11}{5}$$

   Equate constants: $0 = B-2A$.
   Substitute for A:

$$B = -\frac{22}{5}$$

   Therefore:

$$\frac{11}{(x^2+1)(x-2)} \equiv \frac{-11(x+2)}{5(x^2+1)}+\frac{11}{5(x-2)}$$

   Check: put $x = 0$:

$$\text{LHS} = \frac{11}{(1)(-2)} = -\frac{11}{2}$$

$$\text{RHS} = -\frac{22}{5}-\frac{11}{10} = \frac{-44-11}{10} = -\frac{55}{10} = -\frac{11}{2}$$

5. Express $\dfrac{10x-19}{(4x^2+9)(2x-5)} \equiv \dfrac{Ax+B}{4x^2+9}+\dfrac{C}{2x-5}$

i.e. $10x - 19 \equiv (Ax + B)(2x - 5) + C(4x^2 + 9)$

Put $x = \frac{5}{2}$:

$$6 = 0 + C(34)$$

$$C = \frac{3}{17}$$

C may also be obtained by the cover-up rule.
Equate coefficients of x^2: $0 = 2A + 4C$.
Substitute for C:

$$A = -2C = -\frac{6}{17}$$

Equate constants: $-19 = -5B + 9C$.
Substitute for C:

$$5B = 19 + 9C$$

$$= 19 + \frac{27}{17} = \frac{323}{17} + \frac{27}{17} = \frac{350}{17}$$

$$B = \frac{70}{17}$$

Therefore:

$$\frac{10x - 19}{(4x^2 + 9)(2x - 5)} \equiv -\frac{(6x - 70)}{17(4x^2 + 9)} + \frac{3}{17(2x - 5)}$$

Check: put $x = 0$:

$$\text{LHS} = \frac{-19}{-45} = \frac{19}{45}$$

$$\text{RHS} = \frac{70}{17 \times 9} - \frac{3}{85} = \frac{350 - 27}{765} = \frac{323}{765} = \frac{19}{45}$$

6. Express $\dfrac{17 - 14x - 19x^2}{(2x^2 + 3)(3x - 2)}$ in partial fractions.

Assume:

$$\frac{17 - 14x - 19x^2}{(2x^2 + 3)(3x - 2)} \equiv \frac{Ax + B}{2x^2 + 3} + \frac{C}{3x - 2}$$

i.e. $17 - 14x - 19x^2 \equiv (Ax + B)(3x - 2) + C(2x^2 + 3)$

Although the value of C may be found by the cover-up rule or by putting

$x = 2/3$, the calculations in this example by both methods are somewhat complicated. It is probably easier to proceed as follows in this case:

Equate coefficients of x^2:	$-19 = 3A + 2C$	(i)
Equate coefficients of x:	$-14 = -2A + 3B$	(ii)
Equate constants:	$17 = -2B + 3C$	(iii)

Now solve (i), (ii) and (iii) to find the values of A, B and C. Eliminate A from (i) and (ii):

$$2(\text{i}) + 3(\text{ii}) \text{ gives } -38 - 42 = 4C + 9B$$
$$\text{i.e. } -80 = 4C + 9B \qquad \text{(iv)}$$

Eliminate C from (iii) and (iv):

$$4(\text{iii}) - 3(\text{iv}) \text{ gives } 68 + 240 = -8B - 27B$$
$$\text{i.e. } 308 = -35B$$
$$B = -\frac{308}{35} = -\frac{44}{5}$$

Substitute for B in (iv):

$$-80 = 4C - \frac{396}{5}$$

$$4C = 79\frac{1}{5} - 80 = -\frac{4}{5}$$

$$C = -\frac{1}{5}$$

Substitute for C in (i):

$$-19 = 3A - \frac{2}{5}$$

$$3A = -18\frac{3}{5}$$

$$A = -6\frac{1}{5} = -\frac{31}{5}$$

Therefore:

$$\frac{17 - 14x - 19x^2}{(2x^2 + 3)(3x - 2)} \equiv -\frac{(31x + 44)}{5(2x^2 + 3)} - \frac{1}{5(3x - 2)}$$

Check: put $x = 0$:

$$\text{LHS} = \frac{17}{-6} = -\frac{17}{6}$$

$$\text{RHS} = -\frac{44}{15} + \frac{1}{10} = \frac{-88+3}{30} = \frac{-85}{30} = -\frac{17}{6}$$

Exercise 10.5

Express the following in partial fractions.

1. $\dfrac{3}{x(x^2+1)}$

2. $\dfrac{5}{x(x^2+4)}$

3. $-\dfrac{7}{x(x^2+25)}$

4. $\dfrac{11}{(x-1)(x^2+1)}$

5. $\dfrac{9}{x(4x^2+1)}$

6. $\dfrac{2}{x(x^2+1)}$

7. $\dfrac{3}{x(2x^2+1)}$

8. $\dfrac{5}{(x+2)(x^2+3)}$

9. $\dfrac{x+1}{x(x^2+1)}$

10. $\dfrac{2x+3}{x(x^2+1)}$

11. $\dfrac{12-5x}{(x-4)(5x^2+2)}$

12. $\dfrac{-9x}{(x+5)(x^2-7)}$

13. $\dfrac{11x+12}{(2x+1)(3x^2+5)}$

14. $\dfrac{17-4x}{(3x-5)(5x^2+2)}$

15. $\dfrac{x^2+x+1}{(x-1)(x^2+1)}$

16. $\dfrac{1+x-x^2}{(x+1)(x^2+4)}$

17. $\dfrac{2x^2+5x+4}{(x-2)(x^2+4)}$

18. $\dfrac{4x^2-2x-7}{(x-3)(x^2+16)}$

19. $\dfrac{5+3x-4x^2}{(x+2)(x^2+5)}$

20. $\dfrac{8x^2+11x-9}{(2x+3)(3x^2+2)}$

10.5 The procedure when p(x) and q(x) are of the same degree

In the event that p(x) and q(x) are of the same degree then p(x)/q(x) is not a proper algebraic fraction. The first step must be that outlined at the beginning of the chapter.

Example

Express $\dfrac{2x^2+5}{(x-1)(x+2)}$ in a form involving partial fractions.

Divide p(x) by q(x):

$$\begin{array}{r} x^2+x-2)\ 2x^2 \qquad +5(2 \\ \underline{2x^2+2x-4} \\ -2x+9 \end{array}$$

The remainder, $-2x+9$, is of degree less than q(x). Then:

$$\frac{2x^2+5}{(x-1)(x+2)} \equiv 2 - \frac{(2x-9)}{(x-1)(x+2)}$$

Assume:

$$-\frac{(2x-9)}{(x-1)(x+2)} \equiv \frac{A}{x-1} + \frac{B}{x+2}$$

By the cover-up rule:

$$A = -\frac{(2-9)}{(1+2)} = \frac{7}{3}$$

$$B = -\frac{(-4-9)}{(-2-1)} = -\frac{13}{3}$$

Therefore:

$$\frac{2x^2+5}{(x-1)(x+2)} \equiv 2 + \frac{7}{3(x-1)} - \frac{13}{3(x+2)}$$

Check: put $x = 0$:

$$\text{LHS} = \frac{5}{-2} = -2\tfrac{1}{2}$$

$$\text{RHS} = 2 - \tfrac{7}{3} - \tfrac{13}{6} = 2 - 2\tfrac{1}{3} - 2\tfrac{1}{6} = -\tfrac{1}{3} - 2\tfrac{1}{6} = -2\tfrac{1}{2}$$

Exercise 10.6

Express the following in a form involving partial fractions.

1. $\dfrac{x+3}{x-1}$

2. $\dfrac{x-5}{x+1}$

3. $\dfrac{2x+7}{x-2}$

4. $\dfrac{4x-9}{x+3}$

5. $\dfrac{7x+6}{2x+1}$

6. $\dfrac{3-2x}{x+5}$

7. $\dfrac{x^2+1}{x(x+1)}$

8. $\dfrac{x^2-1}{x(x-1)}$

9. $\dfrac{x^2+1}{(x+1)(x+2)}$

10. $\dfrac{x^2+3}{(x-1)(x-2)}$

11. $\dfrac{2x^2+5}{(x+2)(x+3)}$

12. $\dfrac{3x^2+1}{x(3x+1)}$

13. $\dfrac{1-x^2}{x^2+1}$

14. $\dfrac{x^2+x+1}{(x+1)(x+2)}$

15. $\dfrac{x^2-x+1}{(x-1)(x+2)}$

16. $\dfrac{1+x-x^2}{(1-x)(2-x)}$

17. $\dfrac{4x^2+3x+1}{(x+1)(x+3)}$

18. $\dfrac{6x^2+5x+2}{(x+2)(x-5)}$

19. $\dfrac{7-8x^2}{2x^2+3}$

20. $\dfrac{(x+1)(x+2)}{(x-1)(x-2)}$

11

The Binomial Theorem and the Binomial Series

11.1 The expansion of powers of binomial expressions

For various purposes we need to be able to expand expressions of the form $(a+x)^n$ where n is a positive integer. Already we know that:

$$(a+x)^1 \equiv a+x \qquad \text{(i)}$$

And, from Level 1:

$$(a+x)^2 \equiv a^2+2ax+x^2 \qquad \text{(ii)}$$

Multiply (ii) by $(a+x)$:

$$\begin{aligned}(a+x)^3 &\equiv (a+x)(a^2+2ax+x^2)\\ &\equiv a^3+a^2x+2a^2x+2ax^2+ax^2+x^3\\ &\equiv a^3+3a^2x+3ax^2+x^3 \qquad \text{(iii)}\end{aligned}$$

(iii) multiplied by $(a+x)$ produces:

$$(a+x)^4 \equiv a^4+4a^3x+6a^2x^2+4ax^3+x^4 \qquad \text{(iv)}$$

In this way the expansion of $(a+x)^n$ for $n = 5, 6, 7 \ldots$ etc. may be obtained.

If we had to resort to this method to expand, for example, $(a+x)^{10}$, the process would be lengthy and tiresome, so a shorter method is required. By looking at (i), (ii), (iii) and (iv) above we detect the following facts:

1. The first term on the right of the expression is always a^n for the binomial $(a+x)^n$; i.e. the power of a is always the same as the power of $(a+x)$.

2. As we proceed term by term to the right of the expanded form the powers of a decrease by 1 and the powers of x increase by 1; e.g. in (iv) the progression is a^4, a^3x, a^2x^2, ax^3, x^4.
3. In any term the sum of the powers of a and of x is always n; e.g. in (iv) the sum is always 4.
4. The last term in the expanded form is always x^n.
5. The coefficients of the terms are symmetrical about the middle of the expansion. This means there is no need to calculate more than the coefficients up to the middle of the expansion.
6. A rule is needed to determine the coefficients of the various terms. One such rule is called Pascal's triangle. The coefficients alone are extracted and written down in the form below:

Line 1:					1		1				
Line 2:				1		2		1			
Line 3:			1		3		3		1		
Line 4:		1		4		6		4		1	
Line 5:	1		5		10		10		5		1

Note that in Lines 1 to 4 the end numbers are always 1. The remaining numbers in any given line are obtained by adding together the two numbers in the line above which are immediately to each side of it. For Line 5: $5 = 1+4$, $10 = 4+6$, $10 = 6+4$, $5 = 4+1$. Line 6 is 1, 6, 15, 20, 15, 6, 1.
7. The number of terms is always one more than n.

Exercise 11.1

Use points 1 to 7 above to expand the following:

1. $(a+x)^6$
2. $(a+x)^7$
3. $(a+b)^8$
4. $(a+x)^{10}$
5. $(x+y)^{11}$
6. $(a+b)^{12}$

11.2 The binomial theorem

The disadvantage of using Pascal's triangle to calculate the coefficients in the expansion of a binomial is that all the intermediate stages have to be determined in order to arrive at the required result. The binomial theorem does not suffer from this disadvantage. It states that:

$$(a+x)^n \equiv a^n + \binom{n}{1}a^{n-1}.x + \binom{n}{2}a^{n-2}.x^2 + \binom{n}{3}a^{n-3}.x^3 + \ldots$$
$$+ \binom{n}{r}a^{n-r}.x^r + \ldots + x^n \qquad 11.1$$

where $\binom{n}{1} = \frac{n}{1}, \binom{n}{2} = \frac{n(n-1)}{1.2}, \binom{n}{3} = \frac{n(n-1)(n-2)}{1.2.3}$

$$\binom{n}{r} = \frac{n(n-1)(n-2)\ldots(n-r+1)}{1.2.3\ldots r}$$

The expression $1.2.3.4 \ldots r$ is abbreviated to $r!$ (factorial r). Alternative notations to $\binom{n}{1}$ are ${}^nC_r$ and ${}_nC_r$. These are called binomial coefficients.

Examples

1. Determine the following binomial coefficients:

$$\binom{2}{1}, \binom{3}{1}, \binom{3}{2}, \binom{3}{3}, \binom{4}{2}, \binom{5}{3}, \binom{7}{4}, \binom{10}{5}.$$

$$\binom{2}{1} = \frac{2}{1} = 2 \qquad \binom{3}{1} = \frac{3}{1} = 3 \qquad \binom{3}{2} = \frac{3.2}{1.2} = 3$$

$$\binom{3}{3} = \frac{3.2.1}{1.2.3} = 1 \qquad \binom{4}{2} = \frac{4.3}{1.2} = 6 \qquad \binom{5}{3} = \frac{5.4.3}{1.2.3} = 10$$

$$\binom{7}{4} = \frac{7.6.5.4}{1.2.3.4} = 35 \qquad \binom{10}{5} = \frac{10.9.8.7.6}{1.2.3.4.5} = 252$$

2. Expand $(a+x)^{10}$.

$$(a+x)^{10} = a^{10} + \binom{10}{1}a^9x + \binom{10}{2}a^8x^2 + \binom{10}{3}a^7x^3 + \ldots$$
$$+ \binom{10}{r}.a^r.x^{10-r} + \ldots + x^{10}$$
$$= a^{10} + 10a^9x + 45a^8x^2 + 120a^7x^3 + 210a^6x^4$$
$$+ 252a^5x^5 + 210a^4x^6 + 120a^3x^7 + 45a^2x^8 + 10ax^9 + x^{10}$$

3. Write down the fourth term in the expansion of $(2+3x)^7$.
   The fourth term, call it a_4, is:

$$\binom{7}{3}(2)^4(3x)^3 = \frac{7.6.5}{1.2.3}.16.27x^3 = 35.16.27x^3 = 15\,120x^3$$

4. Write down the sixth term in the expansion of $(2x-3y)^8$.
   Express the binomial as $[(2x)+(-3y)]^8$

$$a_6 = \binom{8}{5}(2x)^3(-3y)^5 = \frac{8.7.6.5.4}{1.2.3.4.5}(2^3)(-3)^5 x^3y^5$$

$$= -108\,864x^3y^5$$

5. Write down the rth term of the expansion of $(ax+by)^n$.

$$a_r = \binom{n}{r-1}(ax)^{n-r+1}(by)^{r-1}$$

   Note:

   (a) The rth term has $r-1$ for the lower number in the binomial coefficient.
   (b) (by) is raised to the power $(r-1)$.
   (c) The power of (ax) is $n-(r-1) = n-r+1$.
   (d) Notes (a), (b) and (c) represent the order in which it is advisable to write down the various parts of a_r.

6. Determine the value of the sixth term in the expansion of $(3x-4y)^{10}$ when $x = 2$ and $y = -5$.

$$a_6 = \binom{10}{5}(3x)^5(-4y)^5 = \frac{10.9.8.7.6}{1.2.3.4.5}.3^5(-4)^5x^5y^5$$

   When $x = 2$ and $y = -5$:

$$a_6 = \frac{10.9.8.7.6}{1.2.3.4.5}.3^5.4^5.2^5.5^5$$

$$= \frac{9.4.7}{1} \times 120^5 = 6\,270\,566\,400\,000$$

Exercise 11.2

Determine the expansions of the following expressions:

1. $(x+y)^6$
2. $(x+2y)^4$

3. $(2x+y)^4$
4. $(x+3y)^5$
5. $(x-y)^4$
6. $(x-2y)^4$
7. $(2x-y)^4$
8. $(x-3y)^6$
9. $(x+\frac{1}{2}y)^8$
10. $(x-\frac{1}{2}y)^6$
11. $(2x+\frac{1}{2}y)^4$
12. $(3x+2)^6$
13. $(2-5x)^4$
14. $(2+x)^5$
15. $(1+2x)^4$
16. $(1-2x)^5$

Determine the terms indicated for the expansions of the following expressions:

17. a_5 in $(x+y)^9$
18. a_6 in $(x-y)^{11}$
19. a_4 in $(2x+y)^7$
20. a_3 in $(2+x)^8$
21. a_7 in $(2-x)^{12}$
22. a_4 in $(1-2x)^{10}$
23. a_4 in $(3+2x)^7$
24. a_2 in $(3+2x)^7$ when $x = 0.1$
25. a_3 in $(4+5x)^8$ when $x = 0.2$
26. a_4 in $(10-3x)^6$ when $x = 0.01$

11.3 The binomial series

Formula *11.1* is the form of the expansion of a binomial when n is a positive integer. The expansion always terminates at the $(n+1)$th term. The value of the coefficient of the last term, $\binom{n}{n}$, is 1. If we tried to calculate any term beyond that, for example the $(n+2)$th term, then $\binom{n}{n+1}$ would be:

$$\frac{n(n-1)(n-2)\ldots 1.0}{(n+1)!} = 0$$

All subsequent binomial coefficients contain the factor 0 in the numerator. Consequently they are zero. However, when n is either fractional or a negative integer, 0 will never appear as a factor in the numerator of any binomial coefficient. The series does not terminate for such values of n. The binomial expansion is then called the binomial series. To simplify the form of the expansion we assume the binomial to be expressed in the form $(1+x)$ instead of in the form $(a+x)$. The binomial series is:

$$(1+x)^n = 1+\frac{n}{1}x+\frac{n(n-1)}{2!}x^2+\frac{n(n-1)(n-2)}{3!}x^3+\ldots$$

$$+\frac{n(n-1)(n-2)(n-3)\ldots(n-r+1)}{r!}x^r+\ldots \qquad 11.2$$

This is an infinite series. Because it is, the expansion in the form shown in *11.2* is valid only when the sum of the terms tends to a limit as the number of terms increases indefinitely; i.e., if S_n represents the sum of n terms then S_n must tend to some limit, say S, when $n \to \infty$. In these circumstances the expansion in the form *11.2* is valid. The series is then said to be convergent. The condition that the series *11.2* is convergent is that $|x| < 1$.

Examples

1. Write down the first four terms in the expansion of $(1+2x)^{-5}$ and write down the rth term.

$$(1+2x)^{-5} = 1+\frac{-5}{1}(2x)+\frac{-5.-6}{2!}(2x)^2+\frac{-5.-6.-7}{3!}(2x)^3$$

$$+\ldots+\frac{-5.-6.-7\ldots(-5-r+2)}{(r-1)!}(2x)^{r-1}+\ldots$$

The first four terms are:

$$1-10x+60x^2-280x^3$$

$$a_r = (-1)^{r-1}\frac{5.6.7\ldots(r+3)}{(r-1)!}(2^{r-1}).x^{r-1}$$

$$= (-1)^{r-1}\frac{r(r+1)(r+2)(r+3)}{4!}2^{r-1}.x^{r-1}$$

Note that the expansion is valid when $|2x| < 1$, i.e. $|x| < \frac{1}{2}$ or $-\frac{1}{2} < x < \frac{1}{2}$.

2. Expand $(1-3x)^{\frac{1}{2}}$ as far as the term in x^4.

Use the expansion to determine the value of $\sqrt{0.97}$ correct to three decimal places.

$$(1-3x)^{\frac{1}{2}} = [1+(-3x)]^{\frac{1}{2}}$$

$$= 1 + \frac{\frac{1}{2}}{1}(-3x) + \frac{\frac{1}{2}\cdot -\frac{1}{2}}{2!}(-3x)^2 + \frac{\frac{1}{2}\cdot -\frac{1}{2}\cdot -\frac{3}{2}}{3!}(-3x)^3$$
$$+ \frac{\frac{1}{2}\cdot -\frac{1}{2}\cdot -\frac{3}{2}\cdot -\frac{5}{2}}{4!}(-3x)^4$$

$$= 1 - \frac{3x}{2} - \frac{9x^2}{8} - \frac{27x^3}{16} - \frac{405x^4}{28} \ldots$$

The expansion is valid providing $|3x| < 1$, i.e. $|x| < \frac{1}{3}$.

To determine $\sqrt{0.97}$:
Put $x = 0.01$. This is a value within the range for which the series converges, so the expansion may be used when $x = 0.01$. Then:

$$(1-0.03)^{\frac{1}{2}} = \sqrt{0.97}$$

$$\approx 1 + \frac{\frac{1}{2}}{1}(-0.3) + \frac{\frac{1}{2}\cdot -\frac{1}{2}}{2!}(-0.3)^2 + \frac{\frac{1}{2}\cdot -\frac{1}{2}\cdot -\frac{3}{2}}{3!}(-0.3)^3$$

$$= 1 - 0.015 - 0.0001125 = 0.9848875 \approx 0.985$$

The calculator gives $\sqrt{0.97} = 0.9848858$.

3. Calculate the first four terms in the expansion of $(5+4x^2)^{-\frac{3}{4}}$ and determine the range of values of x for which the expansion is valid.
Write:

$$(5+4x^2)^{-\frac{3}{4}} = \left[5\left(1+\frac{4x^2}{5}\right)\right]^{-\frac{3}{4}} = 5^{-\frac{3}{4}}\left[1+\frac{4x^2}{5}\right]^{-\frac{3}{4}}$$

$$= 5^{-\frac{3}{4}}\left[1 + \frac{-\frac{3}{4}}{1}\left(\frac{4x^2}{5}\right) + \frac{-\frac{3}{4}\cdot -\frac{7}{4}}{1.2}\left(\frac{4x^2}{5}\right)^2\right.$$
$$\left.+ \frac{-\frac{3}{4}\cdot -\frac{7}{4}\cdot -\frac{11}{4}}{1.2.3}\left(\frac{4x^2}{5}\right)^3 + \ldots\right.$$

$$= 5^{-\frac{3}{4}}\left[1 - \frac{3x^2}{5} + \frac{21x^4}{50} - \frac{77x^6}{250} + \ldots\right.$$

$$= 5^{-\frac{3}{4}} - \frac{3}{5}.5^{-\frac{3}{4}}.x^2 + \frac{21}{50}.5^{-\frac{3}{4}}.x^4 - \frac{77}{250}.5^{-\frac{3}{4}}.x^6 + \ldots$$

The expansion is valid providing $\left|\frac{4x^2}{5}\right| < 1$, i.e.

$|x^2| < \frac{5}{4}$ or $-\frac{5}{4} < x^2 < \frac{5}{4}$. However, for real x, $x^2 \nless 0$, therefore the condition for validity becomes $0 < x^2 < \frac{5}{4}$, i.e.

$$-\frac{\sqrt{5}}{2} < x < \frac{\sqrt{5}}{2}, \ |x| < \frac{\sqrt{5}}{2}$$

4. Write down the first four terms and the general term of the expansion of $(a + bx)^n$ where n is not a positive integer.

$$(a + bx)^n = a^n\left(1 + \frac{bx}{a}\right)^n$$

When x is small the expansion is:

$$a^n\left[1 + n\left(\frac{bx}{a}\right) + \frac{n(n-1)}{1.2}\left(\frac{bx}{a}\right)^2 + \frac{n(n-1)(n-2)}{1.2.3}\left(\frac{bx}{a}\right)^3 + \ldots\right]$$

The $(r + 1)$th term is:

$$= a^n . \frac{n(n-1)(n-2) \ldots (n-r+1)}{r!}\left(\frac{bx}{a}\right)^r$$

$$= \frac{n(n-1)(n-2) \ldots (n-r+1)}{r!} a^{n-r}(bx)^r$$

The expansion is valid when $\left|\frac{bx}{a}\right| < 1$, i.e. $|x| < \frac{a}{b}$.

Exercise 11.3

Write down the first four terms and the $(r + 1)$th term of the expansions of the following functions. In each case determine the range of values of the variable for the expansion to be valid.

1. $(1 + x)^{-1}$
2. $(1 - x)^{-1}$
3. $(1 + x)^{-2}$
4. $(1 + x)^{\frac{1}{2}}$
5. $(1 + x)^{-\frac{1}{2}}$
6. $(1 + 2x)^{-2}$
7. $(1 + 2x)^{\frac{1}{2}}$
8. $(1 + 2x)^{-\frac{1}{2}}$
9. $(1 - 2x)^{-2}$

10. $(1-2x)^{-\frac{1}{2}}$
11. $(1-2x)^{\frac{1}{2}}$
12. $\left(1+\frac{x}{2}\right)^{-2}$
13. $\left(1+\frac{x}{2}\right)^{\frac{1}{2}}$
14. $\left(1+\frac{x}{2}\right)^{-\frac{1}{2}}$
15. $\left(1-\frac{x}{2}\right)^{-2}$
16. $\left(1-\frac{x}{2}\right)^{\frac{1}{2}}$
17. $\left(1-\frac{x}{2}\right)^{-\frac{1}{2}}$
18. $(2+x)^{-2}$
19. $(2-x)^{-2}$
20. $(2+x)^{\frac{1}{2}}$
21. $(2+3x)^{\frac{1}{2}}$
22. $(2-3x)^{-2}$
23. $(1+2x)^{\frac{5}{2}}$
24. $(1+x^2)^{-1}$
25. $(1+x^2)^{-2}$
26. $(1+x^2)^{\frac{1}{2}}$
27. $(1+2x^2)^{-1}$
28. $(1-4x^2)^{-1}$
29. $(1-2x^2)^{-1}$
30. $\left(1-\frac{1}{10}x^2\right)^{-1}$

If x is small, and so x^3 and higher powers of x may be neglected, determine approximate expansions of the following:

31. $(1+x)^{\frac{1}{2}}$
32. $(1-x)^{\frac{1}{2}}$
33. $(1-x)^{-\frac{1}{2}}$
34. $(1-x^2)^{\frac{1}{2}}$
35. $(1+x)^{\frac{1}{2}}.(1-x)^{\frac{1}{2}}$ (Hint: use question 34.)
36. $(1+x)^{\frac{1}{2}}.(1-x)^{-\frac{1}{2}}$ (Hint: use the results of question 31 and question 33 and determine the product, terminated at the x^2 term.)

37. $\dfrac{(1+x)^{\frac{1}{2}}}{(1-x)^{\frac{1}{2}}}$ (Hint: rearrange and use question 36.)

38. $(1+2x)^{\frac{1}{2}}.(1+x)^{\frac{1}{2}}$

39. $\sqrt{\dfrac{(1+2x)}{(1+x)}}$

40. $\sqrt{\dfrac{1}{1+x}}$

41. $\sqrt{\dfrac{5}{1+x}}$

42. $\sqrt{\dfrac{5}{5+x}}$

11.4 Errors in formulae

For certain kinds of formulae the binomial expansion provides a useful aid in determining the effect on a calculated value when one or more of the variables in the formula is subjected to a small change or error.

Examples

1. The radius (r) of a circle is calculated from the formula:

$$r = \sqrt{\frac{A}{\pi}}$$

where A is the area of the circle. A is measured to be 100 but it is estimated to be in error by $+2$ per cent. Determine the percentage error in the calculated value of r. Take the true value of A to be 100 and the error to be 2 per cent $\times 100 = 2$.
Strictly speaking the true value of A is less than 100 and such that 2 per cent added to it gives 100. But the error in taking A to be 100 is minimal and it has the advantage of making the calculation easier. Then the real

r is $\sqrt{\dfrac{100}{\pi}}$. Suppose the error in r is δr. Then:

$$r+\delta r = \sqrt{\frac{100(1+0.02)}{\pi}} = \sqrt{\frac{100}{\pi}}[1+0.02]^{\frac{1}{2}}$$

$$\approx r[1+\tfrac{1}{2}(0.02)], \text{ ignoring powers of (0.02) greater than 1}$$

i.e. $r+\delta r \approx r+0.01r$

i.e. $\delta r = 0.01r$

The error in r is a 1 per cent error.

2. The area of a square, of side x, is given by:

$$A = x^2 \qquad \text{(i)}$$

Suppose that x undergoes a change, h, which may be either positive or negative, and this produces a change in A of k. Then:

$$A+k = (x+h)^2$$
$$= x^2+2xh+h^2 \qquad \text{(ii)}$$
$$\text{(ii)} - \text{(i)} \Rightarrow k = 2xh+h^2$$

If h is so small that h^2 and higher powers of h may be ignored, then:

$$k \approx 2x.h \qquad \text{(iii)}$$

Now replace h by δx and k by δA. (iii) becomes:

$$\delta A \approx 2x.\delta x$$

Note the relationship between this result and the calculus approach:

Since $\quad A = x^2$
then $\quad dA/dx = 2x$
or $\quad \delta A/\delta x \approx 2x$
i.e. $\quad \delta A \approx 2x.\delta x$

3. A point starts from rest and moves in a straight line with uniform acceleration, a, for time t. The distance travelled, s, is given by:

$$s = \tfrac{1}{2}at^2$$

If there are errors δa and δt in measuring a and t, determine the error, δs, in calculating s.
Suppose the true values are a, t and s. Then:

$$s = \tfrac{1}{2}at^2$$

Therefore:

$$\begin{aligned} s+\delta s &= \tfrac{1}{2}(a+\delta a)(t+\delta t)^2 \\ &= \tfrac{1}{2}(a+\delta a)(t^2+2t.\delta t+\delta t^2) \\ &= \tfrac{1}{2}at^2+\tfrac{1}{2}a(2t.\delta t+\delta t^2)+\tfrac{1}{2}.\delta a(t^2+2t.\delta t+\delta t^2) \\ \delta s &= at.\delta t+\tfrac{1}{2}t^2.\delta a+\tfrac{1}{2}a.\delta t^2+t.\delta a.\delta t+\tfrac{1}{2}\delta a.\delta t^2 \end{aligned}$$

When δa and δt are so small that $\delta a.\delta t$, δt^2 and $\delta a.\delta t^2$ may be ignored:

$$\delta s \approx at.\delta t+\tfrac{1}{2}t^2.\delta a$$

4. The time of oscillation (T) of a simple pendulum (length L) is given by:

$$T = 2\pi\sqrt{\frac{L}{g}}$$

Suppose the error in calculating g is 1 per cent and the error in measuring L is 2 per cent. Determine the maximum error which may be expected in calculating T.

Suppose the true values are g, L and T. Then:

$$\begin{aligned} T+\delta T &= 2\pi\sqrt{\frac{\left(L\pm\frac{L}{50}\right)}{\left(g\pm\frac{g}{100}\right)}} \\ &= 2\pi\sqrt{\frac{L}{g}\frac{\left(1\pm\frac{1}{50}\right)}{\left(1\pm\frac{1}{100}\right)}} \\ &= T\left(1\pm\frac{1}{50}\right)^{\frac{1}{2}}\left(1\pm\frac{1}{100}\right)^{-\frac{1}{2}} \\ &\approx T\left[1\pm\frac{1}{2}.\frac{1}{50}\right]\left[1\pm\left(-\frac{1}{2}\right)\frac{1}{100}\right] \\ &= T\left[1\pm\frac{1}{100}\right]\left[1\mp\frac{1}{200}\right] \end{aligned}$$

At most $T+\delta T = T(1+1/100)(1+1/200) \approx T(1+3/200)$

Then $\delta T \approx 3T/200$, i.e. $\delta T/T \approx 3/200$.

The percentage error in T is then 1.5 per cent.

At the least $T+\delta T = T(1-1/100)(1-1/200) \approx T(1-3/200)$

Then $\delta T/T \approx -3/200$.

This is a percentage error of -1.5 per cent.

Exercise 11.4

1. When $y = x^3$ determine the approximate change in y when x changes to $x + h$.
2. When $y = x^3$ determine the approximate change in y when x changes from 1 to 1.01.
3. When $y = x^2$ determine the approximate change in y when x changes by 1 per cent.
4. When $W = I^2R$ determine the approximate change in W when I changes from 2 to 2.01 and R changes from 10 to 10.01.
5. Calculate the approximate error in calculating p from the formula $p = 10/v$ when a measurement of $v = 2$ is estimated to be in error by 2 per cent.
6. When using the formula $V = \frac{4}{3}\pi r^3$ for the volume of a sphere, radius r, it is estimated that a measurement of r of 10.6 is in error by $+3$ per cent. Determine the approximate error, as a percentage, for the calculation of V.

12

The Expansion of the Exponential Function

12.1 The expansion of e^x

In Level 2 e^x was defined as that function whose derivative w.r.t. x is itself. Now assume that e^x may be expanded as a series of terms in ascending powers of x, that is:

$$e^x \equiv a_0 + a_1 x + a_2 x^2 + a_3 x^3 + a_4 x^4 + \ldots + a_r x^r + \ldots \quad \text{(i)}$$

Since (i) is an identity, then it is true for all values of x.
Put $x = 0$ in (i):

$$1 = a_0 + 0$$
$$a_0 = 1$$

Differentiate (i) w.r.t. x:

$$e^x \equiv 0 + a_1 + 2a_2 x + 3a_3 x^2 + 4a_4 x^3 + \ldots + r a_r x^{r-1} + \ldots \quad \text{(ii)}$$

In (ii) put $x = 0$:

$$1 = a_1 + 0$$
$$a_1 = 1$$

Differentiate (ii) w.r.t. x:

$$e^x \equiv 0 + 2a_2 + 3.2.a_3 x + 4.3.a_4 x^2 + \ldots r(r-1)a_r x^{r-2} + \ldots \quad \text{(iii)}$$

In (iii) put $x = 0$:

$$1 = 2a_2 + 0$$
$$a_2 = \frac{1}{2} = \frac{1}{2!}$$

Differentiate (iii) w.r.t. x:

$$e^x \equiv 0 + 3.2.1a_3 + 4.3.2a_4x + \ldots r(r-1)(r-2)a_r x^{r-3} + \ldots \quad \text{(iv)}$$

In (iv) put $x = 0$:

$$1 = 3.2.1a_3 + 0$$

$$a_3 = \frac{1}{3!}$$

By repetition of this process, after r successive differentiations we determine that:

$$a_r = \frac{1}{r!}$$

We deduce that:

$$e^x \equiv 1 + x + \frac{1}{2!}x^2 + \frac{1}{3!}x^3 + \frac{1}{4!}x^4 + \ldots + \frac{1}{r!}x^r + \ldots \qquad \textit{12.1}$$

The series *12.1* is an infinite series. The expansion of functions in infinite series is valid only when the sum of terms tends to a limit no matter how many of the terms are taken into account to determine the sum. When the sum does tend to a limit the series is said to be convergent. The expansion of e^x is convergent for all values of x.

12.2 The expansion of e^{-x}

Assume that the expansion of the function e^x in the form *12.1* applies when x is replaced by $-x$. Then:

$$e^{-x} \equiv 1 + (-x) + \frac{1}{2!}(-x)^2 + \frac{1}{3!}(-x)^3 + \frac{1}{4!}(-x)^4 + \ldots + \frac{1}{n!}(-x)^n + \ldots$$

$$\equiv 1 - x + \frac{1}{2!}x^2 - \frac{1}{3!}x^3 + \frac{1}{4!}x^4 - \frac{1}{5!}x^5 + \ldots + (-1)^n\frac{1}{n!}x^n + \ldots \qquad \textit{12.2}$$

The series *12.2* is also convergent for all values of x. From formulae *12.1* and *12.2* other important formulae may be developed.

12.3 Further exponential expansions

By adding formulae *12.1* and *12.2* we obtain:

$$e^x + e^{-x} \equiv 2 + 2.\frac{1}{2!}x^2 + 2.\frac{1}{4!}x^4 + 2.\frac{1}{6!}x^6 + \ldots + 2.\frac{1}{(2n)!}x^{2n} + \ldots$$

$$\equiv 2\left(1 + \frac{1}{2!}x^2 + \frac{1}{4!}x^4 + \frac{1}{6!}x^6 + \ldots + \frac{1}{(2n)!}x^{2n} + \ldots\right) \qquad 12.3$$

12.1 – 12.2 gives:

$$e^x - e^{-x} \equiv 2x + 2.\frac{1}{3!}x^3 + 2.\frac{1}{5!}x^5 + 2.\frac{1}{7!}x^7 + \ldots$$

$$+ 2.\frac{1}{(2r+1)!}x^{2r+1} + \ldots$$

$$\equiv 2\left(x + \frac{1}{3!}x^3 + \frac{1}{5!}x^5 + \frac{1}{7!}x^7 + \ldots + \frac{1}{(2r+1)!}x^{2r+1} + \ldots\right) \qquad 12.4$$

Modifications of *12.3* and *12.4* are used to develop two other important functions. They will not be investigated at this stage.

The series *12.3* and *12.4* are convergent for all x. By substituting px for x in *12.1* we obtain:

$$e^{px} \equiv 1 + (px) + \frac{1}{2!}(px)^2 + \frac{1}{3!}(px)^3 + \frac{1}{4!}(px)^4 + \ldots + \frac{1}{n!}(px)^n + \ldots \qquad 12.5$$

12.5 is true for both positive and negative values of p. It is convergent for all values of x.

Multiply *12.5* by a constant, A:

$$Ae^{px} = A\,(1 + px + \frac{1}{2!}(px)^2 + \frac{1}{3!}(px)^3 + \frac{1}{4!}(px)^4 + \ldots + \frac{1}{n!}(px)^n + \ldots$$

$$= A + Apx + \frac{1}{2!}A\,(px)^2 + \frac{1}{3!}A\,(px)^3 + \ldots + \frac{1}{n!}A\,(px)^n + \ldots \qquad 12.6$$

12.4 Approximations to exponential functions for small x

When x is so small that powers beyond x^4 may be ignored the following approximations may be made:

$$e^x \approx 1 + x + \frac{1}{2!}x^2 + \frac{1}{3!}x^3 + \frac{1}{4!}x^4 = 1 + x + \frac{1}{2}x^2 + \frac{1}{6}x^3 + \frac{1}{24}x^4 \qquad 12.7$$

$$e^{-x} \approx 1 - x + \frac{1}{2!}x^2 - \frac{1}{3!}x^3 + \frac{1}{4!}x^4 = 1 - x + \frac{1}{2}x^2 - \frac{1}{6}x^3 + \frac{1}{24}x^4 \qquad 12.8$$

$$e^x + e^{-x} \approx 2\left(1 + \frac{1}{2!}x^2 + \frac{1}{4!}x^4\right) = 2 + x^2 + \frac{1}{12}x^4 \qquad 12.9$$

$$e^x - e^{-x} \approx 2\left(x + \frac{1}{3!}x^3\right) = 2x + \frac{1}{3}x^3 \qquad 12.10$$

$$e^{px} \approx 1 + px + \frac{1}{2!}p^2x^2 + \frac{1}{3!}p^3x^3 + \frac{1}{4!}p^4x^4$$

$$= 1 + px + \frac{1}{2}p^2x^2 + \frac{1}{6}p^3x^3 + \frac{1}{24}p^4x^4 \qquad 12.11$$

Examples

1. Determine an approximate expansion for $(1+x).e^{\frac{1}{2}x}$ when x is so small that all powers of x beyond x^3 may be ignored.

$$(1+x).e^{\frac{1}{2}x} \approx (1+x)\left[1 + \frac{1}{2}x + \frac{1}{2!}\left(\frac{1}{2}x\right)^2 + \frac{1}{3!}\left(\frac{1}{2}x\right)^3\right]$$

$$= (1+x)\left(1 + \frac{1}{2}x + \frac{1}{8}x^2 + \frac{1}{48}x^3\right)$$

$$\approx 1 + x + \frac{1}{2}x + \frac{1}{2}x^2 + \frac{1}{8}x^2 + \frac{1}{48}x^3 + \frac{1}{8}x^3$$

$$= 1 + \frac{3}{2}x + \frac{5}{8}x^2 + \frac{7}{48}x^3$$

2. Determine an approximate expansion for $\dfrac{e^{-x^2}}{(1-x)}$ when x is so small that all powers of x beyond x^4 may be ignored.

$$\frac{e^{-x^2}}{1-x} = (1-x)^{-1}.e^{-x^2} \approx [1+(-x)]^{-1}.e^{-x^2}$$

$$\approx (1 + x + x^2 + x^3 + x^4)(1 - x^2 + \tfrac{1}{2}x^4)$$

$$\approx 1 + x + x^2 - x^2 + x^3 - x^3 + x^4 - x^4 + \tfrac{1}{2}x^4$$

$$= 1 + x + \tfrac{1}{2}x^4$$

3. Determine an approximate expansion for $2e^x + 3e^{-x/2} - 5e^{x^2/3}$ when all

powers of x beyond x^4 may be ignored.

$$2e^x + 3e^{-x/2} - 5e^{x^2/3}$$

$$\approx 2\left(1 + x + \frac{1}{2!}x^2 + \frac{1}{3!}x^3 + \frac{1}{4!}x^4\right) + 3\left[1 - \frac{x}{2} + \frac{1}{2!}\left(-\frac{x}{2}\right)^2 + \frac{1}{3!}\left(-\frac{x}{2}\right)^3 + \frac{1}{4!}\left(-\frac{x}{2}\right)^4\right]$$

$$-5\left[1 + \frac{x^2}{3} + \frac{1}{2!}\left(\frac{x^2}{3}\right)^2\right]$$

$$= 2 + 2x + x^2 + \frac{1}{3}x^3 + \frac{1}{12}x^4 + 3 - \frac{3}{2}x + \frac{3}{8}x^2 - \frac{1}{16}x^3 + \frac{1}{128}x^4 - 5 - \frac{5}{3}x^2 - \frac{5}{18}x^4$$

$$= (2 + 3 - 5) + \left(2 - \frac{3}{2}\right)x + \left(1 + \frac{3}{8} - \frac{5}{3}\right)x^2 + \left(\frac{1}{3} - \frac{1}{16}\right)x^3 + \left(\frac{1}{12} + \frac{1}{128} - \frac{5}{18}\right)x^4$$

$$= 0 + \frac{1}{2}x - \frac{7}{24}x^2 + \frac{13}{48}x^3 - \frac{215}{1152}x^4$$

4. Calculate the value of $(1 + x + x^2)e^{-2x}$ correct to three decimal places when $x = 0.1$.

$$(1 + x + x^2)e^{-2x} \approx (1 + x + x^2)\left[1 - 2x + \frac{(-2x)^2}{2!} + \frac{(-2x)^3}{3!} + \frac{(-2x)^4}{4!}\right]$$

$$= (1 + x + x^2)\left(1 - 2x + 2x^2 - \frac{4}{3}x^3 + \frac{2}{3}x^4\right)$$

$$= 1 + x + x^2 - 2x - 2x^2 - 2x^3 + 2x^2 + 2x^3 + 2x^4 - \frac{4}{3}x^3 - \frac{4}{3}x^4 - \frac{4}{5}x^5 + \frac{2}{3}x^4 + \frac{2}{3}x^5 + \frac{2}{3}x^6$$

$$\approx 1 - x + x^2 - \frac{4}{3}x^3 + \frac{4}{3}x^4$$

$$= 1 - 0.1 + 0.01 - 0.00133 + 0.000133 \text{ when } x = 0.1$$

$$= 0.90880 \approx 0.909$$

Check: when $x = 0.1$:

$$\begin{aligned}(1+x+x^2)e^{-2x} &= 1.11 \times e^{-0.2} \\ &= 0.9087911 \text{ by calculator} \\ &\approx 0.909\end{aligned}$$

Exercise 12.1

1. Given that $y = (1+x^2)e^x$ and that x is small, find an expansion for y as far as the term in x^3.
2. Expand $(1+x^2)e^{-x}$ as far as the term in x^4.
3. For small x expand $x.e^{-x} + \frac{1}{2}x^2.e^{-2x}$ as far as the term in x^4.
4. For small x expand $1 + x.e^{\frac{1}{2}x}$ as far as the term in x^4.
5. For large x expand $\frac{1}{x}e^{\frac{1}{x}}$ as far as the term in $(1/x)^3$. Hint: substitute $1/x = y$ and expand in ascending powers of y. Finally, substitute back $y = 1/x$.
6. For large x expand $\frac{1}{x-1}e^{-\frac{2}{x}}$ as far as the term in $(1/x)^4$.
7. Use the expansion in question 1 to give a value of $(1+x^2).e^x$ correct to three decimal places when $x = 0.01$.
8. Use the expansion in question 2 to give a value of $(1+x^2).e^{-x}$ correct to three decimal places when $x = 0.05$.
9. By means of an expansion for small x calculate the value $\frac{e^{-x}}{(1-2x)}$ correct to three decimal places when $x = 0.1$.
10. Use the expression in question 5 to calculate the value of $\frac{1}{x}e^{\frac{1}{x}}$ correct to four decimal places when $x = 100$.
11. Use the expansion in question 6 to calculate the value of $\frac{1}{x-1}e^{-\frac{2}{x}}$ correct to four decimal places when $x = 2000$.

12.5. The approximate calculation of e

When x is very small, of the order of 0.01 or 0.001, the series *12.1* and *12.2* converge rapidly. This means that only a few terms in the expansion need to be taken into account to calculate the value of an exponential function correct to a few decimal places. When $x = 1$ the series *12.1* converges only slowly. Many more terms have to be taken into account to give a value of e correct to, for example, four decimal places. By substituting $x = 1$ in *12.1* we obtain:

$$\begin{aligned} e &= 1 + 1 + 1/2! + 1/3! + 1/4! + 1/5! + 1/6! + \dots \\ &\approx 1 + 1 + 0.5 + 0.166667 + 0.41667 + 0.008333 + 0.001389 \\ &\quad + 0.000198 + 0.000025 + 0.00003 \\ &= 2.718282 \approx 2.7183 \text{ correct to four decimal places} \end{aligned}$$

By calculator $e^1 = 2.718282818 \approx 2.7183$

Exercise 12.2

By using the appropriate expansions calculate the values of the following functions correct to the degree indicated for the given value of the variable.

1. e^x to three decimal places when $x = -1$
2. e^x to four decimal places when $x = 0.1$
3. e^{-x} to four decimal places for $x = 0.1$
4. e^x to four decimal places for $x = 0.5$
5. e^{-x} to four decimal places when $x = 0.5$
6. e^{x^2} to four decimal places for $x = 0.1$
7. e^{-x^2} to four decimal places when $x = 0.1$
8. e^{2x} to three decimal places when $x = 1$
9. e^{3x} to three decimal places when $x = 1$
10. e^{-2x} to three decimal places when $x = 1$
11. e^{-3x} to three decimal places when $x = 1$
12. $10e^{\frac{x}{2}}$ to three decimal places when $x = 1$
13. $-\frac{4}{5}e^{-\frac{x}{5}}$ to four decimal places when $x = 1$
14. $0.7e^{\frac{3x}{10}} - 1.3e^{-\frac{2x}{5}}$ to four decimal places when $x = 1$

13

Curves of Exponential Growth and Decay. Determination of Laws

13.1 Simple exponential curves

In Chapter 3 of *Analytical Mathematics 2*, as well as in Chapter 12 of this book, certain properties of e^x were discovered:

1. For all x, $e^x > 0$.
2. When $x = 0$, $e^x = e^{-x} = 1$.
3. For positive x, as $x \to \infty$, $e^x \to \infty$.
4. As $x \to -\infty$, $e^x \to 0$.

Fig. 13.1 is a sketch of the curve $y = e^x$.
The curve $y = e^{-x}$ is symmetrical with $y = e^x$ about Oy. Fig. 13.2 is a sketch of $y = e^{-x}$.

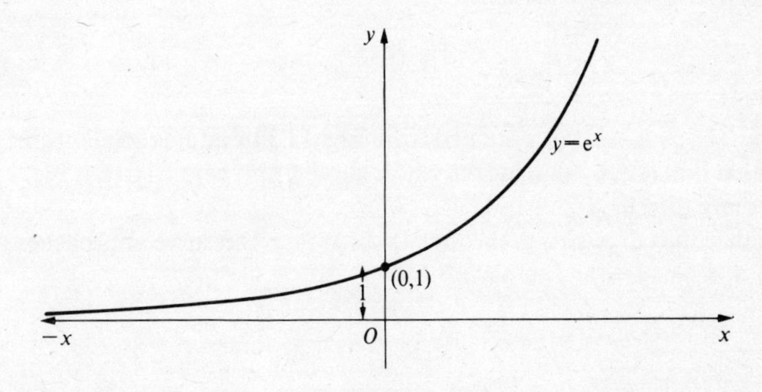

Figure 13.1

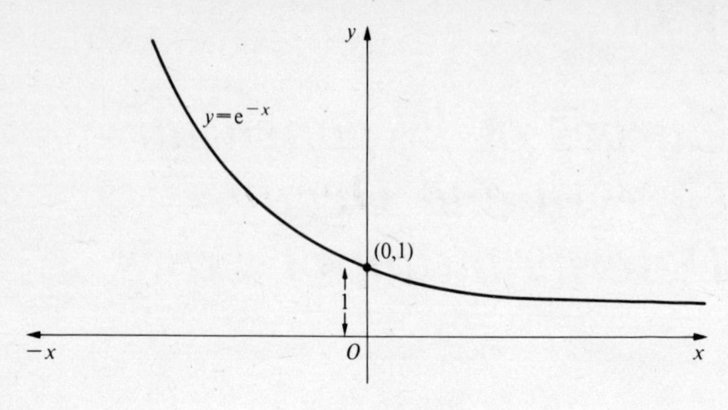

Figure 13.2

Note that both curves pass through the point (0, 1). The curve $y = -e^{-x}$ takes up negative values of y corresponding to the positive values of $y = e^{-x}$. Fig. 13.3 is a sketch of $y = -e^{-x}$.

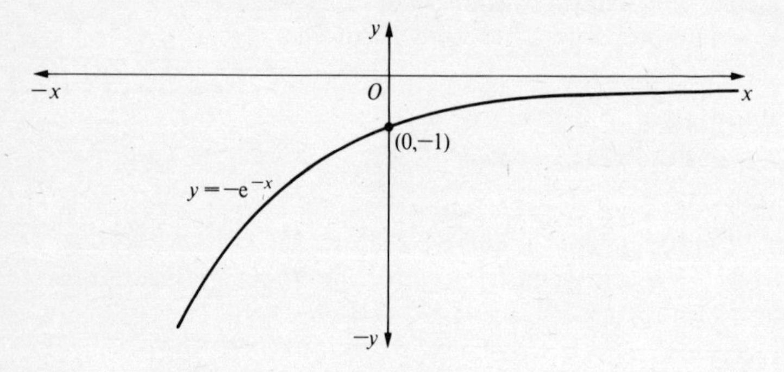

Figure 13.3

The curve $y = 1 - e^{-x}$ is obtained from Fig. 13.3 by adding to all ordinates the value 1, that is, by shifting the whole curve 1 unit parallel to Oy. Fig. 13.4 represents the curve $y = 1 - e^{-x}$.
Note that the curve passes through 0. As $x \to \infty$ the curve approaches the line $y = 1$, i.e. $y = 1$ is an asymptote.

The curve $y = e^{bx}$

Fig. 13.5 illustrates the relationship of the curve $y = e^{bx}$ to the curve $y = e^{x}$ for different values of b:

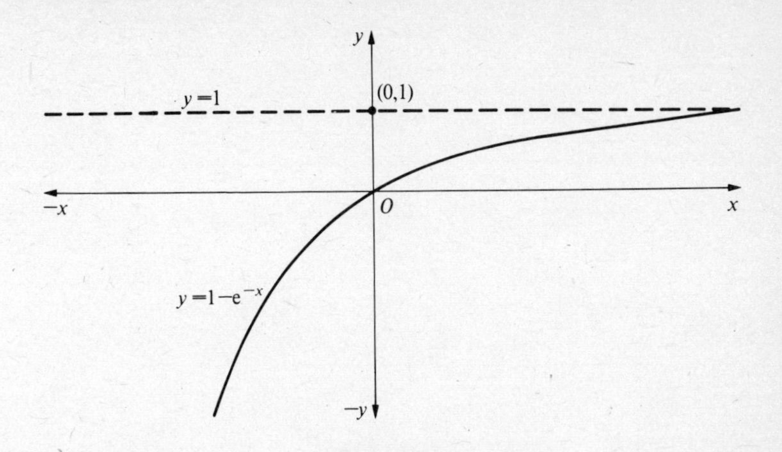

Figure 13.4

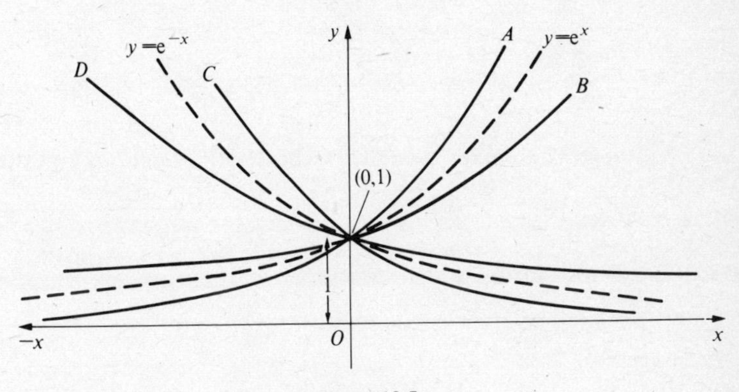

Figure 13.5

1. When b is positive and > 1 the curve is of type A.
2. When b is positive and < 1 the curve is of type B.
3. When b is negative and numerically > 1 the curve is of type C.
4. When b is negative and numerically < 1 the curve is of type D.

Note that all curves $y = e^{bx}$ pass through the point (0, 1).

The curve $y = a\,.\,e^{bx}$

The curve is obtained from the curve $y = e^{bx}$ by multiplying all ordinates by a. We shall suppose that a and b are both positive. Fig. 13.6 is a sketch of the curve $y = a\,.\,e^{bx}$.

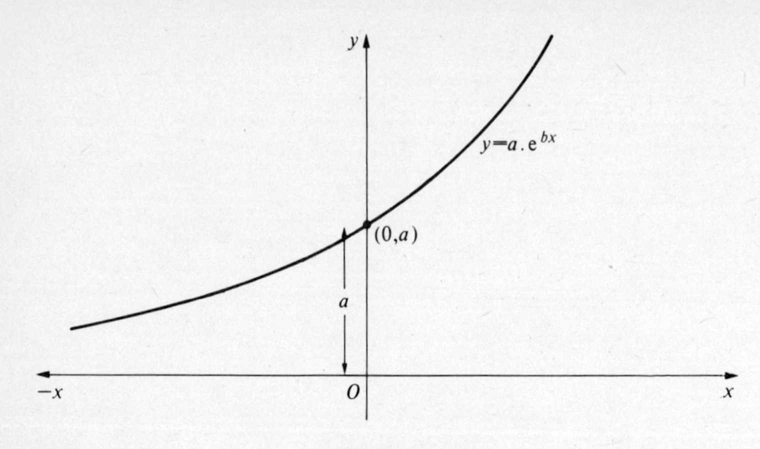

Figure 13.6

All curves $y = a \, . \, e^{bx}$ pass through the point $(0, a)$.

The curve $y = a \, . \, e^{-bx}$

Fig. 13.7 represents a curve $y = a \, . \, e^{-bx}$ when both a and b are positive.

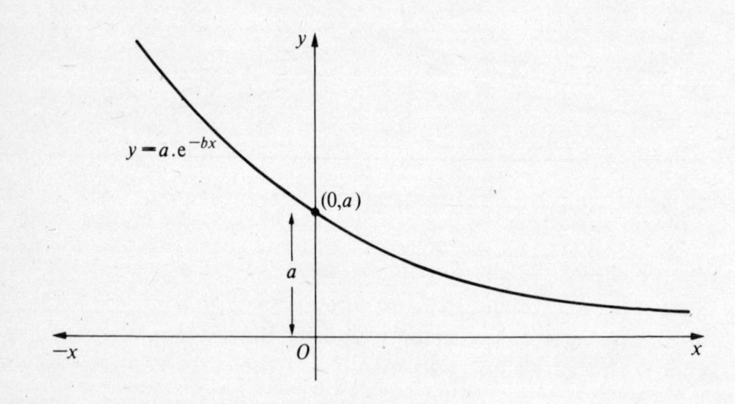

Figure 13.7

The curve $y = -a \, . \, e^{-bx}$

Fig. 13.8 represents the curve $y = -a \, . \, e^{-bx}$ where both a and b are positive. The ordinates are the negatives of those in Fig. 13.7.

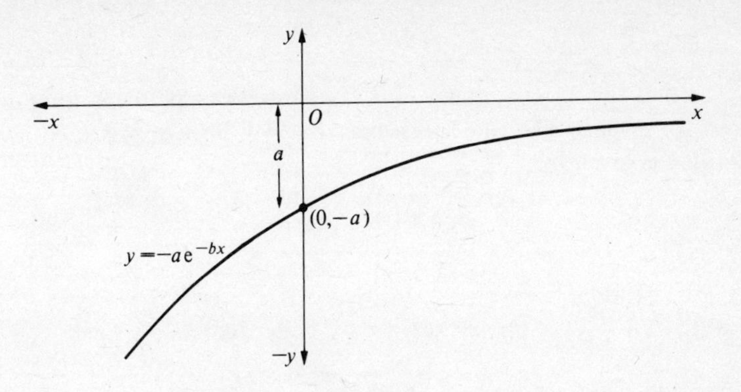

Figure 13.8

The curve $y = a(1 - e^{-bx})$

Fig. 13.9 represents the curve $y = a(1 - e^{-bx})$ when both a and b are positive. It is obtained from Fig. 13.8 by shifting the whole curve parallel to Oy by a units.
It passes through O for all values of a and b. It approaches the line $y = a$ as $x \to \infty$.

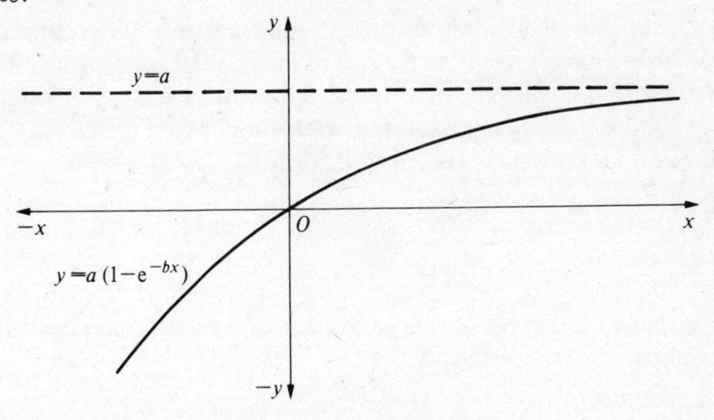

Figure 13.9

13.2 Application of exponential curves to practical problems

In Chapter 3 of *Analytical Mathematics 2* a number of practical problems related to the exponential function were encountered. The following examples represent just a few such applications.

Compound interest

Suppose that the sum of money, £P, is invested at r per cent compound interest for t years. The value to which that will have appreciated in the given time is given by:

$$A = P\left(1 + \frac{r}{100}\right)^t$$
$$= P \, . \, R^t$$

where $R = 1 + \dfrac{r}{100}$. Suppose that R may be written as e^p. Then:

$$A = P \, . \, e^{pt}$$

The graph would be of the form shown in Fig. 13.6.

Population growth

Suppose that a population, initially of magnitude N_0, is of magnitude N after time t. Then:

$$N = N_0 \, . \, e^{at}$$

where a is constant. The value of a depends on the nature of the population. In some cases it is very small, so the size of the population increases slowly, as in the case of human beings. However, in the case of bacteria growing in a suitable culture the value of a is large, so the population increases rapidly. Again, Fig. 13.6 represents such a function.

Radioactive decay

Suppose that a radioactive substance, initially of mass M_0, has mass M after time t. Then:

$$M = M_0 e^{-at}$$

where a is positive. The curve is similar to that shown in Fig. 13.7.

Growth of an electric current

Suppose that the current in a circuit is given by the following formula:

$$i = \frac{E}{R}(1 - e^{-Rt/L})$$

The formula is related to a circuit in which an e.m.f. of E volts, a resistor of R ohms and an inductor of L henry are in series with a switch. The time, t, in seconds, is the time after the switch is closed. The curve obtained is similar to that in Fig. 13.9.

Decay of an electric current

When a charge on a capacitor is discharged slowly through a resistor the formula is:

$$Q = Q_0\,\mathrm{e}^{-t/CR}$$

where Q_0 is the initial charge on the capacitor, Q the charge after time t seconds, C the capacitance of the capacitor, and R the resistance of the resistor. Q is measured in coulombs, C in farads, and R in ohms. The curve obtained is similar to that in Fig. 13.7.

Newton's law of cooling

The temperature, $\theta\,°\mathrm{C}$, of a hot body which is cooling in an environment of steady temperature, is given by:

$$\theta = \theta_0\,\mathrm{e}^{-kt}$$

where k is constant, $\theta_0\,°\mathrm{C}$ is the initial temperature of the body, and $\theta\,°\mathrm{C}$ is the temperature after time t. The value of k will depend on the unit chosen for t. Suppose that the value of k when t is measured in seconds is a. Then the value of k when t is measured in minutes will be $a \times 60$. Fig. 13.7 represents the curve.

The tension in a rope

When a rope is wrapped round a rough surface and supports a load at one end by means of a force applied at the other end then the tension in the rope along the surface varies according to the following formula:

$$T = T_0\,\mathrm{e}^{\mu\theta}$$

where T_0 is the load, and T is the tension at a point of the surface at which the normal makes an angle $\theta°$ with the normal at the last point of contact before the load. The coefficient of friction is μ. Fig. 13.6 represents the curve.

13.3 The use of log-linear graph paper

Although all curves of the type $y = a \, . \, e^{bx}$ have common characteristics, it is not easy to determine from the shape of a particular graph what value of b corresponds to it. It is an easy matter to determine the value of a. For instance, suppose that data are obtained from some source, e.g. an experiment or a series of measurements, and a graph of those data is drawn. Suppose it is known that the data satisfy a relationship of the form:

$$y = ae^{bx} \tag{i}$$

and the values of a and b are to be determined from the graph of y plotted against x. It is not easy to determine b from the graph, although it is easy to calculate a from it.

The one graph whose particular characteristics are easily calculated is the straight line graph. For that reason the relationship (i) is modified by a transformation which converts it into a linear relationship or a straight line form. Take logs to base 10 of (i):

$$\log y = \log a + \log e^{bx}$$
$$\text{i.e.} \quad \log y = \log a + bx \, . \log e$$
$$\text{or} \quad \log y = (b \log e) \, . \, x + \log a \tag{ii}$$

In (ii) put $\log y = Y$. Then:

$$Y = (b \log e) \, . \, x + \log a$$

Now plot Y (or $\log y$) against x to produce a graph of the type shown in Fig. 13.10. From Fig. 13.10 the intercept $c = \log a$ and gradient $m = b \log e$.

The use of special graph paper facilitates the plotting of this graph. Such

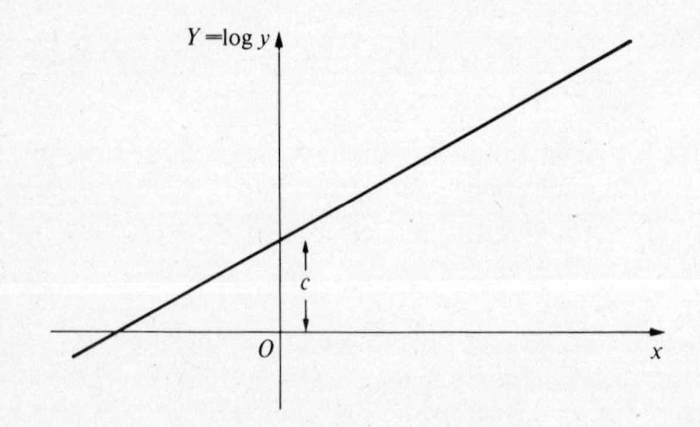

Figure 13.10

graph paper has a scale measured along Ox which is linear (i.e. of the type with which we should all now be familiar). Along Oy it has a scale which is logarithmic (i.e. similar to the markings on a slide rule, scale C). The graph paper is called log-linear graph paper. For other purposes graph paper called log-log graph paper is useful.

Examples

1. Plot the graph of $y = e^x$ on log-linear graph paper.
   Here $\log y = \log e^x = x . \log e$, i.e. $\log y = (\log e) . x + \log 1$. Fig. 13.11 represents the graph.
   Notice that, in Fig. 13.11, because the scale in the vertical direction is a log scale it automatically represents the logarithm of y in length along the axis. Therefore the axis is marked not $\log y$, but y.
   The following table provides the data from which to plot the graph:

x	0	1	2
y	1	$e = 2.718$	$e^2 = 7.389$

   The graph is a straight line.
   From the graph, when $x = 1.5$, $y = 4.5$. By calculator $y = e^{1.5} = 4.4816891$.
   When $y = 2.5$, $x = 0.915$. By calculator $x = \ln 2.5 = 0.9162907$.
2. Plot the graph of $y = e^{-x}$ on log-linear graph paper. The table of values from which to plot the graph is:

x	0	1	2
y	1	$e^{-1} = 0.3678794$	$e^{-2} = 0.1353353$

   Fig. 13.12 represents the graph of $y = e^{-x}$.
   From the graph, when $x = 0.6$, $y = 0.55$. By calculator $y = 0.5488116$.
   When $y = 0.3$, $x = 1.21$. By calculator $x = -\ln 0.3 = \ln (1/0.3) = 1.2039728$.
3. Plot the graph of $y = a . b^x$ on log-linear graph paper. Here, taking logs to base 10 of both sides:

   $$\begin{aligned} \log y &= \log a + \log b^x \\ \text{i.e.} &= x \log b + \log a \\ &= (\log b) . x + \log a \end{aligned}$$

   Fig. 13.13 represents a sketch of the graph.
   In Fig. 13.13 the y axis starts at $y = a$. When $x = 1$, $y = ab$ and when $x = 2$, $y = ab^2$. Since $ab \div a = b$ and $ab^2 \div ab = b$, then the distances

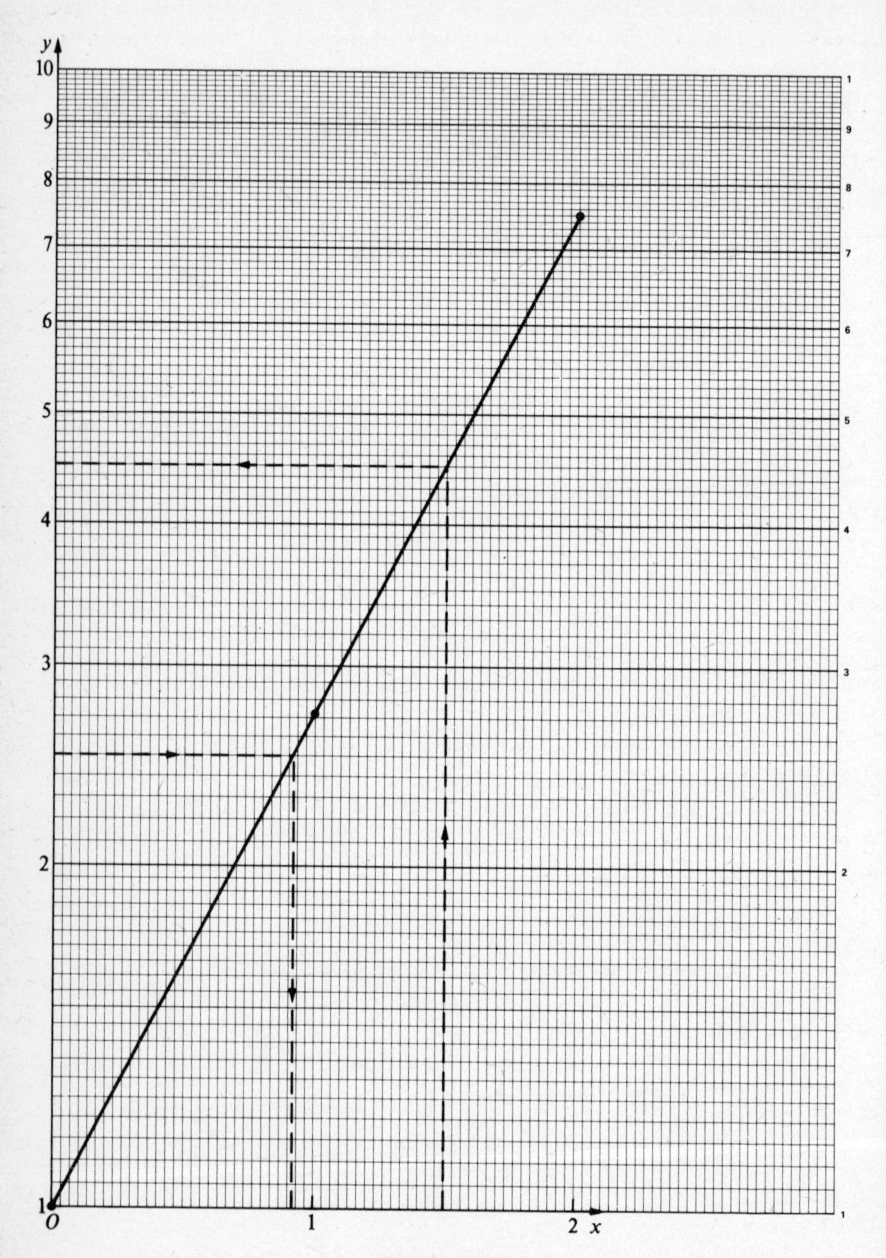

Figure 13.11

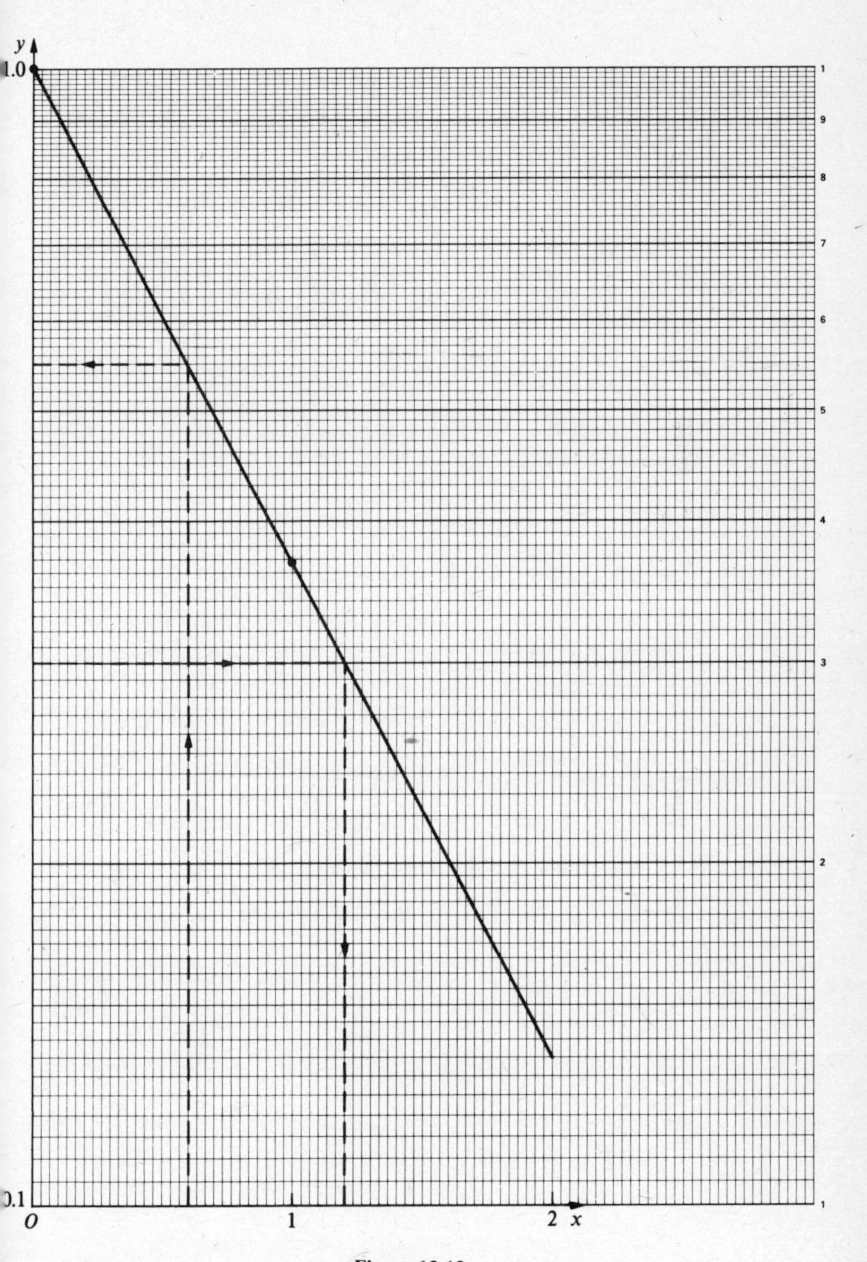

Figure 13.12

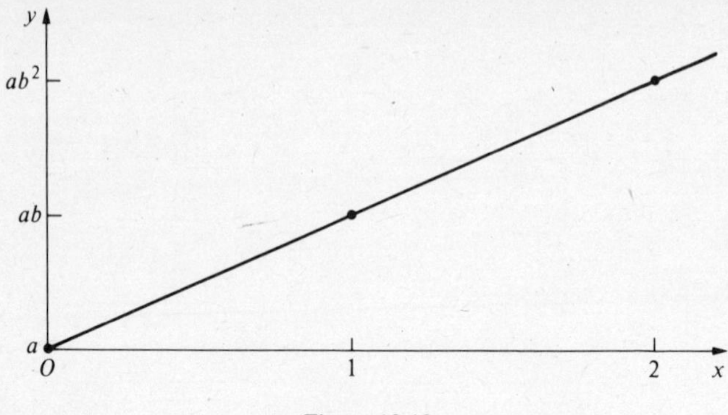

Figure 13.13

on Oy representing the intervals a to ab and ab to ab^2 are equal in length. The graph may be plotted from the following table:

x	0	1	2
y	a	ab	ab^2
$\log y$	$\log a$	$\log a+\log b$	$\log a+2\log b$

In Fig. 13.14 a and b take quite general values. Where a and b have particular numerical values the principle in constructing the graph is the same as that shown in Fig. 13.14.

4. Plot the graph of $y = a\,.\,\mathrm{e}^{bx}$ on log-linear graph paper.

$$\begin{aligned}\log y &= \log (a\mathrm{e}^{bx})\\ &= \log a + \log (\mathrm{e}^{bx})\\ &= \log a + (b \log \mathrm{e})\,.\,x\end{aligned}$$

Fig. 13.15 represents a sketch of the graph.
The graph may be plotted from the following table:

x	0	1	2
y	a	$a\mathrm{e}^b$	$a\mathrm{e}^{2b}$
$\log y$	$\log a$	$\log a+b\log \mathrm{e}$	$\log a+2b\log \mathrm{e}$

Fig. 13.16 represents the graph of $y = a\,.\,\mathrm{e}^{bx}$ where a and b are taken to be general values.

5. Plot the graph of $y = 5\mathrm{e}^{0.7x}$ for values of x from 0 to 3. From the graph determine the value of y when $x = 1.5$ and the value of x when $y = 33$. The following is the table of values from which to plot the graph in Fig. 13.17:

x	0	1	2	3
y	5	$5\mathrm{e}^{0.7}$	$5\mathrm{e}^{1.4}$	$5\mathrm{e}^{2.1}$
	5	10.068764	20.276	40.83085

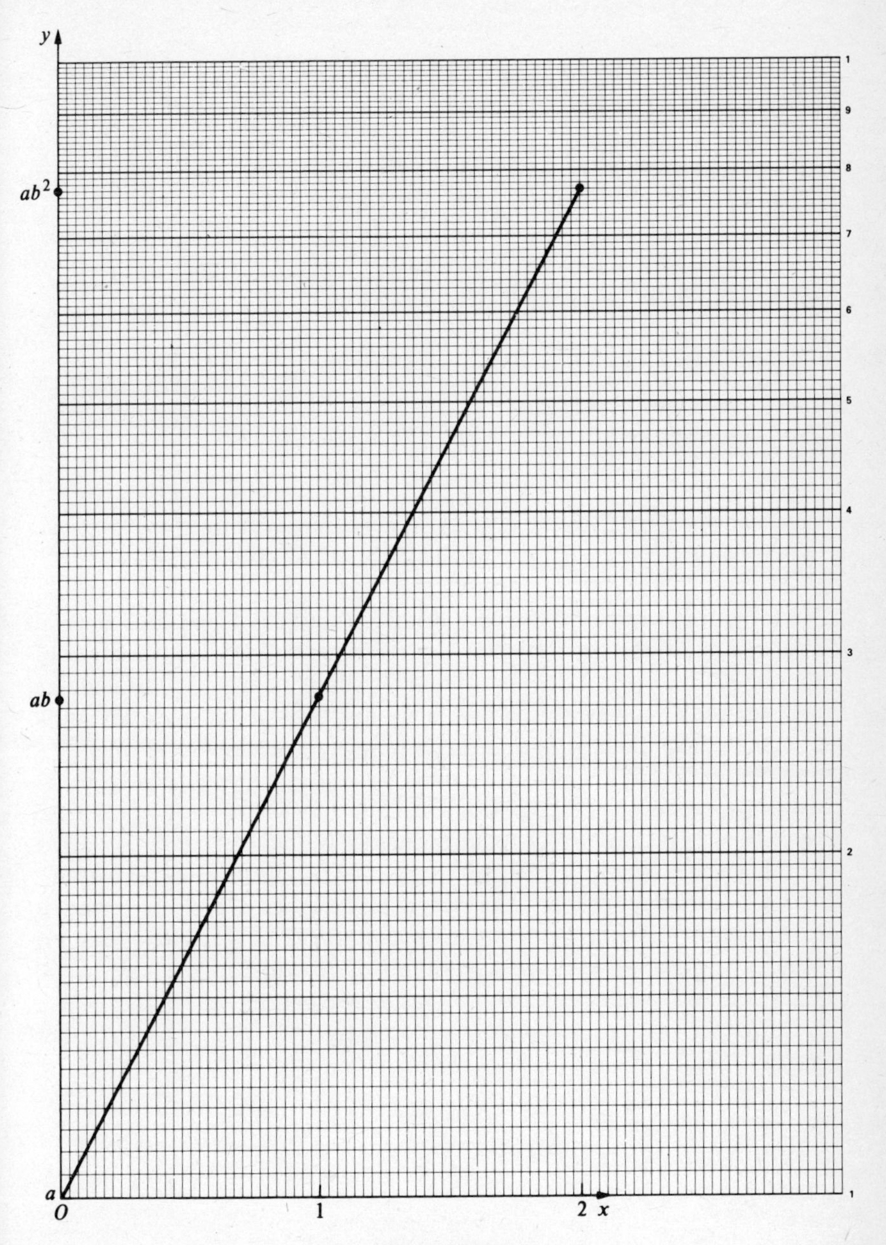

Figure 13.14

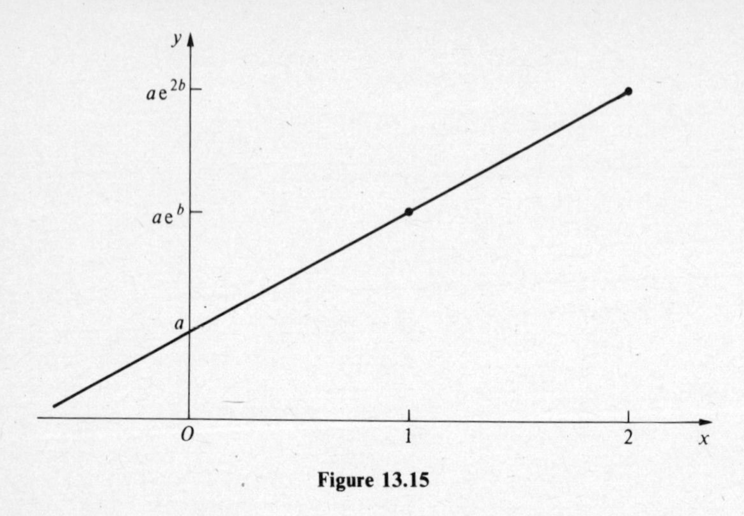

Figure 13.15

From the graph, when $x = 1.5$, $y = 14.25$. By calculator $y = 14.288256$. When $y = 33$, $x = 2.7$. By calculator $x = 2.6958138$.

Note that in example 5 the scale along Oy starts at 5 and each normal interval is multiplied by 5. By that means it is possible to cover the range of values required for y. It does mean that the straight line passes through the intersection of the two axes and does not provide an intercept as in Fig. 13.16. It would have been possible to start the scale along Oy and then represent the point 3 as 10 and the point 9 as 100. That would have given us the range needed. Yet the divisions along Oy are such that it would have been difficult to cater for the required interval of 1 to 10, i.e. 9 units along a length which is subdivided into 40 sections. In fact there are other patterns of log-linear graph paper. The one used here has a cycle of 1 and is subdivided into tenths. With paper which is marked out in two cycles it is quite easy to deal with a range of from 1 to 100. The first cycle is from 1 to 10 and the second cycle from 10 to 100. Of course that itself can be adapted to a first cycle from 0.1 to 1 and a second cycle from 1 to 10, as well as other modifications.

Exercise 13.1

Sketch the graphs of the following functions:

1. $y = 2e^x$
2. $y = 0.7e^x$

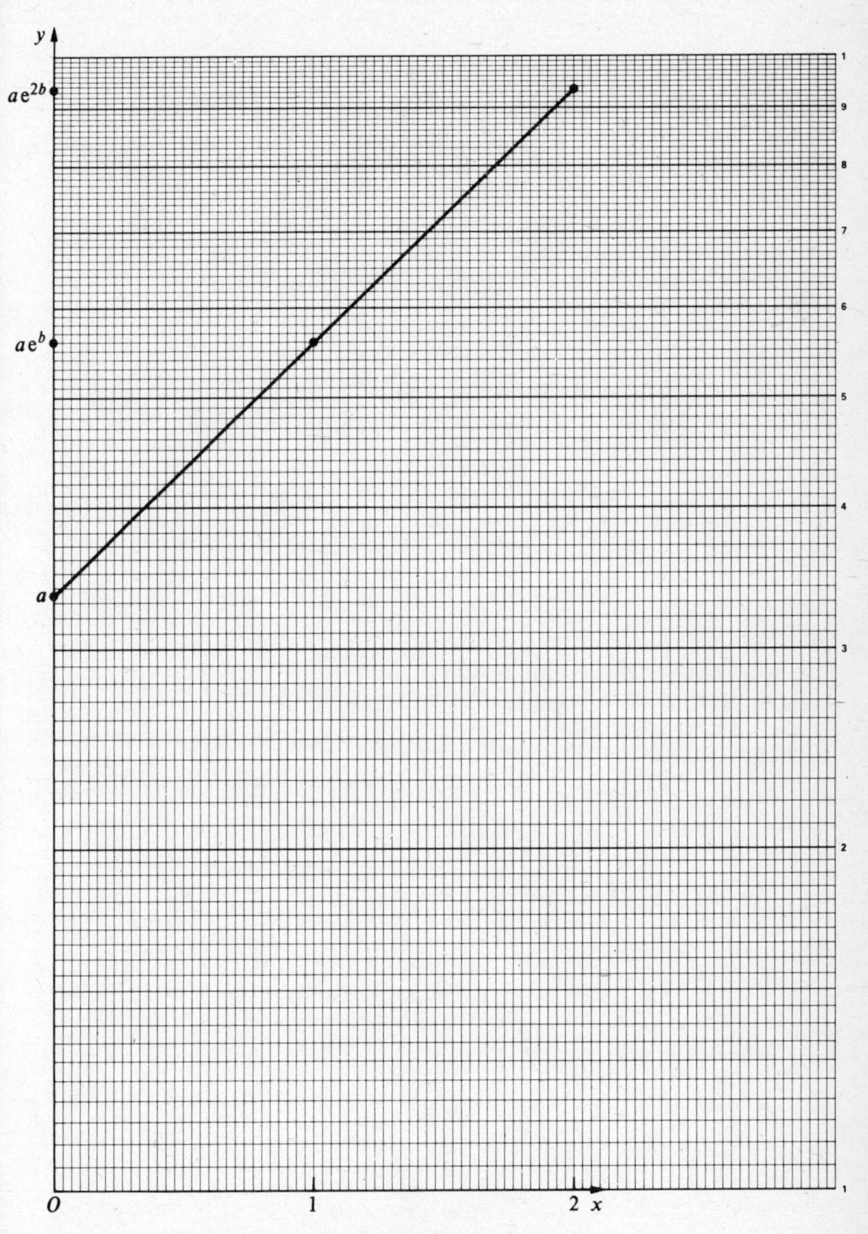

Figure 13.16

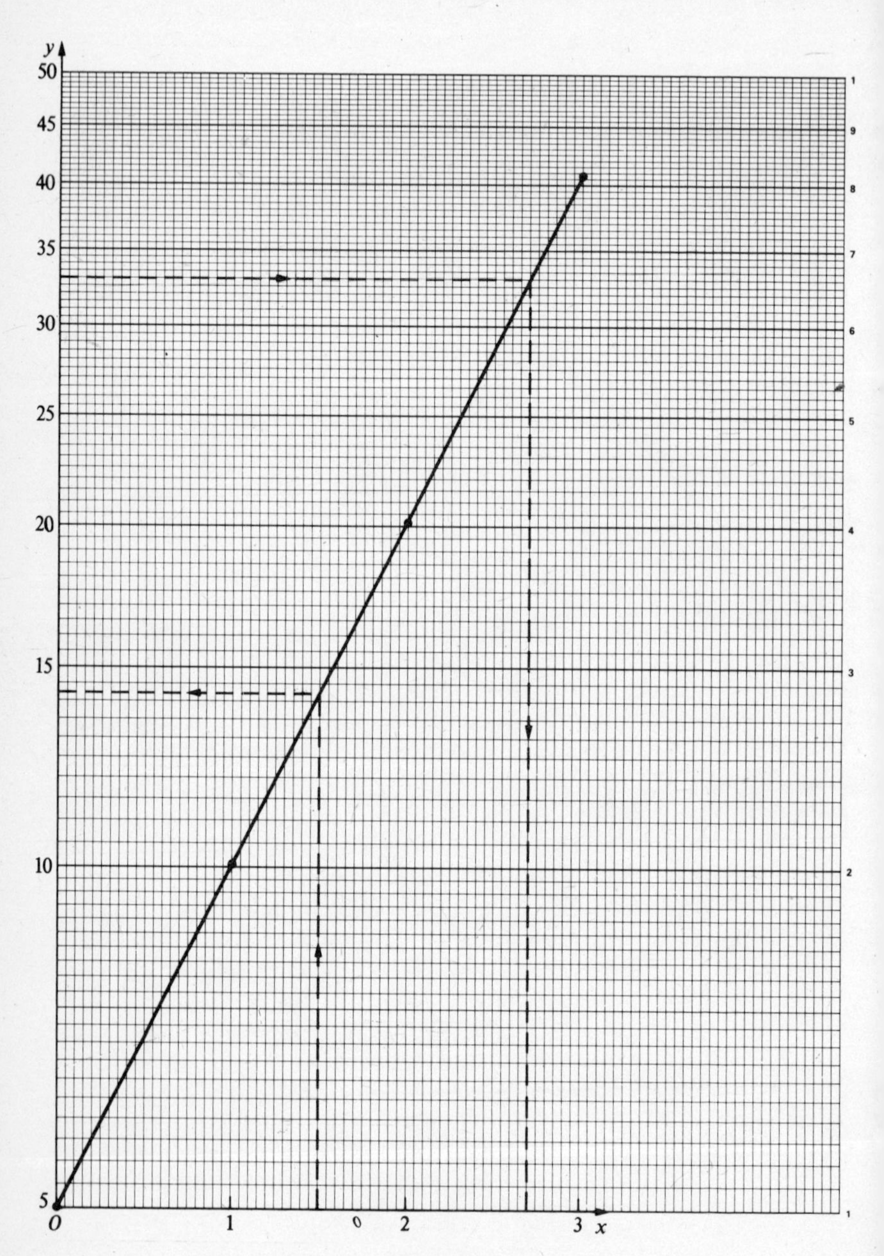

Figure 13.17

3. $y = 10e^{-x}$
4. $y = -2e^{-x}$
5. $y = -10e^{x}$
6. $y = 1 - e^{-x}$
7. $y = 5(1 - e^{-x})$
8. $y = -10(1 - e^{-x})$
9. $y = 2e^{5x}$
10. $y = 50e^{-2x}$
11. $y = -10e^{-2x}$
12. $y = 50(1 - e^{-2x})$
13. $y = -100(1 - e^{-5x})$
14. $y = 20(1 - 10e^{-2x})$

Plot the graphs of the following functions on log-linear graph paper. Check the values obtained from the graphs by calculator, using the original equations.

15. $y = 10e^{x}$ from $x = 0$ to $x = 3$. From the graph determine y when $x = 2.3$ and x when $y = 69$.
16. $y = 200e^{-x}$ from $x = 0$ to $x = 3$. From the graph determine y when $x = 0.7$ and x when $y = 10$.
17. $y = e^{2x}$ between $x = 0$ and $x = 3$. From the graph determine y when $x = 2.5$ and x when $y = 340$.
18. $y = 500e^{-\frac{1}{2}x}$ between $x = 0$ and $x = 6$. From the graph determine y when $x = 1.8$ and x when $y = 108$.
19. $y = 100(1 - e^{-x})$ between $x = 0$ and $x = 3$. From the graph determine y when $x = 2.2$ and x when $y = 84.3$.
20. $y = 250(1 - e^{-\frac{3}{4}t})$ between $t = 0$ and $t = 8$. From the graph determine y when $t = 6.4$ and t when $y = 126$.
21. $y = 50 \times 2^{x}$ between $x = 0$ and $x = 3$. From the graph determine y when $x = 0.9$ and x when $y = 270$.
22. $y = 0.025 \times 10^{x}$ between $x = 0$ and $x = 3$. From the graph determine y when $x = 1.5$ and x when $y = 12$.

13.4 The determination of laws from experimental data

Even when it is known that data collected from some experiment or set of measurements must conform to one or other of the laws discussed earlier in the chapter, it is rare, when those data are plotted on log-linear graph paper, for every point to lie exactly on a straight line.

In that event a line of best fit must be drawn and from that the necessary calculations made to obtain the law to which the data best relate.

Example

It is believed that the data below satisfy a law of the form:

$$m = m_0 e^{-pt}$$

Plot the values in the table on log-linear graph paper. Construct a straight line graph which seems to fit the data. From the graph determine approximate values of m_0 and p, the value of m when $t = 3.5 \times 10^{10}$, and the value of t when $m = 4.8$ kg.

t (seconds)	0	2×10^{10}	4×10^{10}	6×10^{10}
m (kilograms)	10	7.8	5.6	4.4
$\log m$	1	0.892	0.748	0.643

Fig. 13.18 represents a rough sketch of the line.

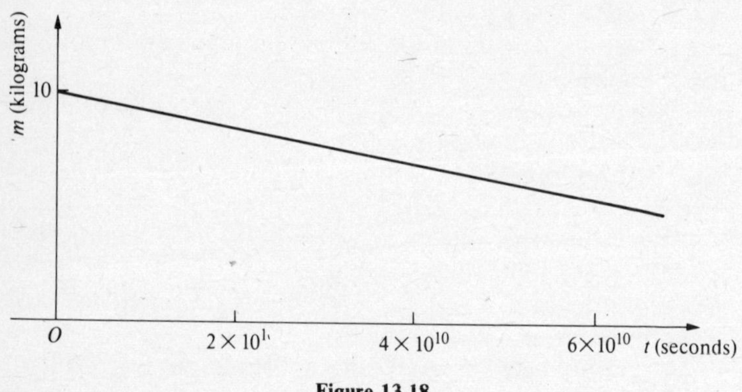

Figure 13.18

From the assumed law:

$$\log m = \log m_0 - pt \log e$$
$$= (-p \log e) . t + \log m_0 \qquad \text{(i)}$$

Fig. 13.19 represents the plotted data on log-linear graph paper. The straight line which is drawn seems to be a reasonable fit.
From the graph, when $t = 3.5 \times 10^{10}$, $m = 6.2$ kg.
When $m = 4.8$ kg, $t = 5.35 \times 10^{10}$ seconds.

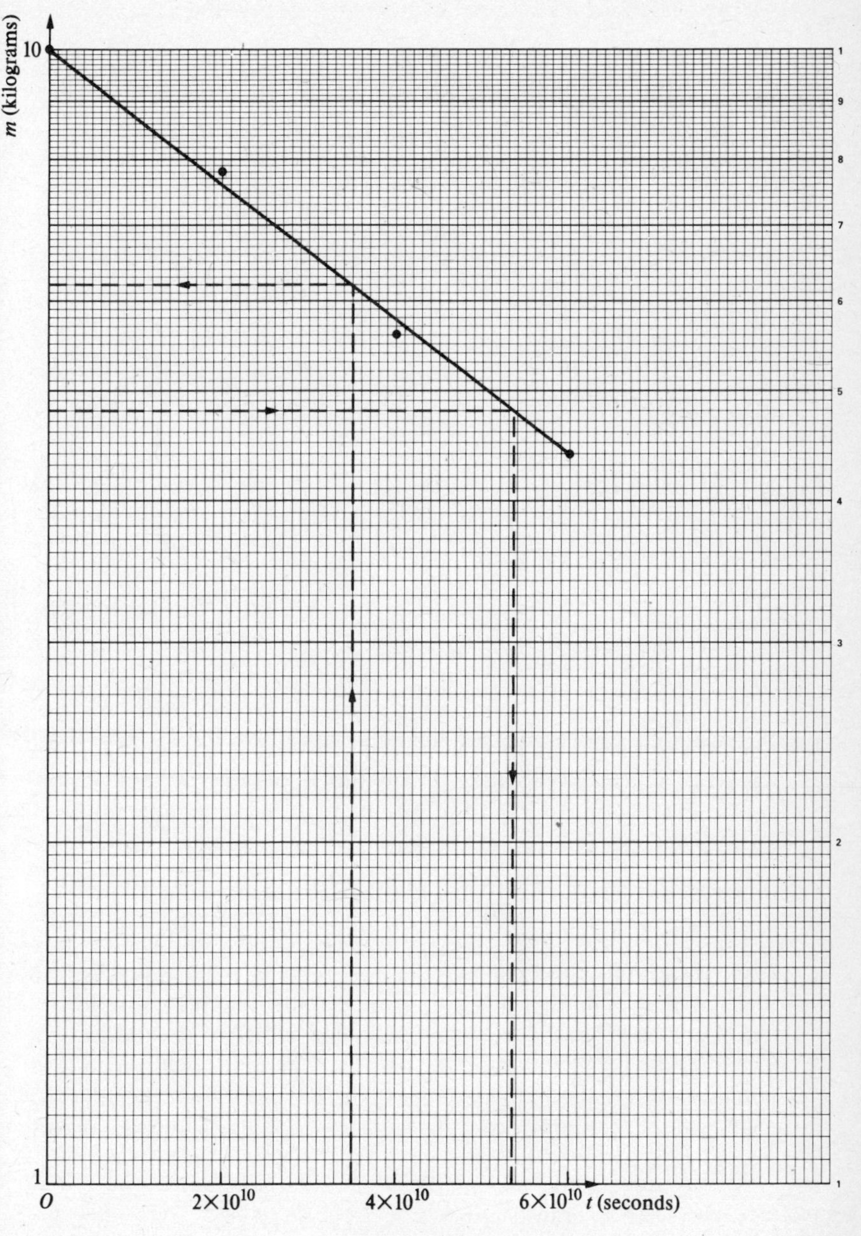

Figure 13.19

To determine the law

From (i) and the data in the table, the gradient of the line is:

$$-p \log e = \frac{0.643 - 1}{6 \times 10^{10}}$$

$$= -0.0595 \times 10^{-10}$$

$$= -5.95 \times 10^{-12}$$

$$p = \frac{5.95 \times 10^{-12}}{\log e}$$

$$= 1.37 \times 10^{-11}$$

The intercept $= m_0 = 10$.
The law is approximately:

$$m = 10e^{-1.37 \times 10^{-11} t}$$

Using this law, when $t = 3.5 \times 10^{10}$, the value of m, by calculator, is 6.1909286 kg.
When $m = 4.8$ kg the calculator determines t to be 5.3574392×10^{10} seconds.

Exercise 13.2

1. The mass, m kilograms, of a radioactive substance is measured at given moments of time, t seconds. From the values in the table below construct a log-linear graph and determine the most suitable values of m_0 and p which fit a law of the form $m = m_0 e^{-pt}$.

t (seconds)	0	2×10^{10}	4×10^{10}	6×10^{10}	8×10^{10}	10^{11}
m (kilograms)	1	0.77	0.58	0.43	0.34	0.26

From the graph determine the values of t when $m = 0.5$ kg and 0.35 kg and the value of m when $t = 5.5 \times 10^{10}$ seconds. Check these values by calculator and the law, using the estimated values of m_0 and p.

2. The formula $T = T_0 e^{\mu\theta}$ relates the tension, T, at points of a belt drive in contact with a pulley to the tension, T_0, in the slack side of the belt. The coefficient of friction between the belt and the pulley is μ. The angle of lap, θ, is measured in radians.

θ (radians)	0.4	0.8	1.6	2.4	2.8
T (newtons)	58	69	95	130	150

From the above data plot on log-linear graph paper the straight line graph which seems to fit the data best. From the graph estimate the values of T_0 and μ. Determine T when $\theta = 2.65$ radians and the value of θ when $T = 100$ newtons.

3. Assume that the data in the table below fit a relationship of the form $Q = Q_0 e^{-pt}$. Draw a graph of Q plotted against t on log-linear graph paper. Use the graph to estimate the values of Q_0 and p.

t (seconds)	1×10^{-3}	3×10^{-3}	5×10^{-3}	8×10^{-3}
Q (coulombs)	2.49×10^{-4}	2.48×10^{-4}	2.47×10^{-4}	2.46×10^{-4}

4. Assume that the data in the table below fit a relationship of the form $I = I_0(1 - e^{-at})$. Draw a suitable straight line graph on log-linear paper to fit the data. From it estimate the values of I_0 and a.

t (seconds)	0.02	0.04	0.06	0.08	0.10
I (amperes)	0.13	0.17	0.18	0.20	0.21

Determine I when $t = 0.07$ seconds and t when $I = 0.15$ amperes.

5. The data in the table below relate the height of a plant, H centimetres, to the time, t, measured in days.

t (days)	4	8	12	16	20
H (centimetres)	13.2	17.5	23	31	40

By drawing a suitable log-linear graph estimate the values of H_0 and k which fit the formula $H = H_0 e^{kt}$. Determine H when $t = 18.5$ and t when $H = 34$.

6. Assume that the data in the table fit a relationship of the form $y = a . b^x$.

x	1	3	5	7
y	99	402	1590	6400

By constructing a log-linear graph estimate the values of a and b. From the graph determine y when $x = 4$ and x when $y = 250$.

14

Simple Curves. Non-linear Physical Laws

Before the graph of a non-linear curve can be drawn a suitable table of values must be constructed.

14.1 The parabola

Examples

1. Plot the graph of $y = \frac{3}{4}x^2$ between $x = -4$ and $x = 4$.

Step 1	x	-4	-3	-2				
Step 2	x^2	16	9	4				
Step 3	$\frac{3}{4}x^2$	12	$6\frac{3}{4}$	3				
	x	-1	0	1	2	3	4	
	x^2	1	0	1	4	9	16	
	$\frac{3}{4}x^2$	$\frac{3}{4}$	0	$\frac{3}{4}$	3	$6\frac{3}{4}$	12	

Step 4 Choose a suitable scale for x.
Step 5 Note the values of y are symmetrical about $x = 0$.
Step 6 Note that y is never negative.
Step 7 Choose a suitable scale for y. (See Fig. 14.1.)
Step 8 When $x = 1.5$, from the graph $y \approx 1.7$. By calculator, when $x = 1.5$, $y = 0.75 \times 1.5^2 = 1.6875$.
Step 9 When $y = 4.8$, from the graph $x \approx \pm 2.5$. By calculator $x = \pm\sqrt{\frac{4}{3}} \times 4.8 = \pm 2.5298221$.

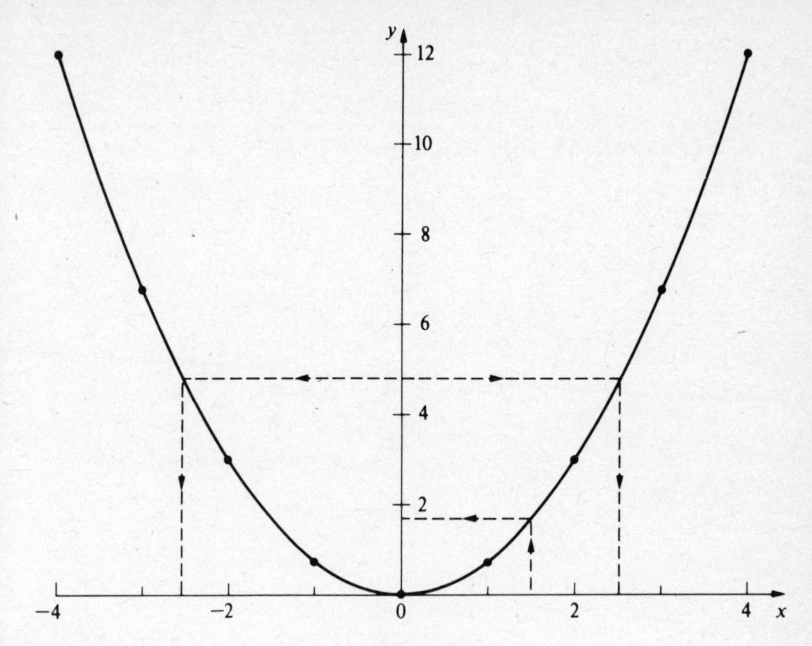

Figure 14.1

2. Plot the graph of $y = -3(x+2)(x-1)$ between $x = -5$ and $x = 3$.

Step 1	x	-5	-4	-3				
Step 2	$(x-1)$	-6	-5	-4				
Step 3	$(x+2)$	-3	-2	-1				
Step 4	$(x+2)(x-1)$	18	10	4				
Step 5	$-3(x+2)(x-1)$	-54	-30	-12				
	x	-2	-1	0	1	2	3	
	$(x-1)$	-3	-2	-1	0	1	2	
	$(x+2)$	0	1	2	3	4	5	
	$(x+2)(x-1)$	0	-2	-2	0	4	10	
	$-3(x+2)(x-1)$	0	6	6	0	-12	-30	

Step 6 Choose a suitable scale for x.
Step 7 Note the values of y are symmetrical about $x = -0.5$.
Step 8 Choose a suitable scale for y. (See Fig. 14.2.)
Step 9 When $x = -3.2$, from the graph, $y \approx -15$. By calculator, $y = -3 \times -1.2 \times -4.2 = -15.12$.
Step 10 When $x = -0.5$, by calculator, $y = -3 \times 1.5 \times -1.5 = 6.75$.
Step 11 When $y = -20$, from the graph, $x \approx 2.5$ or -3.5.

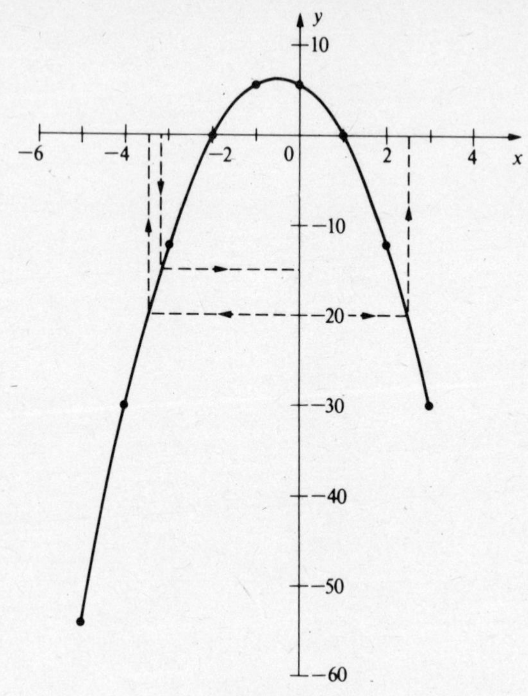

Figure 14.2

3. Plot the graph of $y = 2x^2 - 5x + 11$ between $x = -3$ and $x = 5$.

Step 1	x	-3	-2	-1
Step 2	x^2	9	4	1
Step 3	$2x^2$	18	8	2
Step 4	$-5x$	15	10	5
Step 5	11	11	11	11
Step 6	y	44	29	18

x	0	1	2	3	4	5
x^2	0	1	4	9	16	25
$2x^2$	0	2	8	18	32	50
$-5x$	0	-5	-10	-15	-20	-25
11	11	11	11	11	11	11
y	11	8	9	14	23	36

Step 7 Choose a suitable scale for x.
Step 8 Choose a suitable scale for y. (See Fig. 14.3.)
Step 9 From the above table it does not appear that the curve is symmetrical about any value of x.

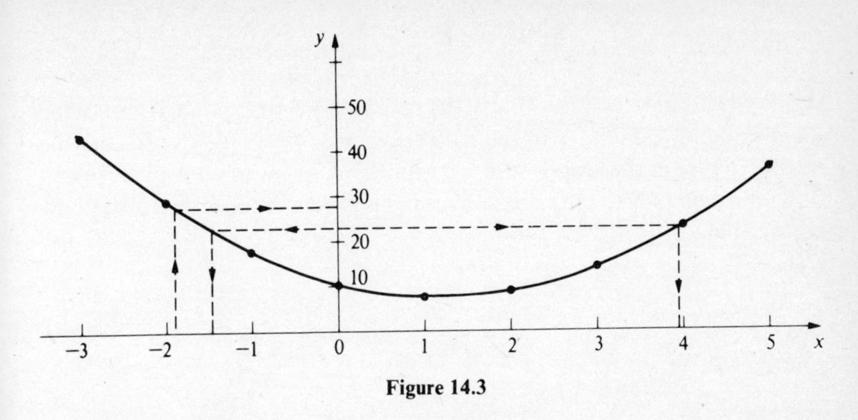

Figure 14.3

Step 10 When $x = -1.8$, from the graph, $y \approx 26$. By calculator, when $x = -1.8$, $y = 26.48$.
Step 11 When $y = 23$, from the graph, $x \approx 3.9$ or -1.4.
Step 12 From the graph the curve is symmetrical about $x = 1.25$.

Each of the above curves is a parabola. The curves in Figs. 14.1 and 14.3 are parabolas with minimum points. The curve in Fig. 4.2 is a parabola with a maximum point. The minimum or maximum point of such a parabola occurs at that value of x about which the curve is symmetrical. The point is often called the vertex.

The basic parabola of this type has an equation $y = x^2$

It has a minimum point (vertex) at (0, 0). y is never negative.

Curves whose equations are $y = ax^2 + bx + c$ are all parabolas. The graphs can all be derived from that of the basic parabola by suitable transformations. Fig. 14.4 is a sketch of $y = x^2$.

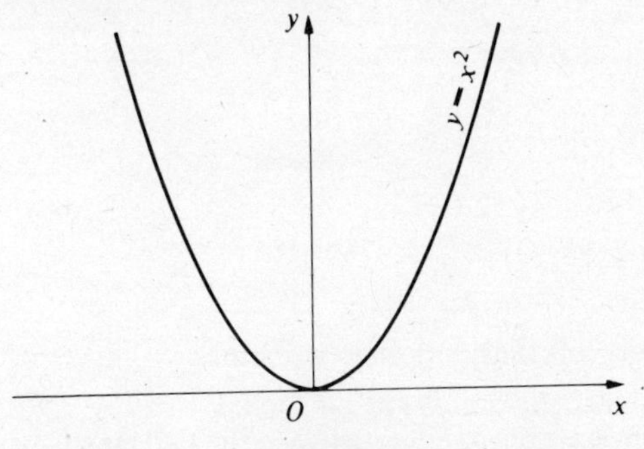

Figure 14.4

The parabola $y = Ax^2$

For any value of x the value of y in the parabola $y = Ax^2$ equals A times the corresponding value of y in the parabola $y = x^2$. Fig. 14.5 represents the relation between the graphs of $y = x^2$ and $y = Ax^2$ when A is positive and $A > 1$. In Fig. 14.5 an ordinate of $y = Ax^2$ is always A times the corresponding ordinate of $y = x^2$.

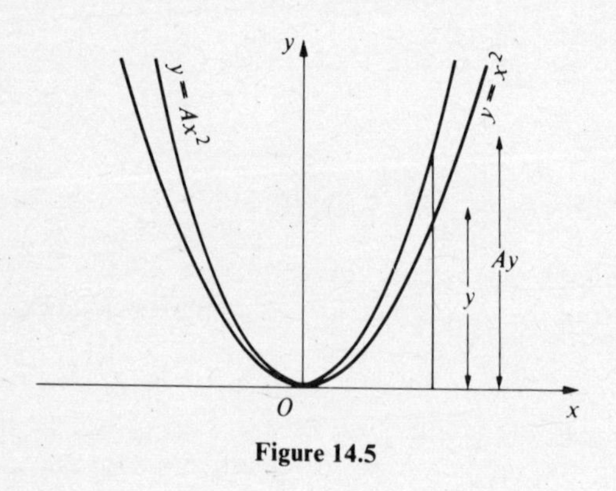

Figure 14.5

Fig. 14.6 represents the relation between the two curves when A is positive and $A < 1$.

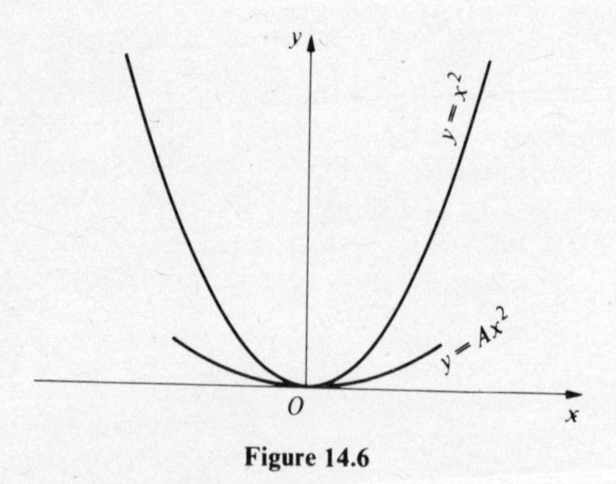

Figure 14.6

Fig. 14.7 represents the relation between the two curves when A is negative and $|A| > 1$.

Fig. 14.8 represents the relation between the two curves when A is negative and $|A| < 1$.

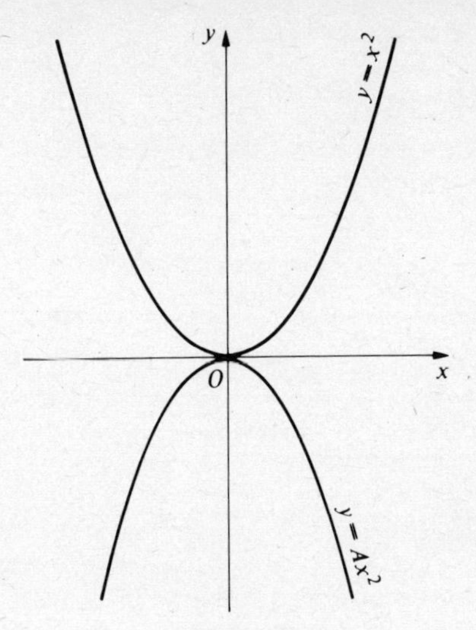

Figure 14.7

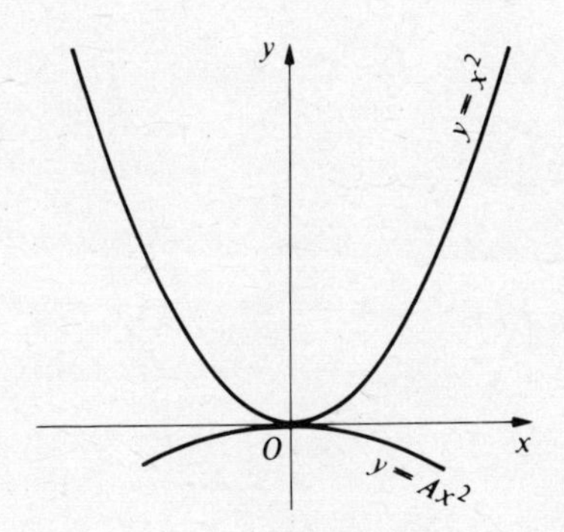

Figure 14.8

The parabola $y = (x - h)^2$

The curve has its minimum point (vertex) when $x = h$, i.e. at the point $(h, 0)$.

The curve has symmetry about $x = h$. Fig. 14.9 represents the relation between the curves $y = (x - h)^2$ and $y = x^2$. The curve $y = (x - h)^2$ is merely $y = x^2$ moved h units parallel to Ox.

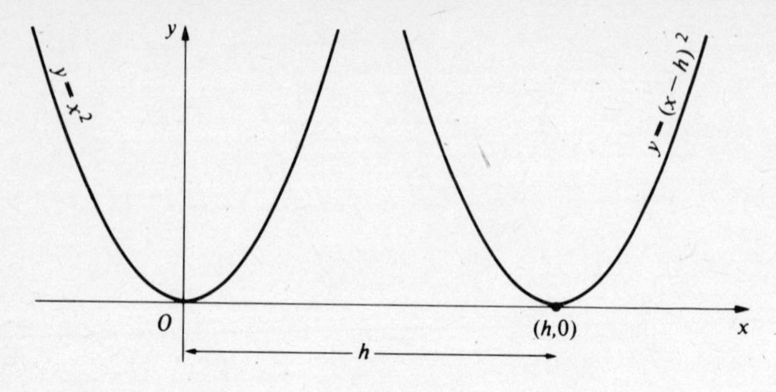

Figure 14.9

The parabola $y = x^2 + k$

The above curve has a rearranged equation $y - k = x^2$. It is symmetrical about $x = 0$. Its minimum point is $(0, k)$.

Fig. 14.10 represents the relation between the curves $y = x^2$ and $y = x^2 + k$. The second parabola is merely the first parabola moved k units parallel to Oy.

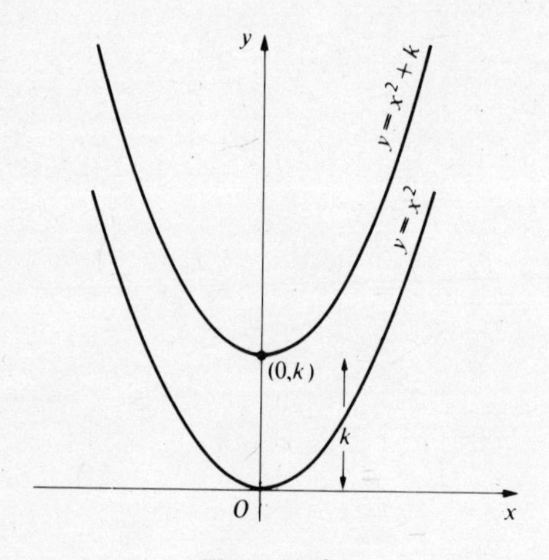

Figure 14.10

The parabola $y = A(x - h)^2 + k$

The above curve has a rearranged equation $y - k = A(x - h)^2$. This curve may be derived from the curve $y = x^2$ by three transformations (see Fig. 14.11).

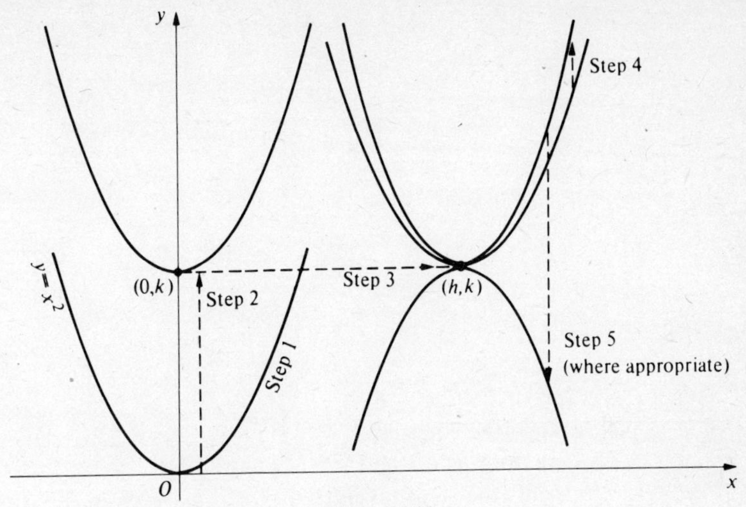

Figure 14.11

Step 1 Draw $y = x^2$.
Step 2 Move the whole curve k units along Oy.
Step 3 Now move the curve h units parallel to Ox.
Step 4 Stretch or compress the curve in the direction of the y axis according to whether $A > 1$ or $A < 1$ and is positive.
Step 5 When A is also negative reflect the curve in the line through its vertex, which is parallel to Ox.

The parabola $y = ax^2 + bx + c$

To determine the relation between the parabola $y = ax^2 + bx + c$ and the basic parabola $y = x^2$: $y = ax^2 + bx + c$ is related to the parabola $y = A(x-h)^2 + k$:

$$y = a\left(x^2 + \frac{b}{a}x\right) + c$$

$$= a\left(x^2 + \frac{b}{a}x + \frac{b^2}{4a^2}\right) + c - \frac{b^2}{4a}$$

$$= a\left(x + \frac{b}{2a}\right)^2 + \left(\frac{4ac - b^2}{4a}\right) \qquad \text{(i)}$$

By relating equation (i) to the form $y = A(x-h)^2 + k$ we get:

$$k = \frac{4ac - b^2}{4a}$$

$$h = -\frac{b}{2a}$$

$$A = a$$

Examples

1. Sketch the graph of $y = 3x^2 - 6x + 3$.
   Step 1 Rearrange the equation: $y = 3(x^2 - 2x + 1) = 3(x-1)^2$.
   Step 2 This equation is of the form $y = A(x-h)^2$.
   Fig. 14.12 represents the steps taken to arrive at a sketch of the curve.

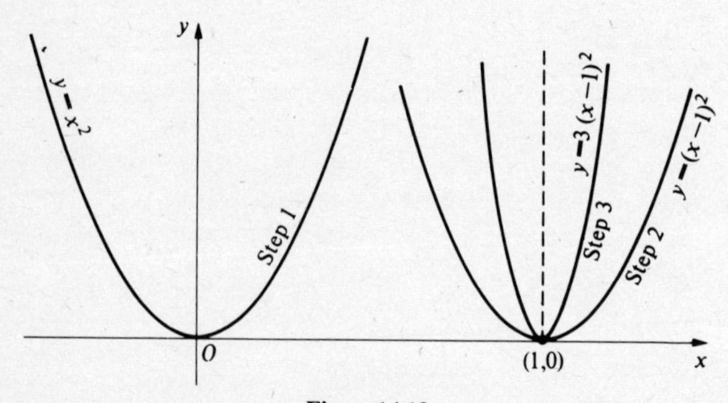

Figure 14.12

2. Sketch the graph of $y = 8 + 5x - 2x^2$.
   Step 1 Rearrange the equation:

$$\begin{aligned} y &= -2(x^2 - \tfrac{5}{2}x) + 8 \\ &= -2(x^2 - \tfrac{5}{2}x + \tfrac{25}{16}) + 8 + \tfrac{25}{8} \\ &= -2(x - \tfrac{5}{4})^2 + \tfrac{89}{8} \end{aligned}$$

Step 2 $\qquad k = 89/8 = 11\tfrac{1}{8}$

Step 3 $\qquad h = 5/4$

Step 4 $\qquad A = -2$

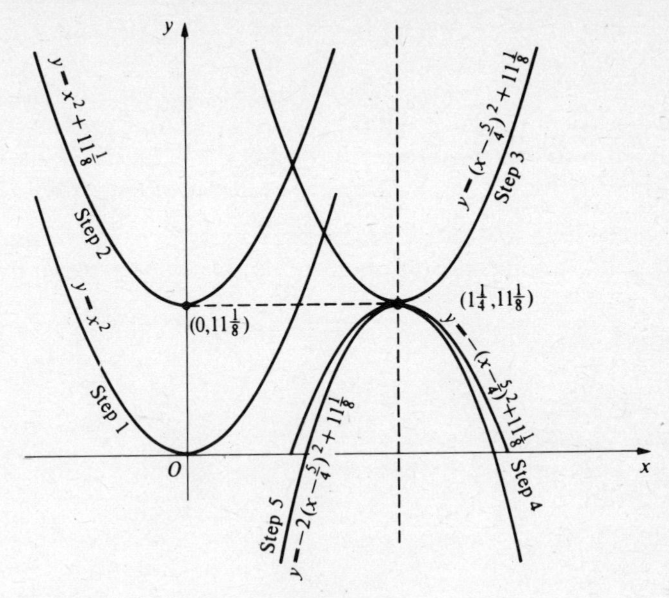

Figure 14.13

Fig. 14.13 represents a sketch of the graph of $y = 8 + 5x - 2x^2$ and the steps taken from the sketch of the basic parabola $y = x^2$ to arrive at the graph.

Exercise 14.1

Construct suitable tables of values and plot the graphs of the following equations between the given values of x. Use the graphs to determine the values of y or x indicated. Wherever possible check those values by calculator.

1. $y = \frac{1}{2}x^2$; between $x = -4$ and $x = 4$. Determine y when $x = 2\frac{1}{2}$ and x when $y = 5\frac{1}{2}$.
2. $y = 4(x-1)(x-2)$; between $x = -5$ and $x = 3$. Determine y when $x = -1.6$ and x when $y = 3.2$.
3. $y = -\frac{1}{5}(x-1)(x-2)$; between $x = -3$ and $x = 4$. Determine y when $x = 1.7$ and x when $y = -1.1$.
4. $y = -\frac{1}{2}x^2 + 2$; between $x = -3$ and $x = 4$. Determine y when $x = -1.3$ and x when $y = 1.5$.
5. $y = x^2 - 5x + 6$; between $x = -1$ and $x = 6$. Determine y when $x = 2.5$ and x when $y = 1.5$.

6. $y = x^2 + 4x + 1$; between $x = -4$ and $x = 2$. Determine where the curve crosses Oy and Ox.
7. $y = 2x^2 - 3x + 2$; between $x = -3$ and $x = 4$. Determine the axis of symmetry of the curve and the co-ordinates of the vertex.
8. $2y + 3 = 6x - 3x^2$; between $x = -1$ and $x = 6$. Determine the axis of symmetry for the curve and the co-ordinates of the vertex.

Sketch the following curves by first rearranging their equations, where necessary, into a suitable form. Show the steps taken to arrive at the final sketch.

9. $y = 3x^2$
10. $y = -\frac{4}{5}x^2$
11. $y = 2(x-1)^2$
12. $y = -\frac{1}{3}(x+2)^2$
13. $y = 2(4-x)^2$
14. $y = x^2 + 3$
15. $y = x^2 - 6$
16. $y = -\frac{3}{4}x^2 + \frac{5}{8}$
17. $y = (x-2)^2 + 5$
18. $3y = -2(x+1)^2 - 4$
19. $y = \frac{4}{3}(x - \frac{2}{5})^2 + \frac{11}{6}$
20. $y = p(x-q)^2 + r$
21. $y = x^2 - 2x + 3$
22. $y = 4x^2 + 12x - 3$
23. $y = x^2 - x + 1$
24. $x = 2(y-1)^2 + 3$

14.2 Further curves

Examples

1. Draw the graph of $y = 2/x$ between $x = -3$ and $x = 3$.

Step 1	x	-3	-2	-1	-0.5	-0.25	0.25	0.5	1	2	3
Step 2	$1/x$	$-\frac{1}{3}$	$-\frac{1}{2}$	-1	-2	-4	4	2	1	$\frac{1}{2}$	$\frac{1}{3}$
Step 3	$2/x$	$-2/3$	-1	-2	-4	-8	8	4	2	1	$2/3$

Step 4 Choose a suitable scale for x.

Step 5 Note the values of y are asymmetrical about $x = 0$, i.e. they are numerically equal in value but opposite in sign.

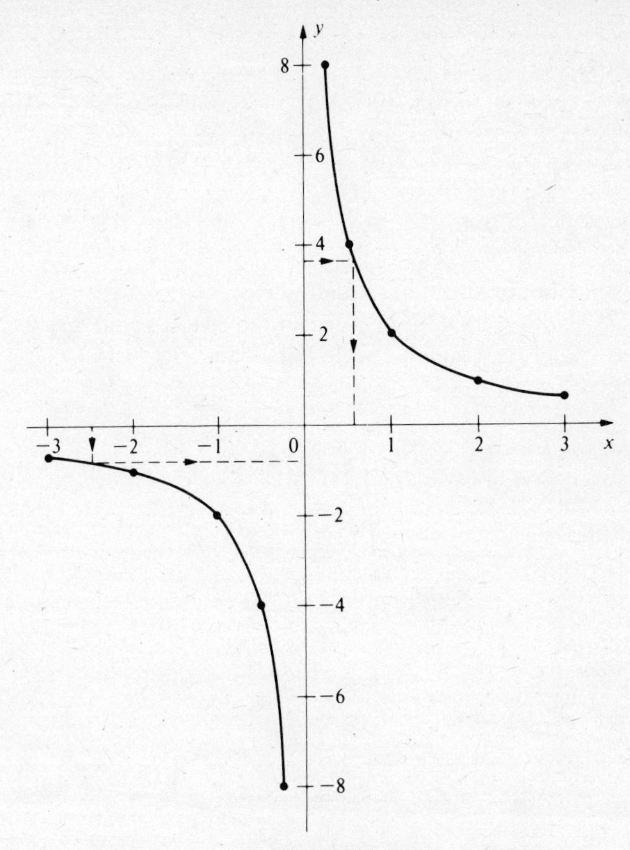

Figure 14.14

Step 6 Choose a suitable scale for y. (See Fig. 14.14.)

Step 7 When $x = -2.5$, from the graph, $y \approx -0.8$. By calculator, $y = -0.8$.

Step 8 When $y = 3.8$, from the graph, $x \approx 0.5$. By calculator, $x = 0.5263158$.

2. Draw the graph of $y = -\dfrac{3}{(2x-5)}$ between $x = -1$ and $x = 6$.

Step 1	x	-1	0	1	1.5	2
Step 2	$2x$	-2	0	2	3	4
Step 3	$(2x-5)$	-7	-5	-3	-2	-1
Step 4	$\dfrac{1}{2x-5}$	$-\frac{1}{7}$	$-\frac{1}{5}$	$-\frac{1}{3}$	$-\frac{1}{2}$	-1
Step 5	$\dfrac{-3}{2x-5}$	$\frac{3}{7}$	$\frac{3}{5}$	1	$\frac{3}{2}$	3

x	2.25	2.75	3	3.5	4	5	6
$2x$	4.5	5.5	6	7	8	10	12
$(2x-5)$	-0.5	0.5	1	2	3	5	7
$\dfrac{1}{2x-5}$	-2	2	1	1/2	1/3	1/5	1/7
$\dfrac{-3}{2x-5}$	6	-6	-3	$-\frac{3}{2}$	-1	$-\frac{3}{5}$	$-\frac{3}{7}$

Step 6 Choose a suitable scale for x.
Step 7 Note the values of y are asymmetrical about $x = 2.5$.
Step 8 Choose a suitable scale for y. (See Fig. 14.15.)

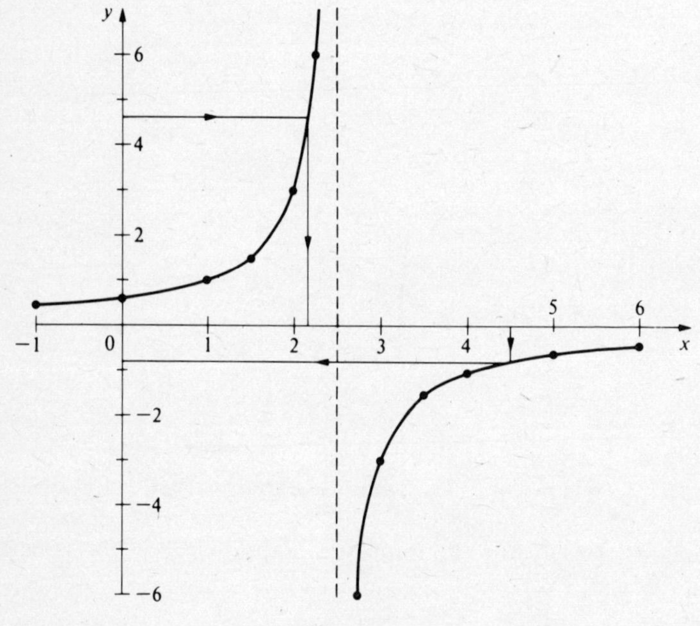

Figure 14.15

Step 9 When $x = 4.5$, from the graph, $y \approx -0.7$. By calculator, $y = -0.75$.

Step 10 When $y = 4.6$, from the graph, $x \approx 2.2$. By calculator, $x = 2.173913$.

3. Draw the graph of $y = 3x^{\frac{1}{2}}$ between $x = 0$ and $x = 10$. Note: y has no real values when x is negative. Consequently no part of the curve lies to the left of the origin.

Step 1	x	0	1	2	3	4	5	6	7	8	9	10
Step 2	$x^{\frac{1}{2}}$	0	1	1.41	1.73	2	2.24	2.45	2.65	2.83	3	3.16
Step 3	$3x^{\frac{1}{2}}$	0	3	4.23	5.19	6	6.72	7.35	7.95	8.49	9	9.48

Step 4 Choose a suitable scale for x.
Step 5 There is no symmetry on this curve.
Step 6 Choose a suitable scale for y. (See Fig. 14.16.)

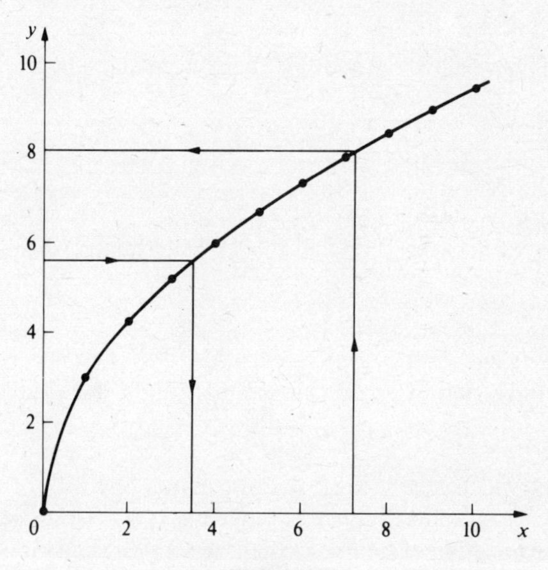

Figure 14.16

Step 7 When $x = 7.2$, from the graph, $y \approx 8.1$. By calculator, $y = 8.0498447$.
Step 8 When $y = 5.6$, from the graph, $x \approx 3.5$. By calculator, $x = 3.4844444$.

Exercise 14.2

Construct suitable tables of values and plot the graphs of the following equations between the indicated values of x. Use the graphs to determine the values of y or x required. Wherever possible check those values by calculator.

1. $y = 4/x$ between $x = -3$ and $x = 3$. Determine y when $x = 2.6$ and x when $y = 8.4$.

2. $y = -1/2x$ between $x = -5$ and $x = 5$. Determine y when $x = -1.4$ and x when $y = -0.9$.
3. $y = \dfrac{48}{2x-1}$ between $x = 2$ and $x = 12$. Determine y when $x = 4.9$ and x when $y = 13.7$.
4. $y = \dfrac{100}{3-2x}$ between $x = -3$ and $x = -10$. Determine y when $x = -5.2$ and x when $y = 6.8$.
5. $y = \dfrac{5}{2}x^{1/2}$ between $x = 0$ and $x = 10$. Determine y when $x = 8.6$ and x when $y = 3.9$.
6. $y = -\dfrac{5}{8}x^{1/2}$ between $x = 10$ and $x = 100$. Determine y when $x = 69$ and x when $y = 4.7$.
7. $y = 3(x-1)^{\frac{1}{2}}$ between $x = 1$ and $x = 11$. Determine y when $x = 8.2$ and x when $y = 5.6$.

14.3 The conversion of certain curves to straight-line graphs

The one graph which is easy to recognize no matter how little of the graph is visible is the straight-line graph. Consequently it is often of advantage to transform the equation of a curve so that the resulting equation is that of a straight line.

The transformation of $y = ax^2 + b$

By substituting $x^2 = X$ the equation becomes $y = aX + b$. This is the equation of a straight line obtained by plotting y against X. The gradient of the line is a and the intercept is b.

Example

The following table of values of x and y indicates a relationship of the form $y = ax^2 + b$. Calculate the values of a and b. From the graph determine y when $x = 6.3$ and x when $y = 16.4$.

x	1	4	7	10
y	10.5	18	34.5	60

Substitute $x^2 = X$ and construct a new table for X and y:

X	1	16	49	100
y	10.5	18	34.5	60

Fig. 14.17 represents the graph of y plotted against X. The four points lie on a straight line, AB. The line crosses the y axis at $y = 10$. Then $b = 10$. The gradient of $AB = 50/100 = 1/2$. The relationship between y and x is:

$$y = \tfrac{1}{2}x^2 + 10$$

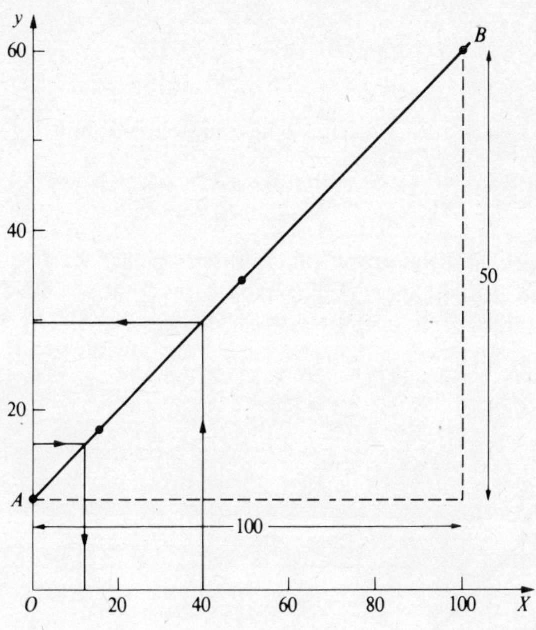

Figure 14.17

When $x = 6.3$, $X = 39.69 \approx 40$. From the graph $y \approx 29.5$. By formula and calculator $y = 29.845$.

When $y = 16.4$, $X \approx 12.5$, $x = \sqrt{12.5} = 3.5$. By the formula and calculator $x = 3.5777088$.

The transformation of $y = p + q/x$

By substituting $1/x = X$ the equation becomes:

$$y = p + qX = qX + p$$

Again this is the equation of a straight line obtained by plotting y against X. The gradient is q and the intercept is p.

Example

The following table of values of x and y refers to a relationship of the form $y = p + q/x$. Calculate the values of p and q. From the graph determine y when $x = \frac{1}{4}$ and x when $y = 35$.

x	1/10	1/5	1/2	1
y	0	25	40	45

Substitute $1/x = X$ and construct a new table for X and y:

X	10	5	2	1
y	0	25	40	45

Fig. 14.18 represents the graph of y plotted against X. The four points plotted lie on a straight line, PQ. The line crosses Oy at $y = 50$. Then $p = 50$. The gradient of $PQ = -50/10 = -5$. Then $q = -5$. The relationship between y and x is:

$$y = 50 - 5/x$$

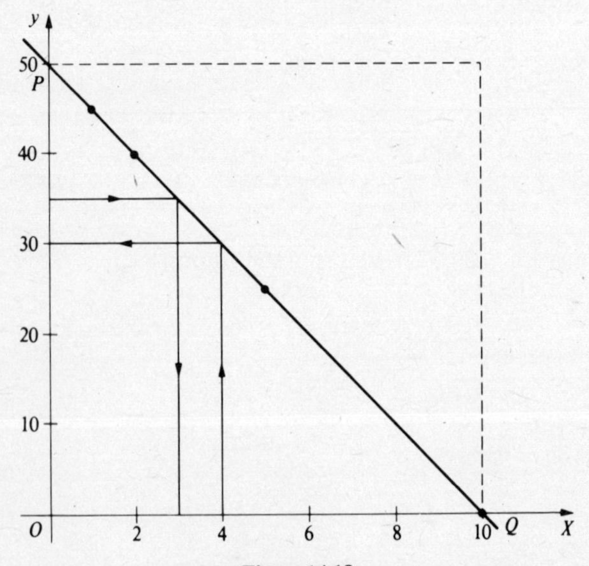

Figure 14.18

When $x = \frac{1}{4}$, $X = 4$. From the graph, $y = 30$. By calculator, $y = 30$.
When $y = 35$, from the graph, $X = 3$, i.e. $x = \frac{1}{3}$. By calculator, $x = \frac{1}{3}$.

Exercise 14.3

Assume that the data in the following tables are consistent with a relationship of the form $y = ax^2 + b$. Make a suitable transformation and plot the revised data on a graph. From the graph calculate the values of a and b.

1.

x	1	3	8	10
y	5.25	7.25	21	30

From the graph determine y when $x = 7.2$ and x when $y = 14.5$.

2.

x	3	5	7	10
y	55.5	47.5	35.5	10

From the graph determine y when $x = 1.6$ and x when $y = 21.9$.

3.

x	2	5	10	11
y	12.4	25	70	82.6

From the graph determine y when $x = 9.2$ and x when $y = 53.5$.

4.

x	5	7	9	10
y	50	2	-62	-100

From the graph determine y when $x = 6.8$ and x when $y = 60$.

5.

x	1/10	1/5	1/2	1
y	20	15	12	11

Assume $y = p + q/x$. Calculate p and q. From the graph determine y when $x = 0.45$ and x when $y = 13.7$.

Assume the data in the following tables are consistent with a relationship of the form $y = p + q/x$. From the graph, obtained after a suitable transformation, calculate the values of p and q.

6.

x	1/5	2/5	4/5	6/5
y	25	20	$17\frac{1}{2}$	$16\frac{2}{3}$

From the graph determine y when $x = 0.5$ and x when $y = 23.2$.

7.

x	1/8	1/6	3/4	1
y	42	44	$48\frac{2}{3}$	49

From the graph determine y when $x = 0.8$ and x when $y = 43.6$.

8.

x	1	4	5	10
y	50	57.5	58	59

From the graph determine y when $x = 7.7$ and x when $y = 53.4$.

14.4 Some standard curves

The parabola is a standard section of a cone. The curve is important in elementary concepts of gravitational motion. Its reflective properties have useful applications in the field of light.

Other curves which are important in many ways have the following standard equations:

$$x^2 + y^2 = a^2$$
$$x^2/a^2 + y^2/b^2 = 1$$
$$x^2/a^2 - y^2/b^2 = 1$$
$$xy = c^2$$

The curve $x^2 + y^2 = a^2$

Fig. 14.19 is a sketch of the curve. It is a circle, centre O, radius a.

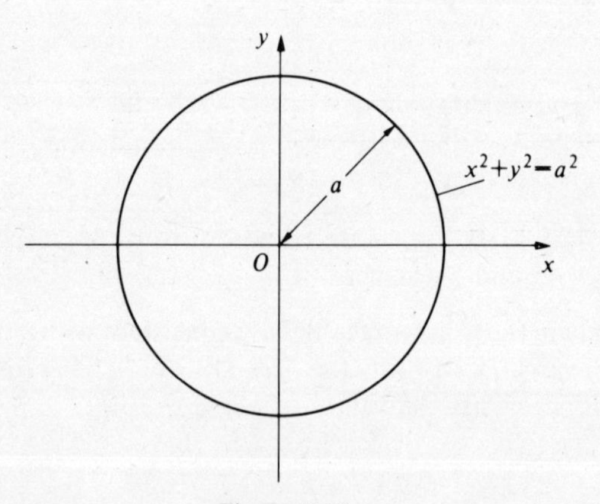

Figure 14.19

It is symmetrical about Ox (detected in the equation because y occurs only as an even power). It is symmetrical about Oy (x occurs only as an even power). It is symmetrical about the axis through O perpendicular to its plane.

The curve $x^2/a^2 + y^2/b^2 = 1$

Fig. 14.20 is a sketch of the curve. It is an ellipse, centre O. Its axes AA^1 and BB^1 are $2a$ and $2b$ respectively. It is symmetrical about Ox. It is symmetrical about Oy. It is symmetrical about the axis through O perpendicular to its plane. When a circle is drawn on AA^1 as diameter then the ratio QN/PN is always a/b for all positions of P on the ellipse.

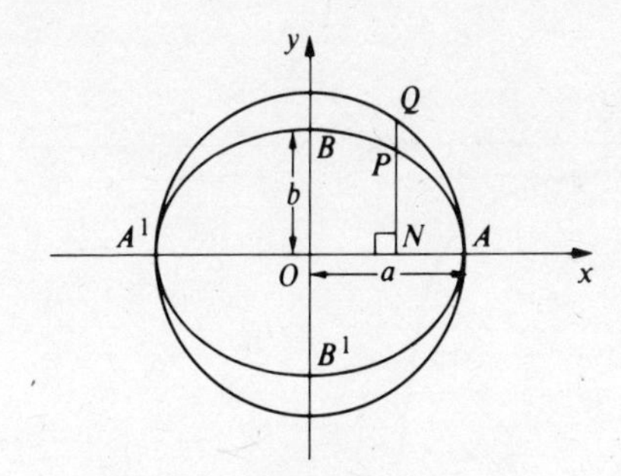

Figure 14.20

The curve $x^2/a^2 - y^2/b^2 = 1$

Fig. 14.21 represents a sketch of the curve. It is a hyperbola with centre O. The curve lies entirely within the angle between two lines, HK and LM. They are the asymptotes. Their gradients are b/a and $-b/a$ respectively. The curve is

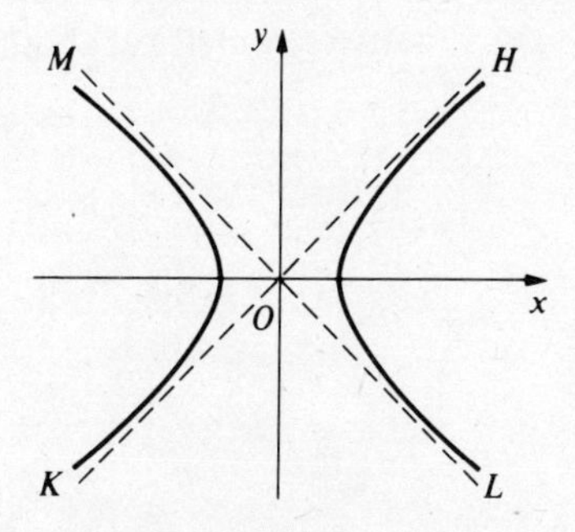

Figure 14.21

symmetrical about Ox. The curve is symmetrical about Oy. The curve is symmetrical about the axis through O perpendicular to its plane.

The curve $xy = c^2$

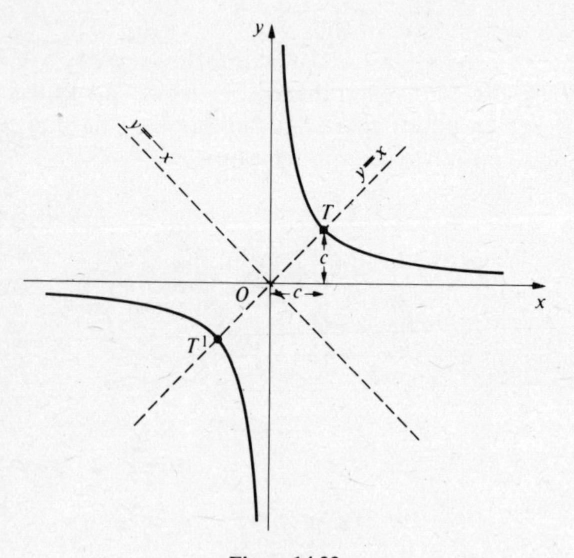

Figure 14.22

Fig. 14.22 represents a sketch of the curve. It is a rectangular hyperbola with centre O. The asymptotes are Ox and Oy. The hyperbola is called rectangular because the asymptotes are perpendicular to each other. The co-ordinates of T, T^1 are (c, c) and $(-c, -c)$ respectively. The curve is not symmetrical about Ox or Oy. It is symmetrical about the bisectors of the angles between Ox and Oy. It is also symmetrical about the axis through O perpendicular to its plane.

15

The Use of Graphs to Determine Values and Established Power Laws

Chapter 2 of *Mathematics Level 2* dealt with graphs of straight lines and the method of determining linear laws from plotted data. Often, though, experimental data and sets of observed measurements are not related by linear laws but by a law which is often called the power law. This is of the form

$$y = ax^n \qquad 15.1$$

where x and y represent the pairs of related values. The values of a and n depend on the nature of the source of the data, e.g. the kind of machine, or the particular technological or scientific principle.

Usually the object in drawing the graph is to use it to determine the values of a and n as well as to use the graph to determine other related values of x and y. If y were plotted against x the shape of the curve would depend on the values of a and n and also on the relative scales of the two axes. The task of recognizing the particular shape which corresponds to particular values of a and n would be almost impossible.

The one shape which is easily recognized is that of the straight line. To achieve this shape of a straight-line graph we carry out a transformation of *15.1*.

15.1 The use of logarithms to reduce laws of the type $y = ax^n$ to straight line form

Take logs to base 10 of *15.1* to obtain:

$$\begin{aligned} \log y &= \log a + \log x^n \\ &= \log a + n \log x \\ \log y &= n \log x + \log a \qquad 15.2 \end{aligned}$$

Put log $x = X$ and log $y = Y$. Then *15.2* becomes:

$$Y = nX + \log a \qquad 15.3$$

By plotting Y against X, *15.3* represents a straight-line graph. Its gradient is n; the intercept is log a. Consequently, by determining the gradient of the line, we obtain n, and by determining the intercept of the line we obtain log a. Fig. 15.1 represents a sketch of such a graph. From it:

$$OA = \log a \qquad 15.4$$

$$a = 10^{OA} \qquad 15.5$$

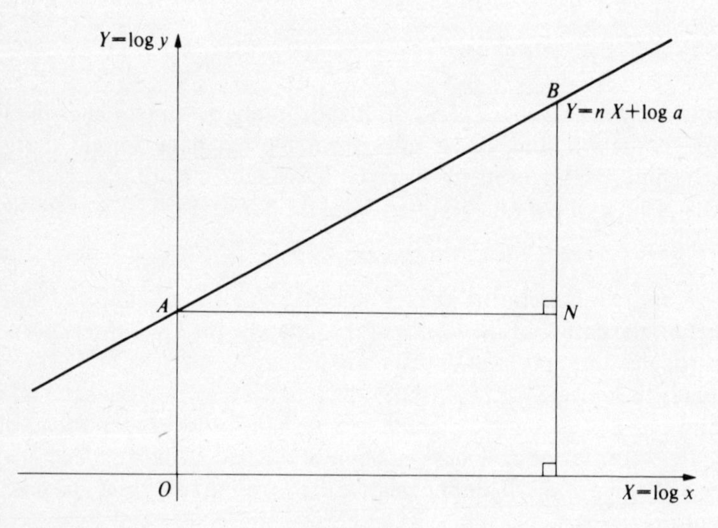

Figure 15.1

Suppose B is a point on the line and that AN and NB are parallel to OX and OY respectively such that AN is a convenient integral value. Then:

$$n = BN/AN \qquad 15.6$$

Particular examples of the power law are:

$s = \frac{1}{2}at^2$, relating distance and time when a is constant
$E = \frac{1}{2}mv^2$, relating kinetic energy and velocity of a given body
$T = 2\pi\sqrt{\frac{L}{g}}$, relating time of oscillation of a pendulum and its length
$p = c/v, = cv^{-1}$, relating pressure and volume of a gas

Examples

1. The energy of a moving body varies with the velocity in the following manner:

Velocity (v) m/s	10	15	20	25	30
Energy (E) joules	2500	5625	10 000	15 625	22 500

Assume:

$$E = a\,.\,v^n$$

Take logs:

$$\begin{aligned}\log E &= \log a + \log v^n \\ &= \log a + n \log v\end{aligned}$$

Now construct a new table:

$\log v$	1	1.1761	1.3010	1.3979	1.4771
$\log E$	3.3979	3.7501	4.0000	4.1939	4.3522

By correcting these figures to two decimal places the graph in Fig. 15.2 is obtained.

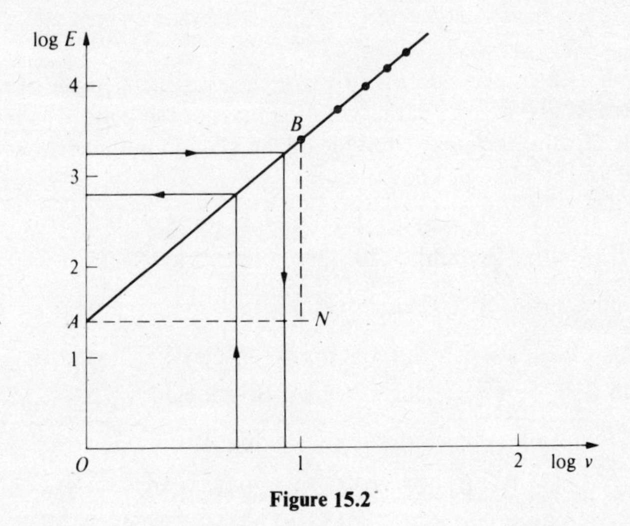

Figure 15.2

From the graph, log $a = 1.4$; $a = 10^{1.4} = 25.118864 \approx 25.12$; $n = BN/AN = (3.4 - 1.4)/1 = 2$. The law appears to be $E = 25.12\,v^2$.
From the graph, when $v = 5$, $\log v = 0.6987$ (by calculator), i.e. log

$v \approx 0.70$. Then $\log E = 2.75$; by calculator, $E = 10^{2.75} = 562.34 \approx 562$.

When $E = 1778$, $\log E = 3.25$. Then $\log v = 0.93$ from the graph, $v = 8.5113804$ (by calculator) ≈ 8.5.

As a check on the graphical method the following is a calculation to obtain the values of a and n. Substituting in $\log E = \log a + n \log v$ from the original table of values:

$$\log 22\,500 = \log a + n \log 30 \qquad \text{(i)}$$

$$\log 2500 = \log a + n \log 10 \qquad \text{(ii)}$$

(i) – (ii):

$$\log 22\,500 - \log 2500 = n \log 30 - n \log 10$$

$$\log \frac{22\,500}{2500} = n \log \frac{30}{10}$$

$$\log 9 = n \log 3$$

$$2 \log 3 = n \log 3, \; n = 2$$

Substitute for n in (ii):

$$\log 2500 = \log a + 2 \log 10$$

$$\log 2500 - \log 100 = \log a$$

$$\log 25 = \log a$$

$$a = 25$$

2. The length of the side of a square sheet of metal plate of uniform thickness and density varies with the mass of the plate. The following table of values relates s, the side of the plate in millimetres, to W, the mass of the plate in kilograms.

W (kg)	4	9	16	25	36
s (mm)	220	330	440	550	660

Suppose $s = a \,.\, W^n$. Then:

$$\log s = \log a + n \log W$$

$$\log s = n \log W + \log a$$

The following table relates $\log s$ and $\log W$:

$\log W$	0.6021	0.9542	1.2041	1.3979	1.5563
$\log s$	2.3424	2.5185	2.6435	2.7404	2.8195

From the graph in Fig. 15.3, $\log a \approx 2.04$; $a = 10^{2.04} = 109.64 \approx 110$. Also, from the graph, $n = BN/AN = (3.04 - 2.04)/2 = 1/2$. Therefore $s = 110 \,.\, W^{\frac{1}{2}} = 110\sqrt{W}$.

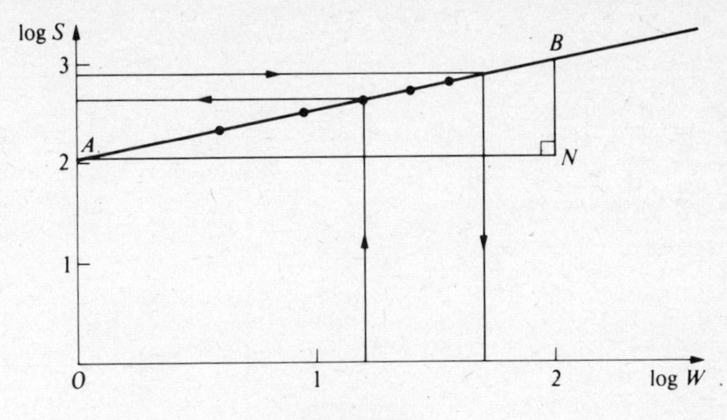

Figure 15.3

From the graph, when log $W = 1.1$, i.e. $W = 12.589254 \approx 12.6$, then log $s = 2.58$, $s = 10^{2.58} = 380.1894 \approx 380$.

From the graph, when log $s = 2.88$, i.e. $s = 10^{2.88} = 758.7758 \approx 759$, then log $W = 2.65$, $W = 10^{2.65} = 446.68359 \approx 447$.

3. A constant potential difference is applied to the terminals of a variable conductor. The current through the conductor and its resistance are connected by a relation represented by the values of R, in ohms, and I, in amperes, in the following table:

R (ohms)	50	100	150	200	250
I (amperes)	5.00	2.50	1.67	1.25	1.00

Assume $I = a \,.\, R^n$. Then:

$$\log I = \log a + n \log R$$

The new table, for log R and log I, is:

log R	1.70	2.00	2.18	2.30	2.40
log I	0.70	0.40	0.22	0.10	0.00

From the graph in Fig. 15.4, log $a = 2.40$, $a = 10^{2.4} = 251.18864 \approx 251$ or, rounded off even more, 250. From the graph, $n = BN/AN = (0.4 - 2.4)/2 = -1$. Therefore $I = 250\,R^{-1}$.

From the graph, when log $R = 1.25$, $R = 17.782794 \approx 17.8$, then log $I = 1.15$, $I = 14.125375 \approx 14.13$.

When log $I = 1.75$, $I = 56.234133 \approx 56.23$, then log $R = 0.65$, $R = 10^{0.65} = 4.4668359 \approx 4.47$.

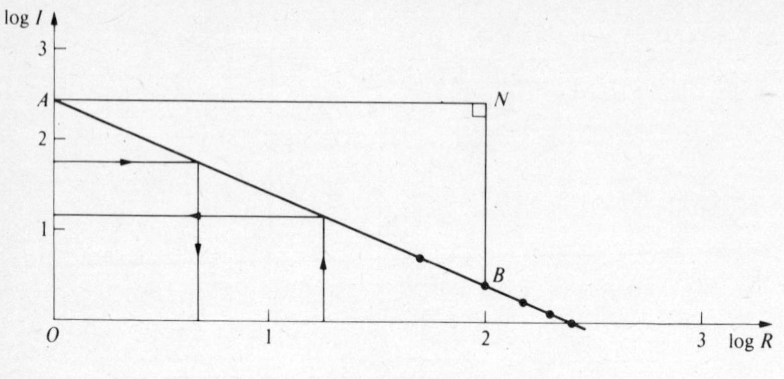

Figure 15.4

Exercise 15.1

1. Draw a graph representing the law $y = 10.x^3$ so as to obtain a straight-line graph. Construct a table by taking the values of x to be 10, 20, 30, 40. Hint: rearrange the equation to $\log y = \log 10 + 3 \log x$. From the graph check that the gradient is 3 and that the intercept is log 10. Determine y when $x = 24$ and x when $y = 12\,000$.
2. Draw a straight-line graph to represent the law $y = 10.x^{-2}$, using the following values of x: 1, 0.8, 0.6, 0.5. From the graph determine y when $x = 0.7$ and x when $y = 200$.
3. Draw a straight-line graph to represent $y = 6.25x^{1.6}$ using the following values of x: 1, 4, 8, 10. From the graph determine y when $x = 5.2$ and x when $y = 63.7$.
4. The masses of circular metal plates vary with the radii of the plates as follows:

Radius (R) mm	200	300	400	500	600
Mass (W) kg	5.00	11.25	20.00	31.25	45.00

Construct a new table relating log R to log W. Use it to plot a graph. From the graph verify that the law relating W to R is of the form $W = k.R^2$. Find an approximate value of k. From the graph determine the value of W when $R = 340$ and the value of R when $W = 33.60$.

5. The velocities which a lead weight attains after falling through varying heights are given in the following table:

Height (h) m	16	25	49	64	100
Velocity (V) m/s	5.6	7.0	9.8	11.2	14.0

Construct a new table relating log V to log h. From it draw a graph. From the graph verify that the law relating V to h is of the form $V = a.h^{\frac{1}{2}}$, where a is a constant. Calculate as accurately as you are able the value of a. From the graph determine the value of V when $h = 37$ and the value of h when $V = 12.3$.

6. The current, I amperes, through a given conductor to which a variable voltage is applied, varies according to the following table:

Voltage (V) volts	50	100	150	200	250	300
Current (I) amperes	0.2	0.4	0.6	0.8	1.0	1.2

Determine the law connecting I and V. When $V = 60.5$ determine I and when $I = 0.65$ determine V.

7. The time of oscillation of a simple pendulum and its length are related by a certain law. The following table gives pairs of values of T and L which are related to one another, where T represents the time of oscillation and L the length of the pendulum.

Length (L) m	1	4	9	16	25
Time (T) s	2	4	6	8	10

Determine the law relating T to L. From your graph determine the value of L when T is 5 and the value of T when L is 2.5.

8. A body subjected to different velocities is timed to cover a given distance at a constant speed. The following pairs of values are recorded:

Velocity (V) m/s	5	10	15	20	25
Time (t) s	8.4	4.2	2.8	2.1	1.68

Determine graphically the relation between t and V. From the graph determine the value of t when $V = 22.5$ and the value of V when $t = 3.0$.

9. A curve with an equation of the form $y = ax^n$ is satisfied by the following data:

x	0.42	0.89	2.08
y	6.2	8.5	12.2

Determine a and n. Calculate y when $x = 1.86$ and x when $y = 7.3$.

10. The following table represents recommended cutting speeds for boring holes of diameter d millimetres in a certain metal. Determine the law relating N to d where N is the speed in rev/min, assuming that it is of the form $N = a.d^n$.

d (mm)	20	25	30	35	40
N (rev/min)	336	265	218	184	160

Calculate N when $d = 25.4$ and 38.1 and d when $N = 260$ and 175.

11. The following table represents the relation between the attractive force, F, of two magnetic poles and their distance apart, d.

d	2	4	6	8
F	49.1	12.7	5.5	3.2

Assume $F = a.d^n$. Determine the values of a and n. Determine F when $d = 5$ and 7 and d when $F = 31.6$ and 5.2.

12. The following data relate the pressure and volume of a gas during a certain kind of expansion:

Volume (v) litre	14.35	11.3	9.55	7.70
Pressure (P) kilopascals	2848.2	3595.6	4221.8	5191.4

Determine the law relating P with v. Calculate from your graph the value of P when $v = 12.2$ and 8.60 and the value of v when $P = 3100.7$ and 4918.7.

13. The following data satisfy a relation of the form $W = a.V^n$, where a and n are constants.

V	2.6	5.8	10.9
W	0.0295858	0.0026651	0.0004015

Calculate a and n. Determine W when $V = 4.7$ and V when $W = 0.00123$.

16

Matrices and Determinants

In *Mathematics Level 1* two methods of solution of simultaneous linear equations in two unknowns were adopted: (a) by elimination, and (b) by substitution. By looking at such problems in a different way a new kind of number emerges: a matrix. As a first step towards that end we look again at a simple, familiar problem.

Examples

1. Solve $3x = 5$.
   Step 1 $\frac{1}{3} \times (3x) = \frac{1}{3} \times 5$
   Step 2 $(\frac{1}{3} \times 3) \times x = 5/3$, the associative law
   Step 3 $1 \times x = 5/3$ ($\frac{1}{3}$ and 3 are inverses under multiplication)
   Step 4 $x = 5/3$
2. Solve $ax = b$, where a and b are constants $\neq 0$.
   Step 1 $\dfrac{1}{a} \times (ax) = \dfrac{1}{a} \times b$, or $a^{-1} \times (ax) = a^{-1} . b$
   Step 2 $\left(\dfrac{1}{a} \times a\right) \times x = \dfrac{b}{a}$, or $(a^{-1} . a) . x = a^{-1} . b$
   Step 3 $1 \times x = b/a$, or $x = a^{-1} . b$
   Step 4 $x = b/a$
3. Solve
   $$\left.\begin{aligned} 2x + y &= 5 \\ x + y &= 3 \end{aligned}\right\}$$
   Rewrite:
   $$\begin{pmatrix} 2 & 1 \\ 1 & 1 \end{pmatrix} \cdot \begin{pmatrix} x \\ y \end{pmatrix} = \begin{pmatrix} 5 \\ 3 \end{pmatrix} \qquad 16.1$$

Put $\begin{pmatrix} 2 & 1 \\ 1 & 1 \end{pmatrix} = A, \quad \begin{pmatrix} x \\ y \end{pmatrix} = X, \quad \begin{pmatrix} 5 \\ 3 \end{pmatrix} = B$

Equation *16.1* becomes $A.X = B$, now an equation like those in Examples (1) and (2) above. To solve this equation we would require:

$$A^{-1}.(A.X) = A^{-1}.B$$
$$X = A^{-1}.B$$

For this to be possible we need to give meanings to:

$$\begin{pmatrix} 2 & 1 \\ 1 & 1 \end{pmatrix} \begin{pmatrix} x \\ y \end{pmatrix} \begin{pmatrix} 5 \\ 3 \end{pmatrix} \text{ and } A^{-1} \text{ when } A = \begin{pmatrix} 2 & 1 \\ 1 & 1 \end{pmatrix}$$

They are all examples of matrices. The solution of the above equations is left until the next chapter.

Definition

A matrix is a rectangular array of numbers arranged in rows and columns. When there are m rows and n columns it is called an $m \times n$ matrix.

For the present we shall restrict ourselves to the study of 2×2 matrices. Examples of 2×2 matrices are:

$$\begin{pmatrix} 1 & 3 \\ 2 & 5 \end{pmatrix} \begin{pmatrix} 4 & -1 \\ 7 & -2 \end{pmatrix} \begin{pmatrix} -1 & -4 \\ -5 & -2 \end{pmatrix} \begin{pmatrix} 1 & 0 \\ 0 & 1 \end{pmatrix} \begin{pmatrix} 2 & 0 \\ 0 & 0 \end{pmatrix} \begin{pmatrix} 0 & -3 \\ 2 & 0 \end{pmatrix}$$

16.1 The sum and difference of two 2 × 2 matrices

Definitions

The sum of two 2×2 matrices is defined by the following rule:

$$\begin{pmatrix} a & b \\ c & d \end{pmatrix} + \begin{pmatrix} A & B \\ C & D \end{pmatrix} = \begin{pmatrix} a+A & b+B \\ c+C & d+D \end{pmatrix} \qquad 16.2$$

The difference between two matrices is defined by the rule:

$$\begin{pmatrix} A & B \\ C & D \end{pmatrix} - \begin{pmatrix} a & b \\ c & d \end{pmatrix} = \begin{pmatrix} A-a & B-b \\ C-c & D-d \end{pmatrix} \qquad 16.3$$

In other words, the sum of two matrices is obtained by adding together

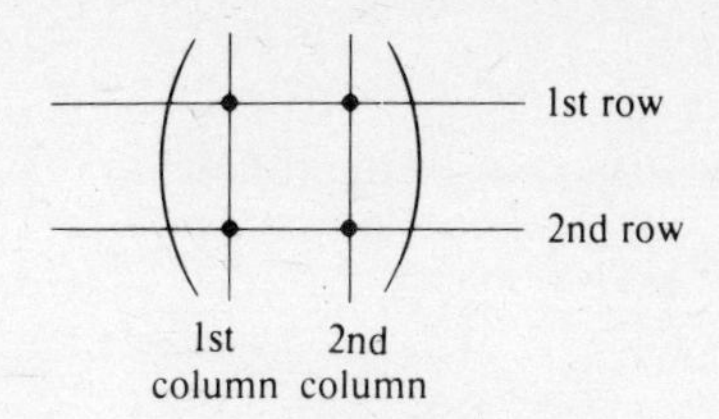

Figure 16.1

corresponding elements of the matrices. The difference between two matrices is obtained by finding the differences between corresponding elements. The term 'corresponding elements' needs clarification. Fig. 16.1 represents a 2×2 matrix in which the dots represent the positions of the four elements. The elements in the matrices which are in the first row and the first column are corresponding elements. The elements in the first row and the second column are corresponding elements, and so on.

$$\text{When } X = \begin{pmatrix} a_{11} & a_{12} \\ a_{21} & a_{22} \end{pmatrix} \text{ and } Y = \begin{pmatrix} b_{11} & b_{12} \\ b_{21} & b_{22} \end{pmatrix}$$

$$\text{then } X + Y = \begin{pmatrix} a_{11} + b_{11} & a_{12} + b_{12} \\ a_{21} + b_{21} & a_{22} + b_{22} \end{pmatrix}$$

The suffixes of the elements in X and Y correspond to the row and column positions of the elements in the matrices: the first suffix for the row and the second suffix for the column.

Examples

1. $\begin{pmatrix} 3 & 2 \\ 1 & 4 \end{pmatrix} + \begin{pmatrix} 9 & 5 \\ 2 & 6 \end{pmatrix} = \begin{pmatrix} 3+9 & 2+5 \\ 1+2 & 4+6 \end{pmatrix} = \begin{pmatrix} 12 & 7 \\ 3 & 10 \end{pmatrix}$

2. $\begin{pmatrix} 4 & 0 \\ -5 & 7 \end{pmatrix} + \begin{pmatrix} -3 & 8 \\ 6 & 2 \end{pmatrix} = \begin{pmatrix} 4-3 & 0+8 \\ -5+6 & 7+2 \end{pmatrix} = \begin{pmatrix} 1 & 8 \\ 1 & 9 \end{pmatrix}$

3. $\begin{pmatrix} 5a & -3a \\ -2a & 4a \end{pmatrix} - \begin{pmatrix} 3a & 4a \\ -4a & -7a \end{pmatrix} = \begin{pmatrix} 8a & a \\ -6a & -3a \end{pmatrix}$

4. $\begin{pmatrix} 2 & 3 \\ 4 & 1 \end{pmatrix} - \begin{pmatrix} 1 & 0 \\ 2 & 0 \end{pmatrix} = \begin{pmatrix} 2-1 & 3-0 \\ 4-2 & 1-0 \end{pmatrix} = \begin{pmatrix} 1 & 3 \\ 2 & 1 \end{pmatrix}$

5. $\begin{pmatrix} a+2b & 6b \\ 3a-4b & 5a \end{pmatrix} + \begin{pmatrix} 2a+3b & a-4b \\ 6a+3b & a+2b \end{pmatrix} = \begin{pmatrix} 3a+5b & a+2b \\ 9a-b & 6a+2b \end{pmatrix}$

Exercise 16.1

Express as a single matrix each of the following:

1. $\begin{pmatrix} 1 & 2 \\ 3 & 4 \end{pmatrix} + \begin{pmatrix} 2 & 4 \\ 3 & 1 \end{pmatrix}$

2. $\begin{pmatrix} 6 & 11 \\ 7 & 4 \end{pmatrix} + \begin{pmatrix} 9 & 13 \\ 14 & 2 \end{pmatrix}$

3. $\begin{pmatrix} 9 & 0 \\ 0 & 8 \end{pmatrix} + \begin{pmatrix} 0 & 7 \\ 5 & 0 \end{pmatrix}$

4. $\begin{pmatrix} 11 & -1 \\ -7 & 4 \end{pmatrix} + \begin{pmatrix} -3 & 15 \\ 12 & -6 \end{pmatrix}$

5. $\begin{pmatrix} 3a & 2a \\ 5a & 6a \end{pmatrix} + \begin{pmatrix} 0 & 9a \\ 4a & 11a \end{pmatrix}$

6. $\begin{pmatrix} -2b & 3b \\ 4b & -7b \end{pmatrix} + \begin{pmatrix} 8b & 2b \\ -6b & -9b \end{pmatrix}$

7. $\begin{pmatrix} a+b & a-b \\ a+2b & b-2a \end{pmatrix} + \begin{pmatrix} a-2b & 2a+3b \\ 4a-3b & 3a+4b \end{pmatrix}$

8. $\begin{pmatrix} 2 & 4 \\ 3 & 1 \end{pmatrix} - \begin{pmatrix} 1 & 3 \\ 2 & 0 \end{pmatrix}$

9. $\begin{pmatrix} 15 & 12 \\ 17 & 23 \end{pmatrix} - \begin{pmatrix} 9 & 11 \\ 14 & 17 \end{pmatrix}$

10. $\begin{pmatrix} 12 & 0 \\ -13 & 2 \end{pmatrix} - \begin{pmatrix} -7 & 5 \\ 0 & 15 \end{pmatrix}$

11. $\begin{pmatrix} 35 & 41 \\ -62 & 53 \end{pmatrix} - \begin{pmatrix} -27 & 34 \\ -93 & -27 \end{pmatrix}$

12. $\begin{pmatrix} 73 & -26 \\ 48 & -53 \end{pmatrix} - \begin{pmatrix} 0 & 28 \\ 36 & 19 \end{pmatrix}$

13. $\begin{pmatrix} 3a & 8a \\ -7a & 15a \end{pmatrix} - \begin{pmatrix} -8a & -3a \\ 15a & -7a \end{pmatrix}$

14. $\begin{pmatrix} 3a+2b & 0 \\ 7a-5b & 3b-2a \end{pmatrix} - \begin{pmatrix} 4a-7b & 2a-5b \\ 3a-4b & 6b+3a \end{pmatrix}$

Examples

1. Write $2\cdot\begin{pmatrix}2 & 1\\ 4 & -3\end{pmatrix}$ as a 2×2 matrix.

   The rule for ordinary numbers is $2x = x + x$, so here, with matrices:

$$2\cdot\begin{pmatrix}2 & 1\\ 4 & -3\end{pmatrix}=\begin{pmatrix}2 & 1\\ 4 & -3\end{pmatrix}+\begin{pmatrix}2 & 1\\ 4 & -3\end{pmatrix}=\begin{pmatrix}4 & 2\\ 8 & -6\end{pmatrix}\text{ by } \mathit{16.2}$$

$$=\begin{pmatrix}2\times 2 & 2\times 1\\ 2\times 4 & 2\times -3\end{pmatrix}$$

2. $$3\begin{pmatrix}a & b\\ c & d\end{pmatrix}=\begin{pmatrix}a & b\\ c & d\end{pmatrix}+\begin{pmatrix}a & b\\ c & d\end{pmatrix}+\begin{pmatrix}a & b\\ c & d\end{pmatrix}$$

$$=\begin{pmatrix}3a & 3b\\ 3c & 3d\end{pmatrix}$$

3. Generally: $$p\begin{pmatrix}a & b\\ c & d\end{pmatrix}=\begin{pmatrix}pa & pb\\ pc & pd\end{pmatrix} \qquad \mathit{16.4}$$

4. $$p.\begin{pmatrix}a & b\\ c & d\end{pmatrix}+q.\begin{pmatrix}x & y\\ z & t\end{pmatrix}=\begin{pmatrix}pa & pb\\ pc & pd\end{pmatrix}+\begin{pmatrix}qx & qy\\ qz & qt\end{pmatrix}$$

$$=\begin{pmatrix}pa+qx & pb+qy\\ pc+qz & pd+qt\end{pmatrix}$$

Exercise 16.2

Simplify:

1. $2.\begin{pmatrix}2 & 1\\ 3 & 4\end{pmatrix}+3.\begin{pmatrix}1 & 2\\ 4 & 3\end{pmatrix}$

2. $3.\begin{pmatrix}6 & 1\\ -2 & 3\end{pmatrix}+5.\begin{pmatrix}-4 & 7\\ 0 & 9\end{pmatrix}$

3. $6.\begin{pmatrix}1 & 0\\ 0 & 1\end{pmatrix}+7.\begin{pmatrix}1 & 0\\ 0 & -1\end{pmatrix}-4.\begin{pmatrix}0 & 1\\ -1 & 0\end{pmatrix}$

4. $2.\begin{pmatrix}a & b\\ c & d\end{pmatrix}+3.\begin{pmatrix}a & -b\\ -c & d\end{pmatrix}$

5. $p.\begin{pmatrix} a & 0 \\ 0 & a \end{pmatrix} + q.\begin{pmatrix} 0 & -a \\ a & 0 \end{pmatrix} + r.\begin{pmatrix} 1 & 1 \\ 1 & 0 \end{pmatrix}$

6. $3.\begin{pmatrix} ax & 0 \\ 0 & by \end{pmatrix} + 7.\begin{pmatrix} 0 & bx \\ -ay & 0 \end{pmatrix}$

7. $\frac{1}{2}.\begin{pmatrix} 2 & 0 \\ 0 & 2 \end{pmatrix}$

8. $\frac{1}{3}.\begin{pmatrix} 3 & 0 \\ 0 & 3 \end{pmatrix}$

9. $\frac{1}{a}.\begin{pmatrix} a & 0 \\ 0 & a \end{pmatrix}$

10. $\frac{1}{2a}.\begin{pmatrix} a & 0 \\ 0 & a \end{pmatrix} + \frac{1}{2b}.\begin{pmatrix} b & 0 \\ 0 & b \end{pmatrix}$

16.2 The product of two matrices

In Example (3) at the beginning of the chapter:

$$\left.\begin{aligned} 2x + y &= 5 \\ x + y &= 3 \end{aligned}\right\} \text{ was replaced by } \begin{pmatrix} 2 & 1 \\ 1 & 1 \end{pmatrix}.\begin{pmatrix} x \\ y \end{pmatrix} = \begin{pmatrix} 5 \\ 3 \end{pmatrix}$$

The operation between the matrices on the LHS is taken to be multiplication. In order that the two statements agree the following statement must be true:

$$\begin{pmatrix} 2 & 1 \\ 1 & 1 \end{pmatrix}.\begin{pmatrix} x \\ y \end{pmatrix} = \begin{pmatrix} 2x + y \\ x + y \end{pmatrix}$$

That is, a 2×2 matrix times a 2×1 matrix $=$ a 2×1 matrix. More generally, it would mean:

$$\begin{pmatrix} \underline{a} & \underline{b} \\ \underline{c} & \underline{d} \end{pmatrix}.\begin{pmatrix} \underline{x} \\ \underline{y} \end{pmatrix} = \begin{pmatrix} \underline{ax} + \underline{by} \\ \underline{cx} + \underline{dy} \end{pmatrix} \qquad 16.5$$

$$\text{or} \quad A.X = Z$$

To obtain any element of Z multiply the elements in the corresponding row of A by the elements in X. The elements underlined similarly are the ones which are multiplied together. The resulting products are added.

Examples

Express as single matrices:

1. $\begin{pmatrix} 3 & 2 \\ 4 & 1 \end{pmatrix} \cdot \begin{pmatrix} 1 \\ 2 \end{pmatrix}$

   The product $= \begin{pmatrix} 3 \times 1 + 2 \times 2 \\ 4 \times 1 + 1 \times 2 \end{pmatrix} = \begin{pmatrix} 3+4 \\ 4+2 \end{pmatrix} = \begin{pmatrix} 7 \\ 6 \end{pmatrix}$

2. $\begin{pmatrix} 6 & -5 \\ -2 & 3 \end{pmatrix} \cdot \begin{pmatrix} 4 \\ -3 \end{pmatrix}$

   The product $= \begin{pmatrix} 6 \times 4 + -5 \times -3 \\ -2 \times 4 + 3 \times -3 \end{pmatrix} = \begin{pmatrix} 39 \\ -17 \end{pmatrix}$

3. $\begin{pmatrix} a & b \\ -b & a \end{pmatrix} \cdot \begin{pmatrix} 1 \\ -1 \end{pmatrix}$

   The product $= \begin{pmatrix} a-b \\ -b-a \end{pmatrix} = \begin{pmatrix} a-b \\ -(a+b) \end{pmatrix}$

4. $\begin{pmatrix} a & b \\ -b & -a \end{pmatrix} \cdot \begin{pmatrix} b \\ a \end{pmatrix}$

   The product $= \begin{pmatrix} ab+ba \\ -b^2-a^2 \end{pmatrix} = \begin{pmatrix} 2ab \\ -(a^2+b^2) \end{pmatrix}$

Exercise 16.3

Express as single matrices:

1. $\begin{pmatrix} 1 & 2 \\ 3 & 4 \end{pmatrix} \cdot \begin{pmatrix} 2 \\ 1 \end{pmatrix}$

2. $\begin{pmatrix} 3 & 4 \\ -2 & 1 \end{pmatrix} \cdot \begin{pmatrix} 1 \\ 5 \end{pmatrix}$

3. $\begin{pmatrix} 6 & -4 \\ 3 & 2 \end{pmatrix} \cdot \begin{pmatrix} -1 \\ 2 \end{pmatrix}$

4. $\begin{pmatrix} -3 & 0 \\ 5 & 2 \end{pmatrix} \cdot \begin{pmatrix} -3 \\ -7 \end{pmatrix}$

5. $\begin{pmatrix} a & b \\ b & a \end{pmatrix} \cdot \begin{pmatrix} 1 \\ 0 \end{pmatrix}$

6. $\begin{pmatrix} a & 2a \\ -a & 2a \end{pmatrix} \cdot \begin{pmatrix} 1 \\ -1 \end{pmatrix}$

7. $\begin{pmatrix} 3a & 4b \\ 5a & -7b \end{pmatrix} \cdot \begin{pmatrix} -a \\ b \end{pmatrix}$

8. $\begin{pmatrix} a & b \\ c & d \end{pmatrix} \cdot \begin{pmatrix} A \\ B \end{pmatrix}$

9. $\begin{pmatrix} a & b \\ c & d \end{pmatrix} \cdot \begin{pmatrix} C \\ D \end{pmatrix}$

Definition of the product of two 2 × 2 matrices

$$\begin{pmatrix} a & b \\ c & d \end{pmatrix} \cdot \begin{pmatrix} A & C \\ B & D \end{pmatrix} = \begin{pmatrix} (aA+bB) & (aC+bD) \\ (cA+dB) & (cC+dD) \end{pmatrix} \qquad 16.6$$

The schematic diagram in Fig. 16.2 is helpful to determine such a product.

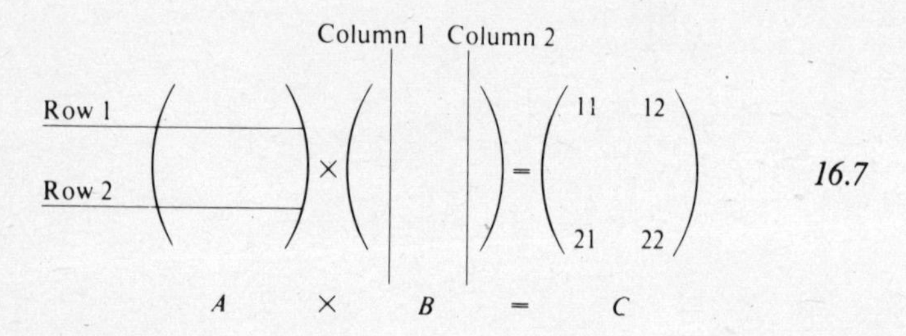

Figure 16.2

Item 1. The elements in a row of A are noted.
Item 2. The elements in a column of B are noted.
Item 3. The position of each element in C is determined by two numbers: the first represents a row of A, the second a column of B.
Item 4. Each element of C is calculated by multiplying together the elements of a row of A by the corresponding elements of a column of B. The individual products are added together.

Examples

1. Calculate $\begin{pmatrix} 1 & 2 \\ 3 & 4 \end{pmatrix} \cdot \begin{pmatrix} 5 & 7 \\ 6 & 8 \end{pmatrix}$

$$\text{The product} = \begin{pmatrix} (1 \times 5 + 2 \times 6) & (1 \times 7 + 2 \times 8) \\ (3 \times 5 + 4 \times 6) & (3 \times 7 + 4 \times 8) \end{pmatrix}$$

$$= \begin{pmatrix} (5+12) & (7+16) \\ (15+24) & (21+32) \end{pmatrix} = \begin{pmatrix} 17 & 23 \\ 39 & 53 \end{pmatrix}$$

2. Calculate $\begin{pmatrix} 5 & -2 \\ -9 & 4 \end{pmatrix} \cdot \begin{pmatrix} 3 & 2 \\ -7 & -6 \end{pmatrix}$

$$\text{The product} = \begin{pmatrix} (15+14) & (10+12) \\ (-27-28) & (-18-24) \end{pmatrix} = \begin{pmatrix} 29 & 22 \\ -55 & -42 \end{pmatrix}$$

3. Calculate $\begin{pmatrix} 3 & 2 \\ -7 & -6 \end{pmatrix} \cdot \begin{pmatrix} 5 & -2 \\ -9 & 4 \end{pmatrix}$

$$\text{The product} = \begin{pmatrix} (15-18) & (-6+8) \\ (-35+54) & (14-24) \end{pmatrix} = \begin{pmatrix} -3 & 2 \\ 19 & -10 \end{pmatrix}$$

Note that, from (2) and (3), the two products are not equal. Therefore, where A and B are two matrices, in general, $A \times B$ is not necessarily equal to $B \times A$. *For matrices, multiplication is not commutative.*

Exercise 16.4

Calculate the following products:

1. $\begin{pmatrix} 1 & 2 \\ 3 & 4 \end{pmatrix} \cdot \begin{pmatrix} 1 & 3 \\ 2 & 4 \end{pmatrix}$
2. $\begin{pmatrix} 1 & 2 \\ 2 & 1 \end{pmatrix} \cdot \begin{pmatrix} 1 & 2 \\ 2 & 1 \end{pmatrix}$
3. $\begin{pmatrix} 2 & -1 \\ 1 & -2 \end{pmatrix} \cdot \begin{pmatrix} 2 & 1 \\ -1 & -2 \end{pmatrix}$
4. $\begin{pmatrix} 5 & 6 \\ 2 & -3 \end{pmatrix} \cdot \begin{pmatrix} 4 & 8 \\ 5 & 7 \end{pmatrix}$
5. $\begin{pmatrix} 4 & 8 \\ 5 & 7 \end{pmatrix} \cdot \begin{pmatrix} 5 & 6 \\ 2 & -3 \end{pmatrix}$
6. $\begin{pmatrix} 4 & -1 \\ -3 & 0 \end{pmatrix} \cdot \begin{pmatrix} -7 & 2 \\ 0 & -5 \end{pmatrix}$

7. $\begin{pmatrix} -7 & 2 \\ 0 & -5 \end{pmatrix} \cdot \begin{pmatrix} 4 & -1 \\ -3 & 0 \end{pmatrix}$

8. $\begin{pmatrix} a & b \\ b & a \end{pmatrix} \cdot \begin{pmatrix} c & d \\ d & c \end{pmatrix}$

9. Calculate $(A \times B) \times C$ and $A \times (B \times C)$

   where $A = \begin{pmatrix} 1 & 2 \\ 3 & 4 \end{pmatrix}$ $B = \begin{pmatrix} 1 & 3 \\ 2 & 4 \end{pmatrix}$ $C = \begin{pmatrix} 3 & 1 \\ 4 & 2 \end{pmatrix}$

10. $\begin{pmatrix} 15 & 6 \\ -12 & 4 \end{pmatrix} \cdot \begin{pmatrix} -13 & 13 \\ 17 & -9 \end{pmatrix}$

11. $\begin{pmatrix} -13 & 13 \\ 17 & -9 \end{pmatrix} \cdot \begin{pmatrix} 15 & 6 \\ -12 & 4 \end{pmatrix}$

12. $\begin{pmatrix} a & b \\ b & a \end{pmatrix} \cdot \begin{pmatrix} c & d \\ d & c \end{pmatrix}$

13. $\begin{pmatrix} c & d \\ d & c \end{pmatrix} \cdot \begin{pmatrix} a & b \\ b & a \end{pmatrix}$

14. $\begin{pmatrix} a & b \\ c & d \end{pmatrix} \cdot \begin{pmatrix} A & C \\ B & D \end{pmatrix}$

15. $\begin{pmatrix} A & C \\ B & D \end{pmatrix} \cdot \begin{pmatrix} a & b \\ c & d \end{pmatrix}$

16. $\begin{pmatrix} 1 & 0 \\ 0 & 1 \end{pmatrix} \cdot \begin{pmatrix} 3 & 5 \\ -2 & 7 \end{pmatrix}$

17. $\begin{pmatrix} 1 & 0 \\ 0 & 1 \end{pmatrix} \cdot \begin{pmatrix} -18 & -35 \\ 29 & -17 \end{pmatrix}$

18. $\begin{pmatrix} 3 & 5 \\ -2 & 7 \end{pmatrix} \cdot \begin{pmatrix} 1 & 0 \\ 0 & 1 \end{pmatrix}$

19. $\begin{pmatrix} -18 & -35 \\ 29 & -17 \end{pmatrix} \cdot \begin{pmatrix} 1 & 0 \\ 0 & 1 \end{pmatrix}$

20. $\begin{pmatrix} 1 & 0 \\ 0 & 1 \end{pmatrix} \cdot \begin{pmatrix} a & b \\ c & d \end{pmatrix}$

21. $\begin{pmatrix} a & b \\ c & d \end{pmatrix} \cdot \begin{pmatrix} 1 & 0 \\ 0 & 1 \end{pmatrix}$

Note the results for questions (4) and (5), (6) and (7), (10) and (11), (12) and

(13), (14) and (15). They endorse the fact that multiplication of matrices is not commutative. The results of questions (16), (17), (18), (19), (20) and (21) indicate that there seems to be something special about the matrix:

$$\begin{pmatrix} 1 & 0 \\ 0 & 1 \end{pmatrix}$$

Call it I. Questions (20) and (21) prove that when we multiply any other matrix by it, in front or behind, then that matrix remains unchanged. It performs a function for matrices similar to that performed by 1 in the multiplication of real numbers. For that reason it is called the *unit matrix*, or the *identity matrix*. Where A represents any 2×2 matrix then:

$$I \times A = A \times I = A \qquad 16.8$$

Or, in full:

$$\begin{pmatrix} 1 & 0 \\ 0 & 1 \end{pmatrix} \cdot \begin{pmatrix} a & b \\ c & d \end{pmatrix} = \begin{pmatrix} a & b \\ c & d \end{pmatrix} \cdot \begin{pmatrix} 1 & 0 \\ 0 & 1 \end{pmatrix} = \begin{pmatrix} a & b \\ c & d \end{pmatrix} \qquad 16.9$$

Examples

Calculate the following:

1. $\begin{pmatrix} 2 & 3 \\ 1 & 2 \end{pmatrix} \cdot \begin{pmatrix} 2 & -3 \\ -1 & 2 \end{pmatrix}$

   The product $= \begin{pmatrix} (4-3) & (-6+6) \\ (2-2) & (-3+4) \end{pmatrix} = \begin{pmatrix} 1 & 0 \\ 0 & 1 \end{pmatrix}$

2. $\begin{pmatrix} -5 & 11 \\ 4 & -9 \end{pmatrix} \cdot \begin{pmatrix} -9 & -11 \\ -4 & -5 \end{pmatrix}$

   The product $= \begin{pmatrix} (45-44) & (55-55) \\ (-36+36) & (-44+45) \end{pmatrix} = \begin{pmatrix} 1 & 0 \\ 0 & 1 \end{pmatrix}$

3. $\begin{pmatrix} 2 & -3 \\ -1 & 2 \end{pmatrix} \cdot \begin{pmatrix} 2 & 3 \\ 1 & 2 \end{pmatrix}$

   The product $= \begin{pmatrix} (4-3) & (6-6) \\ (-2+2) & (-3+4) \end{pmatrix} = \begin{pmatrix} 1 & 0 \\ 0 & 1 \end{pmatrix}$

4. $\begin{pmatrix} -9 & -11 \\ -4 & -5 \end{pmatrix} \cdot \begin{pmatrix} -5 & 11 \\ 4 & -9 \end{pmatrix}$

$$\text{The product} = \begin{pmatrix} (45-44) & (-99+99) \\ (20-20) & (-44+45) \end{pmatrix} = \begin{pmatrix} 1 & 0 \\ 0 & 1 \end{pmatrix}$$

Note that, in each case, the product is the unit matrix.

Exercise 16.5

Calculate the following:

1. $\begin{pmatrix} 4 & 3 \\ 1 & 1 \end{pmatrix} \cdot \begin{pmatrix} 1 & -3 \\ -1 & 4 \end{pmatrix}$
2. $\begin{pmatrix} 1 & -3 \\ -1 & 4 \end{pmatrix} \cdot \begin{pmatrix} 4 & 3 \\ 1 & 1 \end{pmatrix}$
3. $\begin{pmatrix} 5 & 2 \\ 2 & 1 \end{pmatrix} \cdot \begin{pmatrix} 1 & -2 \\ -2 & 5 \end{pmatrix}$
4. $\begin{pmatrix} 1 & -2 \\ -2 & 5 \end{pmatrix} \cdot \begin{pmatrix} 5 & 2 \\ 2 & 1 \end{pmatrix}$
5. $\begin{pmatrix} 1 & 2 \\ 2 & 5 \end{pmatrix} \cdot \begin{pmatrix} 5 & -2 \\ -2 & 1 \end{pmatrix}$
6. $\begin{pmatrix} 5 & -2 \\ -2 & 1 \end{pmatrix} \cdot \begin{pmatrix} 1 & 2 \\ 2 & 5 \end{pmatrix}$
7. $\begin{pmatrix} 4 & 3 \\ 1 & 2 \end{pmatrix} \cdot \begin{pmatrix} 2 & -3 \\ -1 & 4 \end{pmatrix}$
8. $\begin{pmatrix} 2 & -3 \\ -1 & 4 \end{pmatrix} \cdot \begin{pmatrix} 4 & 3 \\ 1 & 2 \end{pmatrix}$
9. $\begin{pmatrix} 5 & 6 \\ 2 & 4 \end{pmatrix} \cdot \begin{pmatrix} 4 & -6 \\ -2 & 5 \end{pmatrix}$
10. $\begin{pmatrix} 4 & -6 \\ -2 & 5 \end{pmatrix} \cdot \begin{pmatrix} 5 & 6 \\ 2 & 4 \end{pmatrix}$
11. $\begin{pmatrix} -5 & -8 \\ -3 & -7 \end{pmatrix} \cdot \begin{pmatrix} -7 & 8 \\ 3 & -5 \end{pmatrix}$
12. $\begin{pmatrix} -7 & 8 \\ 3 & -5 \end{pmatrix} \cdot \begin{pmatrix} -5 & -8 \\ -3 & -7 \end{pmatrix}$

16.3 The determinant of a matrix

The results of the first six questions of exercise 16.5 give a product which is:

$$\begin{pmatrix} 1 & 0 \\ 0 & 1 \end{pmatrix}$$

which is the unit matrix, I.

The results of the last six questions give products which are a little different:

7 and 8 produce the answer $\begin{pmatrix} 5 & 0 \\ 0 & 5 \end{pmatrix}$

9 and 10 produce the answer $\begin{pmatrix} 8 & 0 \\ 0 & 8 \end{pmatrix}$

11 and 12 produce the answer $\begin{pmatrix} 11 & 0 \\ 0 & 11 \end{pmatrix}$

By formula *16.4*:

$$\begin{pmatrix} 5 & 0 \\ 0 & 5 \end{pmatrix} = 5\begin{pmatrix} 1 & 0 \\ 0 & 1 \end{pmatrix}$$

$$\begin{pmatrix} 8 & 0 \\ 0 & 8 \end{pmatrix} = 8\begin{pmatrix} 1 & 0 \\ 0 & 1 \end{pmatrix}$$

$$\begin{pmatrix} 11 & 0 \\ 0 & 11 \end{pmatrix} = 11\begin{pmatrix} 1 & 0 \\ 0 & 1 \end{pmatrix}$$

In other words, they give $5I$, $8I$ and $11I$ respectively. Therefore the products of certain matrices produce either I or a multiple of I. Where this property arises it is of great use in solving equations. This issue will be taken up in the next chapter. One aspect we need to clarify here is what determines whether the product is I or a multiple of I.

Associated with any square matrix is a numerical value which is important. For the matrix:

$$\begin{pmatrix} a & b \\ c & d \end{pmatrix}$$

that value is $ad - cb$, or the difference between the products of the pairs along the diagonals of the matrix, i.e.

$$\begin{matrix} a & \nearrow b \\ c & \searrow d \end{matrix}$$

This value is called the *determinant of the matrix.*

Examples

Calculate the determinants of the following matrices:

1. $\begin{pmatrix} 4 & 5 \\ 2 & 9 \end{pmatrix}$ The determinant $= 4 \times 9 - 2 \times 5 = 36 - 10 = 26$.

2. $\begin{pmatrix} 11 & 4 \\ -7 & 3 \end{pmatrix}$ The determinant $= 11 \times 3 - (-7 \times 4) = 33 + 28 = 61$.

3. $\begin{pmatrix} -3 & -2 \\ 8 & -5 \end{pmatrix}$ The determinant $= (-3 \times -5) - (8 \times -2) = 15 + 16 = 31$.

4. $\begin{pmatrix} 2 & 3 \\ 1 & 2 \end{pmatrix}$ The determinant $= 2 \times 2 - 1 \times 3 = 4 - 3 = 1$.

5. $\begin{pmatrix} -5 & 11 \\ 4 & -9 \end{pmatrix}$ The determinant $= (-5 \times -9) - 4 \times 11 = 45 - 44 = 1$.

6. $\begin{pmatrix} 4 & 3 \\ 1 & 2 \end{pmatrix}$ The determinant $= 4 \times 2 - 1 \times 3 = 8 - 3 = 5$.

Exercise 16.6

Calculate the determinants of the following matrices:

1. $\begin{pmatrix} 4 & 3 \\ 1 & 1 \end{pmatrix}$

2. $\begin{pmatrix} 1 & -3 \\ -1 & 4 \end{pmatrix}$

3. $\begin{pmatrix} 5 & 2 \\ 2 & 1 \end{pmatrix}$

4. $\begin{pmatrix} 4 & 3 \\ 1 & 2 \end{pmatrix}$

5. $\begin{pmatrix} 2 & -3 \\ -1 & 4 \end{pmatrix}$

6. $\begin{pmatrix} -9 & -11 \\ -4 & -5 \end{pmatrix}$

7. $\begin{pmatrix} 5 & 6 \\ 2 & 4 \end{pmatrix}$

8. $\begin{pmatrix} 4 & -6 \\ -2 & 5 \end{pmatrix}$

9. $\begin{pmatrix} 3 & 8 \\ 7 & 4 \end{pmatrix}$

10. $\begin{pmatrix} -8 & 4 \\ -2 & -3 \end{pmatrix}$

11. $\begin{pmatrix} a & -a \\ a & a \end{pmatrix}$

12. $\begin{pmatrix} 2a & 5a \\ a & 3a \end{pmatrix}$

13. $\begin{pmatrix} a & 1/b \\ -b & 1/a \end{pmatrix}$

14. $\begin{pmatrix} (a+b) & b \\ a & a \end{pmatrix}$

15. $\begin{pmatrix} a & 1 \\ -2a & 1 \end{pmatrix}$

In exercise 16.6 note those determinants whose values are 1. Relate them to the questions in exercise 16.5 which produced the unit matrix. Note further the occasions where the determinantal value coincided with the multiple of the unit matrix in exercise 16.5. The determinant itself has a special notation. The determinant of

$$\begin{pmatrix} a & b \\ c & d \end{pmatrix}$$

is variously written as:

det $\begin{pmatrix} a & b \\ c & d \end{pmatrix}$, $\begin{vmatrix} a & b \\ c & d \end{vmatrix}$, and $ad-cb$, or $ad-bc$, and $|A|$, where A is the matrix.

17

The Use of Matrices and Determinants in the Solution of Simultaneous Linear Equations in Two Unknowns

Example

Determine the solution of

$$\begin{aligned} 2x + y &= 5 \\ x + y &= 3 \end{aligned} \tag{i}$$

In matrix notation (i) becomes:

$$\begin{pmatrix} 2 & 1 \\ 1 & 1 \end{pmatrix} \cdot \begin{pmatrix} x \\ y \end{pmatrix} = \begin{pmatrix} 5 \\ 3 \end{pmatrix} \tag{ii}$$

Multiply (ii) by $\begin{pmatrix} 1 & -1 \\ -1 & 2 \end{pmatrix}$

$$\begin{pmatrix} 1 & -1 \\ -1 & 2 \end{pmatrix} \cdot \begin{pmatrix} 2 & 1 \\ 1 & 1 \end{pmatrix} \cdot \begin{pmatrix} x \\ y \end{pmatrix} = \begin{pmatrix} 1 & -1 \\ -1 & 2 \end{pmatrix} \cdot \begin{pmatrix} 5 \\ 3 \end{pmatrix} \tag{iii}$$

$$\begin{pmatrix} 1 & 0 \\ 0 & 1 \end{pmatrix} \cdot \begin{pmatrix} x \\ y \end{pmatrix} = \begin{pmatrix} 2 \\ 1 \end{pmatrix} \tag{iv}$$

$$\begin{pmatrix} x \\ y \end{pmatrix} = \begin{pmatrix} 2 \\ 1 \end{pmatrix} \tag{v}$$

(v) means that $x = 2$, $y = 1$.
Check: substitute back in (i):
LHS of the first equation $= 2 \times 2 + 1 = 5$
LHS of the second equation $= 2 + 1 = 3$

A second check is to solve the equations by either elimination or substitution. The method above solves the equations in (i).

The important step in that solution is the use of the matrix

$$\begin{pmatrix} 1 & -1 \\ -1 & 2 \end{pmatrix}$$

for the multiplication of (ii). We need to know how to determine that matrix which produces the solution of a set of equations when, in fact, the equations are soluble. Suppose we write the matrix

$$\begin{pmatrix} 2 & 1 \\ 1 & 1 \end{pmatrix}$$

as A. Then we write the matrix

$$\begin{pmatrix} 1 & -1 \\ -1 & 2 \end{pmatrix}$$

as A^{-1}, because $A^{-1}.A = I$, the unit matrix. Associate this step in matrices with the step in real numbers:

$$\frac{1}{a} \times a = 1, \text{ or } a^{-1}.a = 1$$

A^{-1} is called the inverse of the matrix A. It is also true that $A.A^{-1} = I$:

$$\begin{pmatrix} 2 & 1 \\ 1 & 1 \end{pmatrix} \cdot \begin{pmatrix} 1 & -1 \\ -1 & 2 \end{pmatrix} = \begin{pmatrix} 1 & 0 \\ 0 & 1 \end{pmatrix}$$

Each matrix is the inverse of the other: A^{-1} is the inverse of A, and A is the inverse of A^{-1}.

17.1 The inverse of a 2×2 matrix

Definition

Matrices which are inverses of each other are such that their product, in either order, is the unit matrix.

Examples

Show that the following pairs of matrices are inverses of each other. Note carefully the relationship of the elements of one to the elements of the other.

1. $\begin{pmatrix} 3 & 5 \\ 1 & 2 \end{pmatrix}, \begin{pmatrix} 2 & -5 \\ -1 & 3 \end{pmatrix}$

   The product $\begin{pmatrix} 3 & 5 \\ 1 & 2 \end{pmatrix} \cdot \begin{pmatrix} 2 & -5 \\ -1 & 3 \end{pmatrix} = \begin{pmatrix} 1 & 0 \\ 0 & 1 \end{pmatrix}$

   The product $\begin{pmatrix} 2 & -5 \\ -1 & 3 \end{pmatrix} \cdot \begin{pmatrix} 3 & 5 \\ 1 & 2 \end{pmatrix} = \begin{pmatrix} 1 & 0 \\ 0 & 1 \end{pmatrix}$

2. $\begin{pmatrix} 8 & 5 \\ 3 & 2 \end{pmatrix}, \begin{pmatrix} 2 & -5 \\ -3 & 8 \end{pmatrix}$

   The product $\begin{pmatrix} 8 & 5 \\ 3 & 2 \end{pmatrix} \cdot \begin{pmatrix} 2 & -5 \\ -3 & 8 \end{pmatrix} = \begin{pmatrix} 1 & 0 \\ 0 & 1 \end{pmatrix}$

   The product $\begin{pmatrix} 2 & -5 \\ -3 & 8 \end{pmatrix} \cdot \begin{pmatrix} 8 & 5 \\ 3 & 2 \end{pmatrix} = \begin{pmatrix} 1 & 0 \\ 0 & 1 \end{pmatrix}$

Exercise 17.1

Show that the following pairs of matrices are inverses of each other. Note carefully the relationship of the elements of one to the corresponding elements of the other.

1. $\begin{pmatrix} 3 & 1 \\ 2 & 1 \end{pmatrix}, \begin{pmatrix} 1 & -1 \\ -2 & 3 \end{pmatrix}$
2. $\begin{pmatrix} 1 & 1 \\ 2 & 3 \end{pmatrix}, \begin{pmatrix} 3 & -1 \\ -2 & 1 \end{pmatrix}$
3. $\begin{pmatrix} 5 & 2 \\ 2 & 1 \end{pmatrix}, \begin{pmatrix} 1 & -2 \\ -2 & 5 \end{pmatrix}$
4. $\begin{pmatrix} 1 & 1 \\ 4 & 5 \end{pmatrix}, \begin{pmatrix} 5 & -1 \\ -4 & 1 \end{pmatrix}$
5. $\begin{pmatrix} 4 & -7 \\ -1 & 2 \end{pmatrix}, \begin{pmatrix} 2 & 7 \\ 1 & 4 \end{pmatrix}$
6. $\begin{pmatrix} -9 & -7 \\ -5 & -4 \end{pmatrix}, \begin{pmatrix} -4 & 7 \\ 5 & -9 \end{pmatrix}$

7. $\begin{pmatrix} 27 & 5 \\ 16 & 3 \end{pmatrix}, \begin{pmatrix} 3 & -5 \\ -16 & 27 \end{pmatrix}$

8. $\begin{pmatrix} -8 & -9 \\ 9 & 10 \end{pmatrix}, \begin{pmatrix} 10 & 9 \\ -9 & -8 \end{pmatrix}$

By observation of the above examples it appears that the inverse of

$$\begin{pmatrix} a & b \\ c & d \end{pmatrix}$$

might be

$$\begin{pmatrix} d & -b \\ -c & a \end{pmatrix}$$

Examples

1. Check whether the matrices

$$\begin{pmatrix} 3 & 1 \\ 1 & 2 \end{pmatrix}, \begin{pmatrix} 2 & -1 \\ -1 & 3 \end{pmatrix}$$

are inverses of each other.

The product $\begin{pmatrix} 3 & 1 \\ 1 & 2 \end{pmatrix} \cdot \begin{pmatrix} 2 & -1 \\ -1 & 3 \end{pmatrix} = \begin{pmatrix} 5 & 0 \\ 0 & 5 \end{pmatrix}$

The product $\begin{pmatrix} 2 & -1 \\ -1 & 3 \end{pmatrix} \cdot \begin{pmatrix} 3 & 1 \\ 1 & 2 \end{pmatrix} = \begin{pmatrix} 5 & 0 \\ 0 & 5 \end{pmatrix}$

The products are not the unit matrix. However, they are closely related to I:

$$\begin{pmatrix} 5 & 0 \\ 0 & 5 \end{pmatrix} = 5\begin{pmatrix} 1 & 0 \\ 0 & 1 \end{pmatrix}$$

Moreover $5 = \det\begin{pmatrix} 3 & 1 \\ 1 & 2 \end{pmatrix}$ i.e. $\begin{vmatrix} 3 & 1 \\ 1 & 2 \end{vmatrix} = 3 \times 2 - 1^2$

2. In view of the result of (1) check whether

$$\frac{1}{5}\begin{pmatrix} 2 & -1 \\ -1 & 3 \end{pmatrix}$$

is the inverse of

$$\begin{pmatrix} 3 & 1 \\ 1 & 2 \end{pmatrix}$$

Step 1 $\frac{1}{5}\begin{pmatrix} 2 & -1 \\ -1 & 3 \end{pmatrix} = \begin{pmatrix} 2/5 & -1/5 \\ -1/5 & 3/5 \end{pmatrix}$

Step 2 $\begin{pmatrix} 3 & 1 \\ 1 & 2 \end{pmatrix} \cdot \begin{pmatrix} 2/5 & -1/5 \\ -1/5 & 3/5 \end{pmatrix} = \begin{pmatrix} 1 & 0 \\ 0 & 1 \end{pmatrix}$

Step 3 $\begin{pmatrix} 2/5 & -1/5 \\ -1/5 & 3/5 \end{pmatrix} \cdot \begin{pmatrix} 3 & 1 \\ 1 & 2 \end{pmatrix} = \begin{pmatrix} 1 & 0 \\ 0 & 1 \end{pmatrix}$

It is the inverse.

Exercise 17.2

Show that the following pairs of matrices are inverses of each other. Check that the denominator of each element in the second matrix equals the determinant of the first matrix.

1. $\begin{pmatrix} 3 & 1 \\ 1 & 1 \end{pmatrix}\begin{pmatrix} 1/2 & -1/2 \\ -1/2 & 3/2 \end{pmatrix}$

2. $\begin{pmatrix} 4 & 1 \\ 1 & 1 \end{pmatrix}\begin{pmatrix} 1/3 & -1/3 \\ -1/3 & 4/3 \end{pmatrix}$

3. $\begin{pmatrix} 5 & 2 \\ 1 & 1 \end{pmatrix}\begin{pmatrix} 1/3 & -2/3 \\ -1/3 & 5/3 \end{pmatrix}$

4. $\begin{pmatrix} 11 & 3 \\ 5 & 2 \end{pmatrix}\begin{pmatrix} 2/7 & -3/7 \\ -5/7 & 11/7 \end{pmatrix}$

5. $\begin{pmatrix} 3 & 1 \\ -1 & 1 \end{pmatrix}\begin{pmatrix} 1/4 & -1/4 \\ 1/4 & 3/4 \end{pmatrix}$

6. $\begin{pmatrix} 7 & 3 \\ -2 & 5 \end{pmatrix}\begin{pmatrix} 5/41 & -3/41 \\ 2/41 & 7/41 \end{pmatrix}$

7. $\begin{pmatrix} a & 0 \\ 1 & 1 \end{pmatrix}\begin{pmatrix} 1/a & 0 \\ -1/a & 1 \end{pmatrix}$

8. $\begin{pmatrix} a & 0 \\ 1 & b \end{pmatrix}\begin{pmatrix} b/ab & 0 \\ -1/ab & a/ab \end{pmatrix}$

9. $\begin{pmatrix} a & 1 \\ c & b \end{pmatrix}\begin{pmatrix} b/(ab-c) & -1/(ab-c) \\ -c/(ab-c) & a/(ab-c) \end{pmatrix}$

10. $\begin{pmatrix} a & b \\ c & d \end{pmatrix} \begin{pmatrix} \dfrac{d}{ad-bc} & \dfrac{-b}{ad-bc} \\ \dfrac{-c}{ad-bc} & \dfrac{a}{ad-bc} \end{pmatrix}$

Question (10) in exercise 17.2 provides a rule for writing down the inverse of any 2×2 matrix when an inverse exists.

Definition

The inverse of the matrix $A = \begin{pmatrix} a & b \\ c & d \end{pmatrix}$ is the matrix $\begin{pmatrix} d/\Delta & -b/\Delta \\ -c/\Delta & a/\Delta \end{pmatrix}$ where $\Delta = ad - bc$, the determinant of A.

Examples

1. Determine the inverse of $A = \begin{pmatrix} 7 & 3 \\ 4 & 5 \end{pmatrix}$.

Step 1 Det $A = 7 \times 5 - 4 \times 3 = 35 - 12 = 23$.

Step 2 $A^{-1} = \begin{pmatrix} 5/23 & -3/23 \\ -4/23 & 7/23 \end{pmatrix}$

Check: $\begin{pmatrix} 7 & 3 \\ 4 & 5 \end{pmatrix} \cdot \begin{pmatrix} 5/23 & -3/23 \\ -4/23 & 7/23 \end{pmatrix} = \begin{pmatrix} 23/23 & 0 \\ 0 & 23/23 \end{pmatrix} = \begin{pmatrix} 1 & 0 \\ 0 & 1 \end{pmatrix}$

2. Determine the inverse of $B = \begin{pmatrix} 11 & 5 \\ -18 & 17 \end{pmatrix}$

Step 1 Det $B = 11 \times 17 - (-18 \times 5) = 187 + 90 = 277$.

Step 2 $B^{-1} = \begin{pmatrix} 17/277 & -5/277 \\ 18/277 & 11/277 \end{pmatrix}$

Check: $\begin{pmatrix} 11 & 5 \\ -18 & 17 \end{pmatrix} \cdot \begin{pmatrix} 17/277 & -5/277 \\ 18/277 & 11/277 \end{pmatrix} = \begin{pmatrix} 277/277 & 0 \\ 0 & 277/277 \end{pmatrix}$

$$= \begin{pmatrix} 1 & 0 \\ 0 & 1 \end{pmatrix}$$

Exercise 17.3

Determine the inverses of the following matrices:

1. $\begin{pmatrix} 3 & 2 \\ 5 & 4 \end{pmatrix}$

2. $\begin{pmatrix} 4 & 1 \\ 3 & 2 \end{pmatrix}$

3. $\begin{pmatrix} 7 & 3 \\ 3 & 2 \end{pmatrix}$

4. $\begin{pmatrix} 9 & 3 \\ 5 & 4 \end{pmatrix}$

5. $\begin{pmatrix} 4 & 5 \\ -2 & 3 \end{pmatrix}$

6. $\begin{pmatrix} -5 & 3 \\ -4 & 2 \end{pmatrix}$

7. $\begin{pmatrix} -1 & 2 \\ -4 & 3 \end{pmatrix}$

8. $\begin{pmatrix} -3 & 2 \\ -7 & -3 \end{pmatrix}$

9. $\begin{pmatrix} a & b \\ -a & b \end{pmatrix}$

10. $\begin{pmatrix} 7 & 12 \\ 11 & 5 \end{pmatrix}$

17.2 The solution of simultaneous linear equations by matrices

Examples

Solve the following pairs of equations:

1.
$$\begin{aligned} 5x + 9y &= 17 \\ 3x + 8y &= 11 \end{aligned} \qquad \text{(i)}$$

Step 1 Rewrite (i):

$$\begin{pmatrix} 5 & 9 \\ 3 & 8 \end{pmatrix} \cdot \begin{pmatrix} x \\ y \end{pmatrix} = \begin{pmatrix} 17 \\ 11 \end{pmatrix} \qquad \text{(ii)}$$

Step 2 $\text{Det}\begin{pmatrix} 5 & 9 \\ 3 & 8 \end{pmatrix} = 40 - 27 = 13.$

Step 3 The inverse of $\begin{pmatrix} 5 & 9 \\ 3 & 8 \end{pmatrix} = \begin{pmatrix} 8/13 & -9/13 \\ -3/13 & 5/13 \end{pmatrix}$

Step 4 Multiply (ii) by this inverse:

$$\begin{pmatrix} 8/13 & -9/13 \\ -3/13 & 5/13 \end{pmatrix} \cdot \begin{pmatrix} 5 & 9 \\ 3 & 8 \end{pmatrix} \cdot \begin{pmatrix} x \\ y \end{pmatrix} = \begin{pmatrix} 8/13 & -9/13 \\ -3/13 & 5/13 \end{pmatrix} \cdot \begin{pmatrix} 17 \\ 11 \end{pmatrix}$$

$$\begin{pmatrix} 1 & 0 \\ 0 & 1 \end{pmatrix} \cdot \begin{pmatrix} x \\ y \end{pmatrix} = \begin{pmatrix} 37/13 \\ 4/13 \end{pmatrix}$$

Step 5 $\begin{pmatrix} x \\ y \end{pmatrix} = \begin{pmatrix} 37/13 \\ 4/13 \end{pmatrix}$

Step 6 $x = 37/13; \ y = 4/13.$

Check: Substitute in (i):

$$5 \times \frac{37}{13} + 9 \times \frac{4}{13} = \frac{185}{13} + \frac{36}{13} = 221/13 = 17$$

$$3 \times \frac{37}{13} + 8 \times \frac{4}{13} = \frac{111}{13} + \frac{32}{13} = \frac{143}{13} = 11$$

2.

$$\begin{aligned} 23x - 12y &= 37 \\ 4x + \ 5y &= -5 \end{aligned} \qquad \text{(i)}$$

Step 1 Rewrite (i):

$$\begin{pmatrix} 23 & -12 \\ 4 & 5 \end{pmatrix} \cdot \begin{pmatrix} x \\ y \end{pmatrix} = \begin{pmatrix} 37 \\ -5 \end{pmatrix} \qquad \text{(ii)}$$

Step 2 $\text{Det}\begin{pmatrix} 23 & -12 \\ 4 & 5 \end{pmatrix} = 115 + 48 = 163$

Step 3 The inverse of $\begin{pmatrix} 23 & -12 \\ 4 & 5 \end{pmatrix} = \begin{pmatrix} 5/163 & 12/163 \\ -4/163 & 23/163 \end{pmatrix}$

Step 4 Multiply (ii) by this inverse:

$$\begin{pmatrix} 5/163 & 12/163 \\ -4/163 & 23/163 \end{pmatrix} \cdot \begin{pmatrix} 23 & -12 \\ 4 & 5 \end{pmatrix} \cdot \begin{pmatrix} x \\ y \end{pmatrix} = \begin{pmatrix} 5/163 & 12/163 \\ -4/163 & 23/163 \end{pmatrix} \cdot \begin{pmatrix} 37 \\ -5 \end{pmatrix}$$

$$\begin{pmatrix} 1 & 0 \\ 0 & 1 \end{pmatrix} \cdot \begin{pmatrix} x \\ y \end{pmatrix} = \begin{pmatrix} 125/163 \\ -263/163 \end{pmatrix}$$

Step 5 $\begin{pmatrix} x \\ y \end{pmatrix} = \begin{pmatrix} 125/163 \\ -263/163 \end{pmatrix}$

Step 6 $x = 125/163$; $y = -263/163$.

Check: Substitute in (i):

$$23 \times \frac{125}{163} - 12\left(-\frac{263}{163}\right) = \frac{2875}{163} + \frac{3156}{163} = \frac{6031}{163} = 37$$

$$4 \times \frac{125}{163} + 5\left(-\frac{263}{163}\right) = \frac{500}{163} - \frac{1315}{163} = -\frac{815}{163} = -5$$

3.

$$\begin{aligned} ax + by &= p \\ cx + dy &= q \end{aligned} \qquad \text{(i)}$$

Step 1 Rewrite (i):

$$\begin{pmatrix} a & b \\ c & d \end{pmatrix} \cdot \begin{pmatrix} x \\ y \end{pmatrix} = \begin{pmatrix} p \\ q \end{pmatrix} \qquad \text{(ii)}$$

Step 2 Det $\begin{pmatrix} a & b \\ c & d \end{pmatrix} = ad - bc = \Delta$

Step 3 The inverse of $\begin{pmatrix} a & b \\ c & d \end{pmatrix} = \begin{pmatrix} d/\Delta & -b/\Delta \\ -c/\Delta & a/\Delta \end{pmatrix}$

Step 4 Multiply (ii) by this inverse:

$$\begin{pmatrix} d/\Delta & -b/\Delta \\ -c/\Delta & a/\Delta \end{pmatrix} \cdot \begin{pmatrix} a & b \\ c & d \end{pmatrix} \cdot \begin{pmatrix} x \\ y \end{pmatrix} = \begin{pmatrix} d/\Delta & -b/\Delta \\ -c/\Delta & a/\Delta \end{pmatrix} \cdot \begin{pmatrix} p \\ q \end{pmatrix}$$

$$\begin{pmatrix} 1 & 0 \\ 0 & 1 \end{pmatrix} \cdot \begin{pmatrix} x \\ y \end{pmatrix} = \begin{pmatrix} (pd - bq)/\Delta \\ (aq - cp)/\Delta \end{pmatrix}$$

Step 5 $\begin{pmatrix} x \\ y \end{pmatrix} = \begin{pmatrix} (dp - bq)/\Delta \\ (aq - cp)/\Delta \end{pmatrix}$

Step 6 $x = (dp - bq)/\Delta$; $y = (aq - cp)/\Delta$ *17.1*

Exercise 17.4

Solve the following pairs of equations by the matrix method:

1. $x + 2y = 3$
   $2x + y = 12$
2. $5x - 2y = 13$
   $3x + 2y = 11$

3. $3x - 2y = -11$
   $3x - 7y = -14$
4. $3a + 2b = 11$
   $2a - b = 5$
5. $9x + 5y = 17$
   $3x + 2y = -1$
6. $6p - 9q = -7$
   $9p - 7q = -8$
7. $3x - 4y = 0$
   $2x + 4y = 13$

17.3 The solution of simultaneous linear equations by determinants

An alternative to the matrix method of solution of simultaneous linear equations is the determinant method. Example (3) in section 17.2 provides the basis for the alternative solution. From (i) the arrangement of the coefficients only in the two equations may be written in the form of an array, i.e. a 2×3 matrix, as follows:

$$\begin{pmatrix} a & b & p \\ c & d & q \end{pmatrix}$$

From this array we may extract three 2×2 matrices by excluding, in turn, one of the columns. These three matrices are:

$$\begin{pmatrix} a & b \\ c & d \end{pmatrix} \quad \text{omitting column 3}$$

$$\begin{pmatrix} b & p \\ d & q \end{pmatrix} \quad \text{omitting column 1}$$

$$\begin{pmatrix} a & p \\ c & q \end{pmatrix} \quad \text{omitting column 2}$$

Their determinants are, respectively, $(ad - cb) = \Delta$; $(bq - dp)$, call it Δ_1; $(aq - cp)$, call it Δ_2. Using this final notation the solution to Example (3) may be expressed:

$$x = -\Delta_1/\Delta; \; y = \Delta_2/\Delta \qquad \textit{17.2}$$

The following examples illustrate the method. As a check the data of examples (1) and (2) in section 17.2 will be used again.

Examples

1. Solve

$$5p+9q=17$$
$$3p+8q=11$$

Step 1 Write down the array:

$$\begin{pmatrix} 5 & 9 & 17 \\ 3 & 8 & 11 \end{pmatrix}$$

Step 2

$$\Delta = \begin{vmatrix} 5 & 9 \\ 3 & 8 \end{vmatrix} = 40-27 = 13$$

$$\Delta_1 = \begin{vmatrix} 9 & 17 \\ 8 & 11 \end{vmatrix} = 99-136 = -37$$

$$\Delta_2 = \begin{vmatrix} 5 & 17 \\ 3 & 11 \end{vmatrix} = 55-51 = 4$$

Step 3 Substitute in $x = -\Delta_1/\Delta$; $y = \Delta_2/\Delta$:

$$x = 37/13; \quad y = 4/13$$

2. Solve

$$23a-12b = 37$$
$$4a+5b = -5$$

Step 1 Write down the array:

$$\begin{pmatrix} 23 & -12 & 37 \\ 4 & 5 & -5 \end{pmatrix}$$

Step 2

$$\Delta = \begin{vmatrix} 23 & -12 \\ 4 & 5 \end{vmatrix} = 115+48 = 163$$

$$\Delta_1 = \begin{vmatrix} -12 & 37 \\ 5 & -5 \end{vmatrix} = 60-185 = -125$$

$$\Delta_2 = \begin{vmatrix} 23 & 37 \\ 4 & -5 \end{vmatrix} = -115-148 = -263$$

Step 3 Substitute in $x = -\Delta_1/\Delta$; $y = \Delta_2/\Delta$:

$$x = 125/163; y = -263/163$$

Exercise 17.5

Solve the following pairs of equations by the determinant method. Check the answers to questions (1) to (7) by those obtained in Exercise 17.4.

1. $p+2q=3$
   $2p+q=12$
2. $5a-2b=13$
   $3a+2b=11$
3. $3p-5q=-11$
   $3p-7q=-14$
4. $3x+2y=11$
   $2x-y=5$
5. $9x+5y=17$
   $3x+2y=-1$
6. $6a-9b=-7$
   $9a-7b=-8$
7. $3x-4y=0$
   $2x+4y=13$
8. $7x-4y=44$
   $6x+3y=57$
9. $4x-6y=3/2$
   $7x-5y=27/4$
10. $14x+3y=7$
    $28x-2y=6$
11. $\frac{2}{3}x+\frac{5}{9}y=11$
    $6x+5y=46$
12. $\frac{1}{5}a-\frac{3}{10}b=10$
    $6a-9b=20$

Now try to solve the following equations by both the matrix method and the determinant method:

13. $4x+6y=19$
    $2x+3y=17$
14. $4x+6y=30$
    $2x+3y=15$
15. $ax+by=c$
    $pax+pby=d$
16. $ax+by=c$
    $pax+pby=pc$

Both the matrix method and the determinant method fail when we try to apply them to questions (11) to (16) in the last exercise. The determinant of the matrix on the left of the set of equations in each case is 0. Consequently the inverse does not exist; each element is indeterminate. For example, when we try to write down the inverse of

$$\begin{pmatrix} 4 & 6 \\ 2 & 3 \end{pmatrix} \quad \text{we produce} \quad \begin{pmatrix} 3/0 & -6/0 \\ -2/0 & 4/0 \end{pmatrix}$$

Division by 0 is meaningless: none of the four elements can be determined.

17.4 A singular matrix

Definition

Any matrix whose determinant is zero is called a *singular matrix.*

Case 1

In both question 14 and question 16 in Exercise 17.5 the two equations are really identical, e.g. $4x + 6y = 30$ and $2x + 3y = 15$ both represent the same straight line:

$$y = -\tfrac{2}{3}x + 5$$

The number of solutions must be infinitely large.

Case 2

In both question 13 and question 15 in Exercise 17.5 the two equations represent lines which are parallel, e.g. $pax + pby = d$ is $ax + by = d/p$, which is a line parallel to:

$$ax + by = c$$

Two such lines will never meet. An alternative view is that they do meet but at an infinitely great distance away. Either view leads to the conclusion that there is no solution to the equations.

Examples

Show that the following matrices are singular:

1. $\begin{pmatrix} 5 & 1 \\ 10 & 2 \end{pmatrix}$ The determinant $= 5 \times 2 - 10 \times 1 = 10 - 10 = 0$

2. $\begin{pmatrix} -6 & -2 \\ 9 & 3 \end{pmatrix}$ The determinant $= (-6)(3) - (-2)(9) = -18 + 18 = 0$

3. $\begin{pmatrix} a & a^2 \\ b & ab \end{pmatrix}$ The determinant $= a(ab) - b(a^2) = a^2b - a^2b = 0$

4. Determine the condition for the following matrix to be singular:

$$\begin{pmatrix} a & b \\ b & a \end{pmatrix}$$

The determinant $= a^2 - b^2 = (a-b)(a+b)$. When $a - b = 0$, or $a + b = 0$, the determinant $= 0$, i.e. either $a = b$, or $a = -b$ for the matrix to be singular.

5. What values of a will make the following matrix singular?

$$\begin{pmatrix} a & 1 \\ 1 & a^2 \end{pmatrix}$$

The determinant $= a^3 - 1 = (a-1)(a^2 + a + 1)$. When $a = 1$ the determinant $= 0$. No other real value of a will make the determinant vanish. The matrix is singular only when $a = 1$.

6. Determine the conditions for the solution of the following pair of equations to be unique:

$$ax + by = c$$
$$bx + ay = d$$

For the solution to be unique:

$$\begin{vmatrix} a & b \\ b & a \end{vmatrix} \neq 0$$
$$a^2 - b^2 \neq 0$$
$$a \neq \pm b$$

Exercise 17.6

Determine which of the following matrices are singular:

1. $\begin{pmatrix} 4 & 2 \\ 2 & 1 \end{pmatrix}$
2. $\begin{pmatrix} -2 & 4 \\ 1 & 2 \end{pmatrix}$

3. $\begin{pmatrix} 6 & 3 \\ 8 & 4 \end{pmatrix}$
4. $\begin{pmatrix} 6 & 3 \\ -8 & -4 \end{pmatrix}$
5. $\begin{pmatrix} 7 & -9 \\ -21 & 27 \end{pmatrix}$
6. $\begin{pmatrix} 1/2 & 1/3 \\ 1/3 & 1/2 \end{pmatrix}$
7. $\begin{pmatrix} 2a & 2b \\ -3a & -3b \end{pmatrix}$
8. $\begin{pmatrix} a/b & 1 \\ -1 & d/a \end{pmatrix}$

Determine the conditions for the following matrices to be singular:

9. $\begin{pmatrix} p & 1 \\ 1 & p \end{pmatrix}$
10. $\begin{pmatrix} q & 3 \\ 5 & 4 \end{pmatrix}$
11. $\begin{pmatrix} 7 & -p \\ 4 & 5 \end{pmatrix}$
12. $\begin{pmatrix} p & 3 \\ 27 & p \end{pmatrix}$
13. $\begin{pmatrix} a^2 & 5 \\ 25 & a \end{pmatrix}$
14. $\begin{pmatrix} a & 5 \\ -25 & a^2 \end{pmatrix}$
15. $\begin{pmatrix} 2x & 3y \\ 4 & -2 \end{pmatrix}$

Determine the conditions in which the following pairs of equations have a unique solution:

16. $px + 3y = 7$
    $4x - 2y = 9$
17. $3x - ay = 12$
    $5x + 2y = 17$
18. $ax - by = c$
    $bx + ay = d.$

Exercise 17.7

1. Currents i_1 and i_2 in a network are related by the equations:

$$i_1 + 2i_2 = 4.8$$
$$3i_1 - i_2 = 4.3$$

Determine the values of i_1 and i_2.

2. The potential differences V_1 and V_2 along two parts of a circuit are given by:

$$1.2\,V_1 + 0.5\,V_2 = 72.2$$
$$0.8\,V_1 + 1.1\,V_2 = 74.2$$

Determine V_1 and V_2.

3. Solve the following equations for i_1 and i_2:

$$0.1\,i_1 + 2(i_1 + i_2) = 4.1$$
$$0.04\,i_2 + (i_1 + i_2) = 2.15$$

4. The law relating effort, P, to load, W, for a particular machine is of the form $P = aW + b$. Given that when W is 1000 N P is 150 N and that when W is 10 000 N P is 750 N, determine the values of a and b.
5. The wind resistance, R, on a particular vehicle is related to the velocity, v, of the vehicle by the formula $R = p + qv^3$. Determine the values of p and q given that R is 5000 N when v is 20 m/s and R is 10 000 N when v is 30 m/s.
6. Network voltages V_1 and V_2 are given by:

$$3\,V_1 = 4\,V_2 + 185$$
$$5\,V_1 + 3\,V_2 = 647$$

Determine the values of V_1 and V_2.

7. The resistances r_1 and r_2 in a network satisfy the relations:

$$1.4r_1 + 1.3r_2 = 14.8$$
$$3r_1 - 2.7r_2 = 3.8$$

Determine r_1 and r_2.

18

Introduction to Complex Numbers

Some quadratic equations which are apparently simple cannot be solved by the methods we have previously employed.

Examples

1. Solve the equation:

$$x^2 + 1 = 0 \qquad \text{(i)}$$

Rearrange (i):

$$x^2 = -1$$

Take the square root of both sides:

$$x = \pm\sqrt{-1}$$

No real number fits the description $\sqrt{-1}$, because if such a number were positive, e.g. $+p$, then $(+p)^2$ would be positive, and if such a number were negative, e.g. $-q$, then $(-q)^2$ would also be positive. Again, if such a number were zero, then $0^2 = 0$. Consequently no number we have yet encountered has a square which is negative.

When we cannot discover a number whose square is negative then we must invent one which does fit the description, or else we should never be able to solve an equation such as (i). We assume that there is a number which satisfies (i) and we call it j. We define j to be $\sqrt{-1}$, i.e.

$$j = \sqrt{-1} \qquad \textit{18.1}$$

From this we obtain:

$$j^2 + 1 = 0 \qquad 18.2$$

$$\text{or} \quad j^2 = -1$$

There is no point in our trying to identify j with some real number because j cannot be real. It has been called, and is still referred to as, an imaginary number.

2. Solve the equation:

$$x^2 + 4 = 0 \qquad \text{(ii)}$$

Rearrange:

$$\begin{aligned} x^2 &= -4 \\ &= -1 \times 4 \\ &= j^2 \times 2^2 \end{aligned}$$

Take the square root of both sides:

$$\begin{aligned} x &= \pm j \times 2 \\ \text{or} \quad x &= \pm j.2 \end{aligned}$$

3. Solve the equation:

$$x^2 = -5 \qquad \text{(iii)}$$

Then:

$$x^2 = -1 \times 5 = j^2 \times 5$$

Take the square root:

$$x = \pm j.\sqrt{5}$$

4. Solve the equation:

$$x^2 + c = 0 \qquad \text{(iv)}$$

where c is positive.

$$\begin{aligned} x^2 &= -c \\ &= -1 \times c = j^2 \times c \end{aligned}$$

Take the square root:

$$x = \pm j.\sqrt{c}$$

5. Solve the equation:

$$2x^2 + 3 = 0$$

Rearrange:

$$2x^2 = -3$$

Divide by 2:

$$x^2 = -\frac{3}{2}$$

$$= j^2 \times \left(\frac{3}{2}\right)$$

Take the square root:

$$x = \pm j\sqrt{\frac{3}{2}}$$

6. Solve $ax^2 + b = 0$, where a and b are positive.
Then:

$$ax^2 = -b = j^2 . b$$

$$x^2 = j^2 . \frac{b}{a}$$

Take the square root:

$$x = \pm j . \sqrt{\frac{b}{a}}$$

Exercise 18.1

Solve the following equations:

1. $x^2 + 9 = 0$
2. $x^2 + 25 = 0$
3. $x^2 + 121 = 0$
4. $x^2 + a^2 = 0$
5. $x^2 + 2 = 0$
6. $x^2 + 7 = 0$
7. $2x^2 + 5 = 0$
8. $1/x^2 + 1 = 0$
9. $1/(x^2 + 1) + 1 = 0$
10. $3x^2 + 8 = 0$
11. $p^2x^2 + q^2 = 0$
12. $px^2 + q = 0$, where p and q are positive.

In fact all the roots in the above examples and exercises are imaginary, i.e. real multiples of j.

Examples

1. Solve $x^2+4x+5=0$.
   This equation cannot be solved by factorization. By the formula:

$$x=\frac{-4\pm\sqrt{4^2-4\times1\times5}}{2}$$
$$=\frac{-4\pm\sqrt{16-20}}{2}$$
$$=\frac{-4\pm\sqrt{-4}}{2}$$
$$=\frac{-4\pm\sqrt{j^2.2^2}}{2}$$
$$=\frac{-4-j.2}{2}\quad\text{or}\quad\frac{-4+j.2}{2}$$
$$\text{i.e.}\quad -2+j\quad\text{or}\quad -2-j$$

   Each root contains two terms. The first (-2) is real; the second $(\pm j)$ is imaginary. *Such a number is called a complex number.*
   Definition: A complex number is of the form $a+j.b$, where a and b are real (they may be positive or negative).
2. Solve the equation $3x^2+2x+5=0$.
   By the formula:

$$x=\frac{-2\pm\sqrt{2^2-4\times3\times5}}{6}$$
$$=\frac{-2\pm\sqrt{4-60}}{6}$$
$$=\frac{-2\pm\sqrt{-56}}{6}$$
$$=\frac{-2\pm\sqrt{j^2.56}}{6}$$
$$=\frac{-2\pm j.\sqrt{56}}{6}$$
$$=-\frac{1}{3}\pm j\cdot\frac{\sqrt{56}}{6}$$

This again is of the type $a \pm j.b$ where a and b are real. In fact $\frac{\sqrt{56}}{6}$ may be simplified to $\frac{\sqrt{4 \times 14}}{6} = \frac{2\sqrt{14}}{6} = \frac{\sqrt{14}}{3}$.

Examples

The following are examples of complex numbers.

1. $3+j.2$
2. $4+j.5$
3. $3-j.2$ or $3+j(-2)$
4. $4-j.5$ or $4+j.(-5)$
5. $-4+j.5$
6. $\frac{4}{3}+j.\frac{2}{5}$
7. $-\frac{3}{5}+j.\frac{2}{7}$
8. $-\frac{5}{6}-j.\frac{7}{8}$
9. $0+j.3$ (or, written as $j.3$, is purely imaginary)
10. $5+j.0$ (or, when written as 5, is purely real)

Examples 9 and 10 illustrate that $a+j.b$, which is complex when a and b are real and not zero, can be either purely real or purely imaginary in special circumstances. For example:

$a+j.b$ is complex when a and b are real and not zero
$a+j.b$ is real when a is real and $b=0$
$a+j.b$ is imaginary when b is real and $a=0$

Exercise 18.2

Solve the equations below, expressing the answers as complex numbers.

1. $x^2+2x+10=0$
2. $x^2-2x+17=0$
3. $x^2+4x+8=0$
4. $x^2-4x+8=0$

5. $x^2+x+1=0$
6. $x^2-x+1=0$
7. $x^2+x+3=0$
8. $4x^2+5x+3=0$

18.1 Representation of complex numbers

It is often helpful to be able to represent a number in a diagrammatic form. For example, real numbers may be represented by the points on a line. To arrive at a representation of complex numbers it is essential first to agree on a convention, that is, a complex number, for example z, is written:

$$x+j.y$$

$$\text{i.e.}\quad z=x+j.y \qquad 18.3$$

in which the real part, x, is always quoted before the imaginary part, jy. Therefore, since all complex numbers are to be written in the form *18.3*, to distinguish one such number from any other we need only quote the values of x and y, in the order *first* x, *second* y; or x, y; or (x, y); and that reminds us of the co-ordinates of a point in a plane. *In fact* (x, y) *is an ordered pair.*

Another ordered pair is $\begin{pmatrix} x \\ y \end{pmatrix}$, and that reminds us of the vector in a plane from the origin to the point whose co-ordinates are (x, y). Fig. 18.1 illustrates the idea. It is called the Argand diagram.

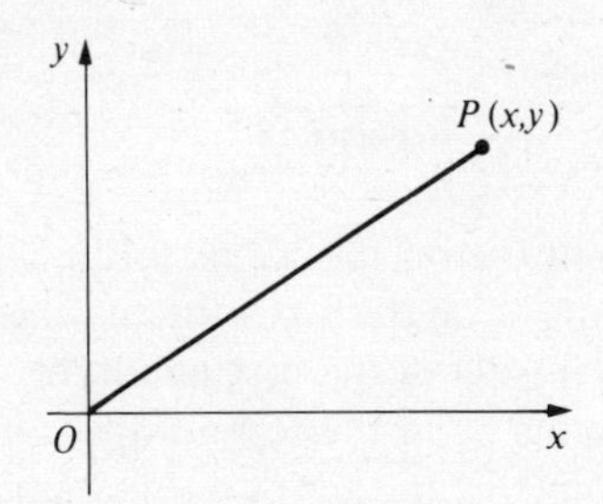

Figure 18.1

The complex number $z = x+j.y$ may be represented either by the point $P \equiv (x, y)$, or by the vector $OP \equiv \begin{pmatrix} x \\ y \end{pmatrix}$.

Examples

Represent the following complex numbers on an Argand diagram.

1. $2+j.3$
2. $3-j.2$
3. $-4+j.2$
4. $-3-j.4$
5. $2+j.0=2$
6. $0+j.2=j.2$

The above complex numbers are represented in Fig. 18.2. The xy plane may be subdivided into six parts to correspond to the special cases of complex numbers which may arise. Where $a+j.b$ represents the complex number in general:

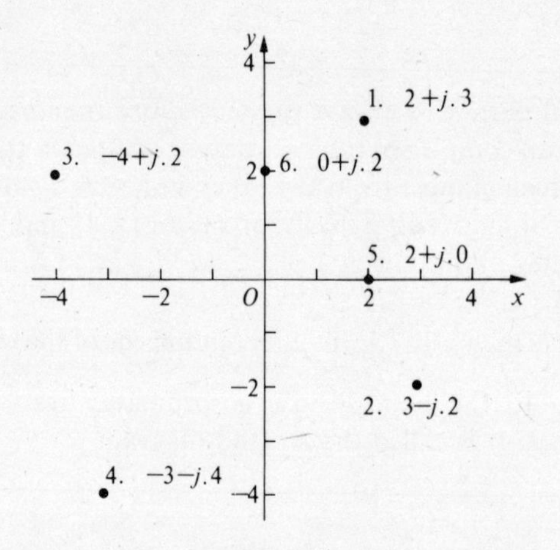

Figure 18.2

1. when $b=0$, all points lie on the x axis (often called the real axis);
2. when $a=0$, all points lie on the y axis (often called the imaginary axis);
3. $a>0$, $b>0$, all points lie in the first quadrant;
4. $a<0$, $b>0$, all points lie in the second quadrant;
5. $a<0$, $b<0$, all points lie in the third quadrant;
6. $a>0$, $b<0$, all points lie in the fourth quadrant.

Exercise 18.3

Plot the points which represent the following complex numbers on an Argand diagram:

1. $1+j.1$
2. $2+j.1$
3. $1+j.2$
4. $1+j.3$
5. $-1+j.2$
6. $-2-j.3$
7. $-\frac{3}{2}-j.\frac{5}{2}$
8. $-2+j.3$
9. $0+j.3$
10. $0-j.2$
11. $2+j.0$
12. $-3+j.0$

With the data in the exercise above the vector representing question (1) is $\begin{pmatrix}1\\1\end{pmatrix}$, that representing question (2) is $\begin{pmatrix}2\\1\end{pmatrix}$, and that representing question (3) is $\begin{pmatrix}1\\2\end{pmatrix}$.

Exercise 18.4

Write down the vectors representing the following complex numbers:

1. $1+j.3$
2. $-1+j.2$
3. $-2-j.3$
4. $-\frac{3}{2}-j.\frac{5}{2}$
5. $-2+j.3$
6. $a+j.b$
7. $2a+j.2b$
8. $3a+j.3b$
9. $-a-j.b$
10. $pa+j.pb$
11. $(a+c)+j.(b+d)$

18.2 Addition of complex numbers

Now that we have introduced a new kind of number, a complex number, we need to know how to add, subtract, multiply and divide using such numbers. For addition we may proceed in either of two ways.

Method 1

Suppose two complex numbers are:

$$z_1 = x_1 + j.y_1 \qquad 18.4$$

and

$$z_2 = x_2 + j.y_2 \qquad 18.5$$

They may be represented either by points $P_1(x_1, y_1)$ and $P_2(x_2, y_2)$ or by vectors $\boldsymbol{OP}_1\begin{pmatrix} x_1 \\ y_1 \end{pmatrix}$ and $\boldsymbol{OP}_2\begin{pmatrix} x_2 \\ y_2 \end{pmatrix}$. We know how to add vectors together:

$$\boldsymbol{OP}_1 + \boldsymbol{OP}_2 = \begin{pmatrix} x_1 \\ y_1 \end{pmatrix} + \begin{pmatrix} x_2 \\ y_2 \end{pmatrix} = \begin{pmatrix} x_1 + x_2 \\ y_1 + y_2 \end{pmatrix}$$

which is a vector that must represent the complex number:

$$(x_1 + x_2) + j.(y_1 + y_2)$$

And that must mean:

$$(x_1 + j.y_1) + (x_2 + j.y_2) = (x_1 + x_2) + j.(y_1 + y_2) \qquad 18.6$$

Method 2

We simply define:

$$(x_1 + j.y_1) + (x_2 + j.y_2) = (x_1 + x_2) + j.(y_1 + y_2)$$

Examples

Add together the following sets of complex numbers:

1. $(2 + j.3) + (3 + j.4) = (2 + 3) + j.(3 + 4) = 5 + j.7$
2. $(-4 + j.3) + (2 - j.7) = (-4 + 2) + j.(3 - 7) = -2 - j.4$
3. $(1 + j.3) + (-2 + j.4) + (3 - j.6) = (1 - 2 + 3) + j.(3 + 4 - 6) = 2 + j$

Note that in all cases we are really supposing that the associative law operates, i.e. that the order in which we add terms together may be changed, even when some terms are real and others imaginary.

4. $(1 + j.3) + (-1 - j.3) = (1 - 1) + j.(3 - 3) = 0 + j.0 = 0$
5. $(a + j.b) + (-a - j.b) = (a - a) + j.(b - b) = 0 + j.0 = 0$
6. If $z = a + j.b$, then $(a + j.b) + (a + j.b) + (a + j.b) = 3a + j.3b$ i.e. $z + z + z = 3a + j.3b$, or $3z = 3a + j.3b$
7. If $z = a + j.b$, then $pz = pa + j.pb$

Exercise 18.5

Express each of the following as a single complex number:

1. $(1+j.2)+(2+j.1)$
2. $(3+j.2)+(3+j.4)$
3. $(5+j.6)+(2+j.3)$
4. $(-2+j.1)+(1+j.2)$
5. $(-2+j.1)+(1-j.2)$
6. $(-2+j.1)+(-1-j.2)$
7. $(-4+j.3)+(6-j.7)$
8. $(3+j.2)+(-2+j.3)$
9. If $z_1 = a+j.b$ and $z_2 = c+j.d$ express $2z_1+3z_2$ in terms of a, b, c and d.
10. If $z = x+j.y$ write az in terms of a, x and y where a is real.
11. If $z = x+j.y$ write $-z$ in terms of x and y.
12. If $z = x+j.y$ simplify $z+(-z)$.
13. If $z_1 = x_1+j.y_1$, $z_2 = x_2+j.y_2$, write $z_1+(-z_2)$ in terms of x_1, y_1, x_2 and y_2.

Question (13) in Exercise 18.5 produces:

$$\begin{aligned} z_1+(-z_2) &= x_1+j.y_1+(-x_2-j.y_2) \\ &= (x_1-x_2)+j.(y_1-y_2) \end{aligned}$$

But $z_1+(-z_2)$ must be z_1-z_2. Therefore here we have the rule for the difference between two complex numbers.

18.3 The difference between two complex numbers

The rule is:

$$(x_1+j.y_1)-(x_2+j.y_2) = (x_1-x_2)+j.(y_1-y_2) \qquad 18.7$$

Exercise 18.6

Express each of the following as a single complex number:

1. $(4+j.3)-(2+j)$
2. $(5+j.7)-(4+j.3)$
3. $(4+j.3)-(-2+j)$

4. $(4+j.3)-(-2-j)$
5. $(5-j.7)-(4+j.3)$
6. $(-2-j.3)-(4+j.5)$
7. $(-2+j.3)-(4-j.5)$
8. $(-8-j.10)-(12-j.9)$
9. $(6+j.5)-(-4+j.2)+(3-j.5)$
10. $2(2+j)+3(j+2)-4(1+j)$
11. $3(3-j.2)-2(-4-j.3)+5(2+j.4)$

18.4 Representation of sums and differences on an Argand diagram

The results and data of Examples 1 and 2 in section 18.2 are plotted in Fig. 18.3. They illustrate that, to find the sum of two complex numbers graphically, a parallelogram is drawn and the diagonal of the parallelogram represents that complex number which is the sum of the given complex numbers.

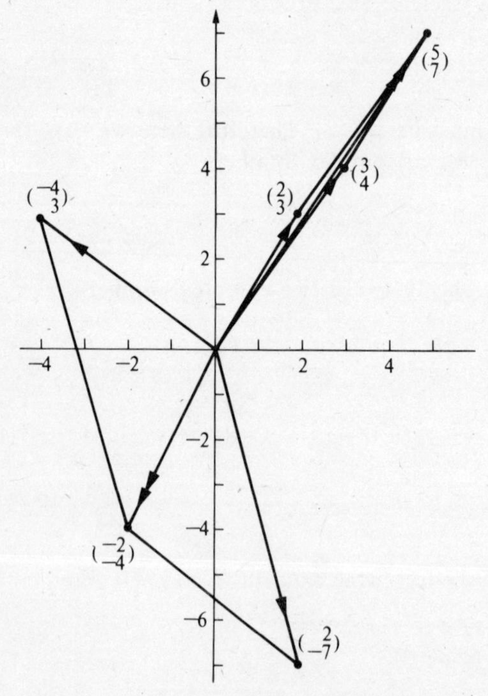

Figure 18.3

Fig. 18.4 represents the sum of two general complex numbers, i.e.

where $z_1 = x_1 + j.y_1$ and $z_2 = x_2 + j.y_2$

then $z_1 + z_2 = (x_1 + x_2) + j(y_1 + y_2)$

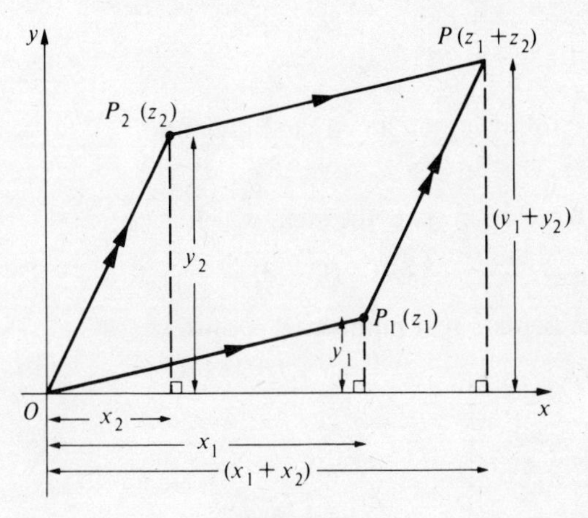

Figure 18.4

In Fig. 18.4 P_1 represents z_1 and P_2 represents z_2. OP_1 and OP_2 are taken to be adjacent sides of a parallelogram. To find P (representing $z_1 + z_2$), draw P_2P parallel to OP_1 and P_1P parallel to OP_2.

Fig. 18.5 represents the difference between two complex numbers.

where $z_1 = x_1 + j.y_1$ and $z_2 = x_2 + j.y_2$

then $z_1 - z_2 = (x_1 - x_2) + j.(y_1 - y_2)$.

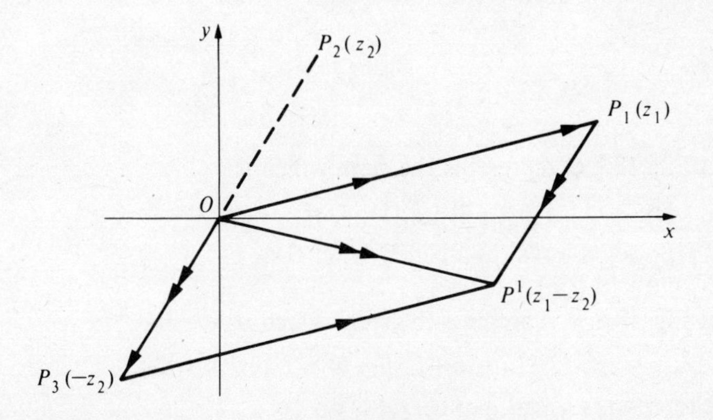

Figure 18.5

P_1 represents z_1, P_2 represents z_2, P_3 represents $-z_2$, where OP_3 is equal and opposite to OP_2. Write $z_1 - z_2$ as $z_1 + (-z_2)$. Construct a parallelogram with OP_1 and OP_3 adjacent sides. The fourth vertex of the parallelogram is P^1, which represents $z_1 - z_2$.

Examples

Represent the following on an Argand diagram:

1. $(2+j.3)+(2-j.3)$
   In Fig. 18.6 P represents the sum, which is:

$$(2+2)+j(3-3) = 4+j.0 = 4$$

   i.e. P represents a real number. P is on Ox.

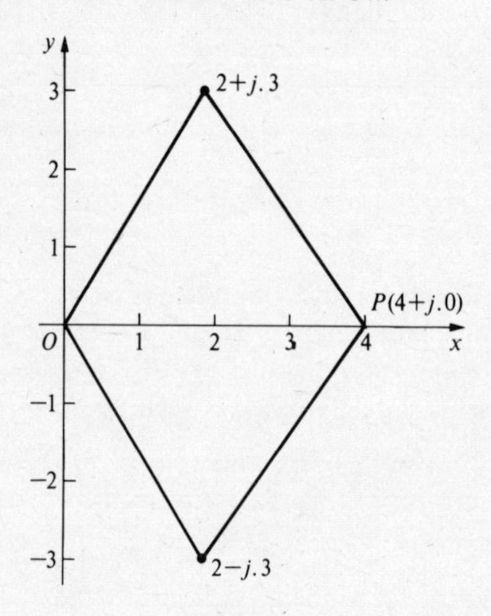

Figure 18.6

2. $(-4+j)+(-4-j)$
   In Fig. 18.7 Q represents the sum, which is:

$$(-4-4)+j.(1-1) = -8+j.0 = -8$$

   Q represents a real number. Q is on Ox.
3. $(a+jb)+(a-jb)$
   In Fig. 18.8 R represents the sum, which is:

$$(a+a)+j.(b-b) = 2a+j.0 = 2a$$

   R represents a real number. R is on Ox.

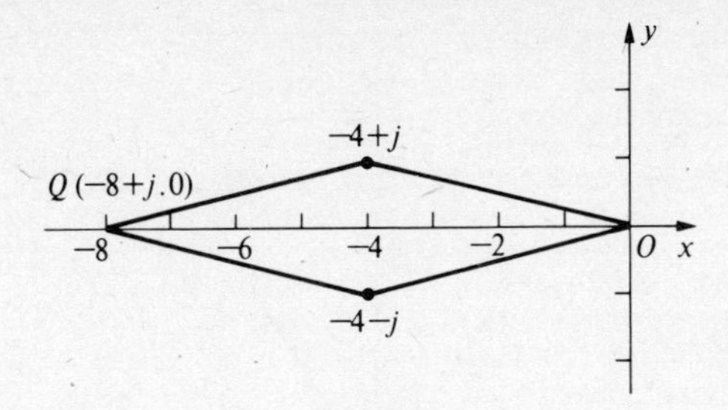

Figure 18.7

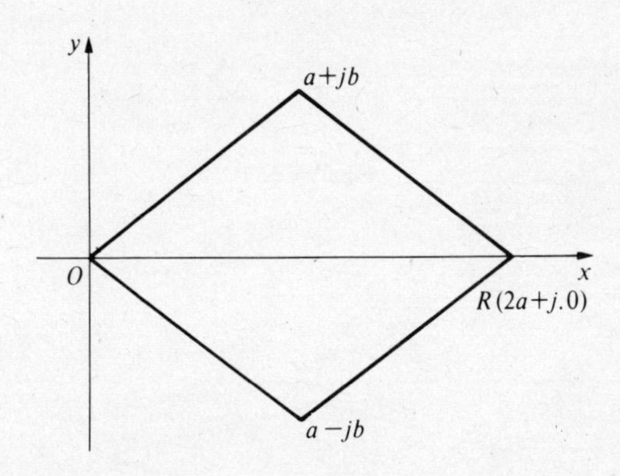

Figure 18.8

The above examples, particularly Example 3, represent sums of pairs of complex numbers whose sums are real. *$a+j.b$ and $a-j.b$ are called conjugate complex numbers.* Note that the points which represent them on the Argand diagram are images of each other in Ox.

4. $(2+j.3)-(2-j.3)$
   In Fig. 18.9 P^1 represents the difference, which is:

   $$(2-2)+j(3+3) = 0+j.6 = j.6$$

   P^1 represents an imaginary number. P^1 is on Oy.
5. $(-4+j)-(-4-j)$
   In Fig. 18.10 Q^1 represents the difference, which is:

   $$(-4+4)+j.(1+1) = j.2$$

   Q^1 represents an imaginary number. Q^1 is on Oy.

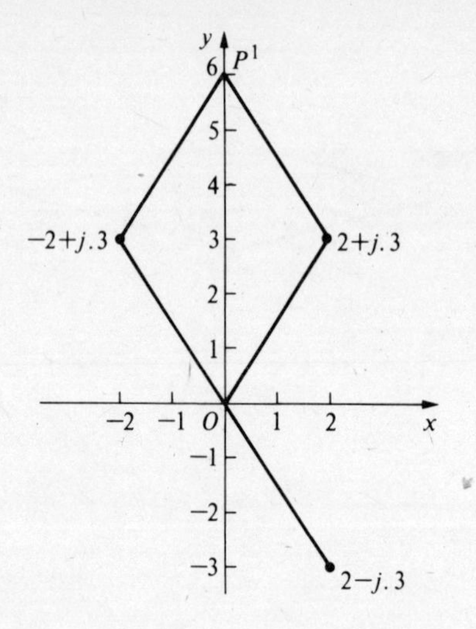

Figure 18.9

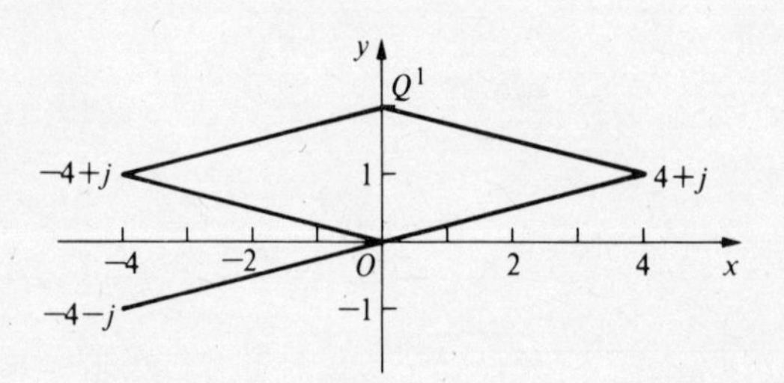

Figure 18.10

6. $(a+j.b)-(a-j.b)$
   In Fig. 18.11 R^1 represents the difference, which is:

$$(a-a)+j.(b+b)=j.2b$$

   R^1 represents an imaginary number. R^1 is on Oy.

The difference between the conjugate complex numbers is always an imaginary number.

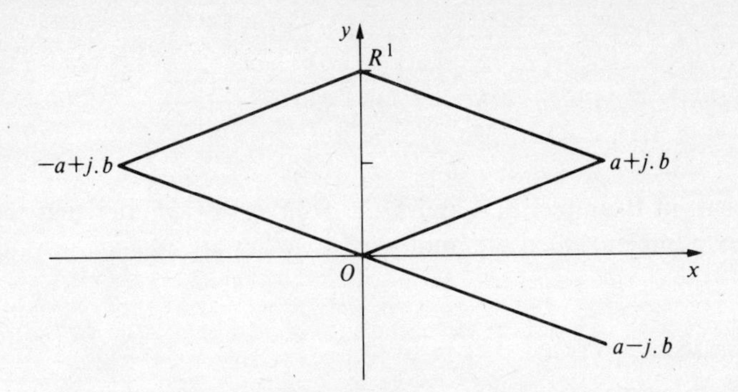

Figure 18.11

18.5 The product of two complex numbers

Where $z_1 = x_1 + j.y_1$ and $z_2 = x_2 + j.y_2$ we define:

$$z_1.z_2 = (x_1x_2 - y_1y_2) + j.(x_1y_2 + x_2y_1) \qquad \textit{18.8}$$

In fact if we assumed that we could calculate $(x_1+jy_1)(x_2+jy_2)$ by the rules of real algebra we would arrive at *18.8*:

$$\begin{aligned}(x_1+jy_1)(x_2+jy_2) &= x_1x_2 + j^2y_1y_2 + jx_1y_2 + jy_1x_2\\ &= (x_1x_2 - y_1y_2) + j.(x_1y_2 + y_1x_2)\end{aligned}$$

It is probably easier to work in that way for multiplication rather than try to remember rule *18.8*.

Examples

Express the following as single complex numbers.

1. $(2+j.3)(3+j.2)$
   By *18.8* the product $= (2\times 3 - 3\times 2) + j.(3\times 3 + 2\times 2)$
   $= (6-6)+j\ (9+4) = 0 + j.13$
2. $(5+j.2)(-4+j.7) = (-20-14)+j.(-8+35)$
   $= -34 + j.27$
3. $\left(\frac{1}{2}+j.\frac{\sqrt{3}}{2}\right)\left(\frac{1}{2}-j.\frac{\sqrt{3}}{2}\right) = \left(\frac{1}{4}+\frac{3}{4}\right)+j.\left(\frac{\sqrt{3}}{4}-\frac{\sqrt{3}}{4}\right)$
   $= 1 + j.0 = 1$

4. $(-2+j.3)(-2-j.3) = (4+9)+j.(-6+6)$
$= 13+j.0 = 13$
5. $(a+jb)(a-jb) = (a^2+b^2)+j.(ab-ab)$
$= a^2+b^2$

Note that, in Examples 3, 4 and 5 the *product is real*, and that the two complex numbers which are multiplied together are *conjugates.*

Exercise 18.7

Express each of the following as a single complex number.

1. $(2+j.3)(3+j.4)$
2. $(3+j.2)(4+j.3)$
3. $(5+j.2)(2-j.5)$
4. $(2-j.3)(4-j.3)$
5. $(-7-j.6)(4+j.5)$
6. $(1+j)(-1-j)$
7. $(1+j)(1-j)$
8. $(2+j)(2-j)$
9. $(-1+j)(-1-j)$
10. $\left(-\frac{1}{2}+j.\frac{\sqrt{3}}{2}\right)\left(-\frac{1}{2}-j.\frac{\sqrt{3}}{2}\right)$
11. $\left(\frac{1}{\sqrt{2}}+j.\frac{1}{\sqrt{2}}\right)\left(\frac{1}{\sqrt{2}}-j.\frac{1}{\sqrt{2}}\right)$
12. $(1+j)(1+j.2)(1+j.3)$ (Hint: first calculate $(1+j)(1+j.2)$, then multiply the result by $(1+j.3)$.)
13. $(1+j)(1+j.2)(1-j.2)(1-j)$ (Hint: rearrange the order before multiplying.)
14. $(a+jb)(b+ja)(a-jb)$

18.6 Division of complex numbers

We might proceed by formulating a definition for division as was done in formula *18.8*. It is simpler to avoid involved rules.

As a first step we will show how to write down $1/z$ as a single complex number when z is known. Suppose $z = x+j.y$. The conjugate complex number is $x - j \cdot y$, called $\bar{z}$, or z^1. Then:

$$z \cdot \bar{z} = (x + j \cdot y)(x - j \cdot y) = x^2 + y^2 = OP^2 \text{ (Fig. 18.12)}$$

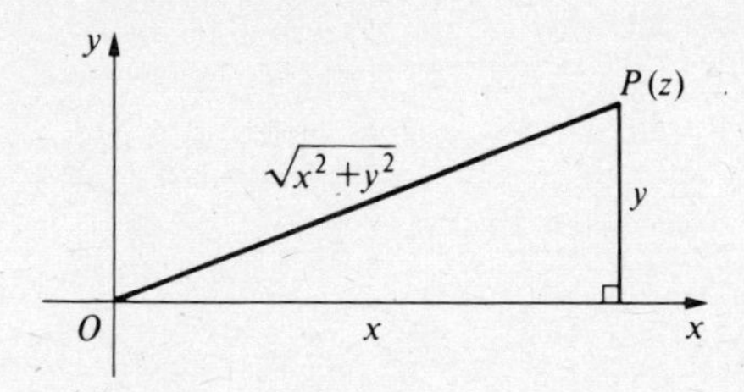

Figure 18.12

OP^2 is called:

$$(\text{mod } z)^2 \text{ or } |z|^2$$

Therefore:

$$z \cdot \bar{z} = |z|^2 \qquad 18.9$$

18.9 is one of the most important formulae we shall encounter. By rearranging it we obtain:

$$\frac{1}{z} = \frac{\bar{z}}{|z|^2} \qquad 18.10$$

Alternatively, *18.10* may be written:

$$\frac{1}{x + j \cdot y} = \frac{x - j \cdot y}{x^2 + y^2} \qquad 18.11$$

$$\text{or} \quad \frac{x}{x^2 + y^2} - j \cdot \frac{y}{x^2 + y^2} \qquad 18.12$$

The RHS of *18.12* is a complex number. In other words, *the reciprocal of a complex number may be expressed as a different complex number.*

Examples

Express the following as complex numbers:

1. $\dfrac{1}{1+j}$: By *18.11* $\dfrac{1}{1+j} = \dfrac{1-j}{1^2+1^2} = \dfrac{1-j}{2} = \dfrac{1}{2} - j \cdot \dfrac{1}{2}$

2. $\dfrac{1}{2+j.3}$: By *18.11* $\dfrac{1}{2+j.3}=\dfrac{2-j.3}{2^2+3^2}=\dfrac{2-j.3}{13}=\dfrac{2}{13}-j.\dfrac{3}{13}$

3. $\dfrac{3}{4+j.5}$: By *18.11* $\dfrac{3}{4+j.5}=3\times\dfrac{1}{4+j.5}=3\times\dfrac{4-j.5}{4^2+5^2}$

$$=3\frac{(4-j.5)}{41}=\frac{12-j.15}{41}=\frac{12}{41}-j.\frac{15}{41}$$

4. $\dfrac{4}{2-j.3}=\dfrac{4(2+j.3)}{2^2+3^2}=\dfrac{8+j.12}{13}=\dfrac{8}{13}+j.\dfrac{12}{13}$

Exercise 18.8

Express the following as complex numbers:

1. $\dfrac{1}{1+j.2}$
2. $\dfrac{1}{1-j}$
3. $\dfrac{1}{3+j.2}$
4. $\dfrac{1}{5+j.4}$
5. $\dfrac{1}{5-j.6}$
6. $\dfrac{1}{-6-j.5}$
7. $\dfrac{3}{1+j.4}$
8. $\dfrac{8}{12-j.5}$
9. $\dfrac{1}{3+j.4}$
10. $\dfrac{17}{15-j.8}$
11. $\dfrac{25}{-7-j.24}$
12. $\dfrac{41}{40+j.9}$

The above ideas are developed to establish a method of carrying out division with complex numbers. Suppose we wish to determine $\frac{z_1}{z_2}$ where $z_1 = x_1 + j.y_1$ and $z_2 = x_2 + j.y_2$. We write:

$$\frac{z_1}{z_2} = z_1 \times \frac{1}{z_2} = z_1 \times \frac{\bar{z}_2}{|z_2|^2} \text{ using } \mathit{18.10}$$

$$= \frac{z_1 \times \bar{z}_2}{|z_2|^2} \qquad \mathit{18.13}$$

In *18.13* $|z_2|^2$ is real and $z_1 \times \bar{z}_2$ can be evaluated by the methods discussed in section 18.5.

Examples

Express the following as complex numbers.

1. $\dfrac{1+j}{1+j.2}$

   Rewrite as:

$$(1+j) \times \frac{1}{(1+j.2)}$$

$$= (1+j)\frac{(1-j.2)}{1^2+2^2} = \frac{(1+j)(1-j.2)}{1^2+2^2} = \frac{(1+2)+j(1-2)}{5}$$

$$= \frac{3-j}{5} = \frac{3}{5} - j.\frac{1}{5}$$

   An alternative approach is to multiply numerator and denominator by the conjugate of the denominator. The two methods amount to the same in the end.

$$\frac{(1+j)(1-j.2)}{(1+j.2)(1-j.2)} = \frac{3-j}{1+4} = \frac{3}{5} - j.\frac{1}{5}$$

   Another alternative is the following. Suppose:

$$\frac{1+j}{1+j.2} = a + j.b$$

   Multiply both sides by $(1+j.2)$. Then:

$$1 + j = (a + j.b)(1 + j.2)$$
$$= a - 2b + j(b + 2a)$$

Equate real and imaginary parts:

$$a - 2b = 1 \qquad \text{(i)}$$

and

$$2a + b = 1 \qquad \text{(ii)}$$

(i) + 2 × (ii) gives:

$$5a = 3;\ a = 3/5$$

Substitute for a in (ii):

$$b = 1 - \frac{6}{5} = -\frac{1}{5}$$

2. $$\frac{3+j.5}{6-j.7} = \frac{(3+j.5)(6+j.7)}{6^2+7^2} = \frac{(18-35)+j(30+21)}{85}$$

$$= \frac{-17+j.51}{85} = -\frac{17}{85} + j.\frac{51}{85}$$

3. $$\frac{(1+j)(2-j.3)}{(3+j.2)} = \frac{(1+j)(2-j.3)(3-j.2)}{3^2+2^2}$$

$$= \frac{(5-j)(3-j.2)}{3^2+2^2} = \frac{13-j.13}{13} = 1-j$$

4. $$\frac{(a+j.b)}{(c+j.d)} = \frac{(a+j.b)(c-j.d)}{(c^2+d^2)} = \frac{(ac+bd)+j(bc-ad)}{(c^2+d^2)}$$

$$= \left(\frac{ac+bd}{c^2+d^2}\right) + j.\left(\frac{bc-ad}{c^2+d^2}\right)$$

The result in Example 4 could be used as the definition for division of complex numbers. The RHS is a difficult expression to memorize. It is easier to adopt the method illustrated.

Exercise 18.9

Express the following as complex numbers.

1. $\dfrac{1+j}{2+j}$

2. $\dfrac{1+j}{3+j.2}$

3. $\dfrac{2-j}{2+j.3}$

4. $\dfrac{3+j.4}{4+j.3}$

5. $\dfrac{3-j.4}{4+j.3}$

6. $\dfrac{5+j.8}{2+j.7}$

7. $\dfrac{(1-j)(2+j)}{(3+j.2)}$

8. $\dfrac{(4+j.3)(5-j.2)}{(12+j.5)}$

9. $\dfrac{(2+j.3)}{(1+j)(2-j.3)}$

10. $(1+j)^2$

11. $(2+j.3)^2$

12. $(1+j)^3$

13. $(1+j)^2+(1+j)$

14. $2(1+j)^2+3(1+j)$

15. $\dfrac{(1+j)}{(2+j)}+(2-j.3)$

16. $\dfrac{5}{(3+j.4)}+\dfrac{(3-j.4)}{5}$

17. $\dfrac{(1+j.3)}{(1-j.3)}+\dfrac{(1-j.3)}{(1+j.3)}$

18. $(3-j.4)^3$

19

The Modulus-argument Form of Complex Numbers

19.1 Polar co-ordinates

Chapter 18 introduced the representation of a complex number using an Argand diagram. The complex number

$$z = x + j.y$$

which is really an ordered pair, may be represented either by the point P, co-ordinates (x, y), or by the vector $\boldsymbol{OP}\begin{pmatrix} x \\ y \end{pmatrix}$.

In fact any system of co-ordinates which determines the position of a point in a plane may be used to represent complex numbers. One such is the system of polar co-ordinates. For polar co-ordinates a fixed reference point, O, is chosen, together with an axis of reference or initial line, Ox. Fig. 19.1 represents such a system and its relationship to the cartesian system.

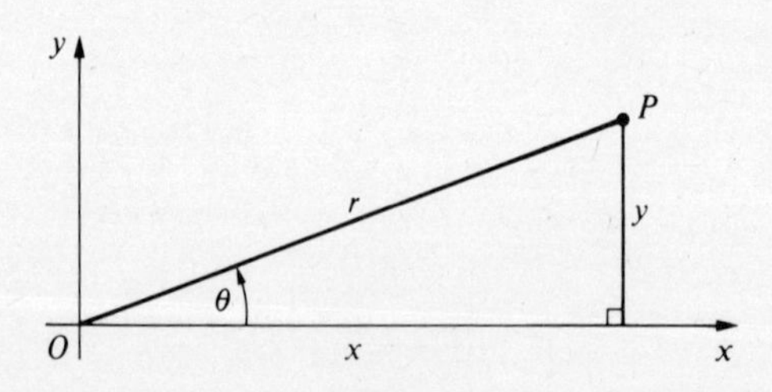

Figure 19.1

In polar co-ordinates the position of P is determined by its distance from O, $OP = r$, and the angle of rotation of OP from the initial line Ox in an anticlockwise direction, θ. Therefore P may be determined either by the co-ordinates (x, y) or by $[r, \theta]$, where θ is usually measured in radians. Note the square brackets, which are used to avoid confusion between the two systems. From Fig. 19.1:

$$x = r\cos\theta \qquad 19.1$$
$$y = r\sin\theta \qquad 19.2$$

Square and add *19.1* and *19.2*:

$$\begin{aligned} x^2 + y^2 &= r^2(\cos^2\theta + \sin^2\theta) \\ &= r^2 \end{aligned} \qquad 19.3$$

Take the square root of *19.3*:

$$r = \sqrt{x^2 + y^2} \qquad 19.4$$

The $\pm$ is not introduced because r is taken to be positive. *19.2* $\div$ *19.1* gives:

$$\tan\theta = y/x \qquad 19.5$$

Formulae *19.1* and *19.2* enable us to convert from polar to the cartesian form. Formulae *19.4* and *19.5* enable us to convert from the cartesian form to the polar form.

19.2 Conversion from cartesian to polar form of a complex number and vice versa

Where a complex number, z, may be written:

$$z = x + j.y$$

by substitution from *19.1* and *19.2* we may rewrite:

$$\begin{aligned} z &= r\cos\theta + j.r\sin\theta \\ &= r(\cos\theta + j\sin\theta) \end{aligned} \qquad 19.6$$

19.6 is said to be the polar form of the complex number.

Examples

Express the following complex numbers in polar form.

1. $1+j$

Although *19.5* is often regarded as the method of obtaining the angle θ from the values of x and y, it is sometimes more convenient to use *19.1* and *19.2* instead. Both methods will be illustrated.

From *19.1*:

$$r \cos \theta = 1 \quad \text{(i)}$$

From *19.2*:

$$r \sin \theta = 1 \quad \text{(ii)}$$

From *19.4*:

$$r = \sqrt{1^2 + 1^2} = \sqrt{2}$$

Substitute for r in (i):

$$\cos \theta = 1/r = 1/\sqrt{2}$$

Cos θ is positive so θ may be in either the first or the fourth quadrant. (iii)

Substitute for r in (ii):

$$\sin \theta = 1/r = 1/\sqrt{2}$$

Sin θ is positive so θ may be in either the first or the second quadrant. (iv)

From (iii) and (iv) together θ must be in the first quadrant. Therefore:

$$\theta = \pi/4$$

Then:

$$1+j = \sqrt{2}(\cos \pi/4 + j \sin \pi/4)$$

Alternatively, by using *19.5*:

$$\tan \theta = 1/1 = 1$$

θ may be in either the first or the third quadrant. However, $1+j$ is represented by the point (1, 1), which is in the first quadrant. So θ must be in the first quadrant, and $\theta = \pi/4$. Again:

$$1+j = \sqrt{2}(\cos \pi/4 + j \sin \pi/4)$$

2. $3-j.4$

From *19.1*:

$$r \cos \theta = 3 \quad \text{(i)}$$

From *19.2*:

$$r \sin\theta = -4 \qquad \text{(ii)}$$

From *19.4*:

$$r = \sqrt{3^2 + (-4)^2} = 5$$

(i) becomes $\cos\theta = 3/5$; θ is in either the first or the fourth quadrant. (iii)

(ii) becomes $\sin\theta = -4/5$; θ is in either the third or the fourth quadrant. (iv)

(iii) and (iv) together give θ in the fourth quadrant.
$\theta = 2\pi - 0.9272952 = 5.3558901 \approx 5.36$ radians
Or, in degrees, $\theta = 360° - 53.130102° = 306.8699° \approx 306.87°$
$3 - j.4 = 5(\cos 306.87° + j \sin 306.87°)$
Or $= 5(\cos 5.36 + j \sin 5.36)$

3. Convert $5(\cos \pi/3 + j \sin \pi/3)$ to cartesian form.
   19.1 becomes:

$$x = 5 \cos \pi/3 = 5 \times \tfrac{1}{2} = \tfrac{5}{2}$$

   19.2 becomes:

$$y = 5 \sin \pi/3 = 5 \times \frac{\sqrt{3}}{2}$$

   The cartesian form of the complex number is:

$$\frac{5}{2} + j.\frac{5\sqrt{3}}{2}$$

4. Convert $11(\cos 224° + j \sin 224°)$ to cartesian form.
   19.1 becomes:

$$x = 11 \cos 224° = -7.9127378 \approx -7.91$$

   19.2 becomes:

$$y = 11 \sin 224° = -7.6412421 \approx -7.64$$

   giving the cartesian form:

$$z = -7.91 - j.7.64$$

5. Convert $14(\cos 4.52 + j.\sin 4.52)$ to cartesian form where the angle is measured in radians.
   19.1 becomes:

$$x = 14 \cos 4.52 = 14(-0.1912043) = -2.6768608$$

19.2 becomes:

$$y = 14\sin 4.52 = 14(-0.9815503) = -13.741704$$

$$x \approx -2.677,\ y \approx -13.742$$

The cartesian form is:

$$-2.677 - j.13.742$$

Exercise 19.1

Convert the following to polar form:

1. $2+j.2$
2. $4+j.3$
3. $1+j.\sqrt{3}$
4. $3-j.5$
5. $5+j.12$
6. $24-j.7$
7. $-8-j.15$
8. $6+j.7$
9. $a+j.a$
10. $\frac{4}{5}-j.\frac{3}{5}$

Convert the following to cartesian form:

11. $4(\cos\pi/4+j.\sin\pi/4)$
12. $5(\cos\pi/6-j.\sin\pi/6)$
13. $10(\cos 7\pi/4+j\sin 7\pi/4)$
14. $5(\cos 5\pi/4+j\sin 5\pi/4)$
15. $2(\cos 5\pi/6+j\sin 5\pi/6)$
16. $8(\cos 2\pi/3+j\sin 2\pi/3)$
17. $16(\cos 5\pi/3+j\sin 5\pi/3)$
18. $2(\cos 15^\circ+j\sin 15^\circ)$
19. $5(\cos 103^\circ+j\sin 103^\circ)$
20. $6(\cos 249.6^\circ+j\sin 249.6^\circ)$
21. $17(\cos 306^\circ+j\sin 306^\circ)$

19.3 The modulus and argument of a complex number

Fig. 19.1 shows $OP = r = \sqrt{x^2+y^2}$. This is the *modulus of* z, or *mod* z, or $|z|$. It also shows θ where $\tan\theta = y/x$, or $\cos\theta = x/r$ and $\sin\theta = y/r$. θ is the *argument of z*, or *arg z*. In other words, where $z = x + j.y$:

$$\text{mod}\, z, \text{ or } |z| = \sqrt{x^2 + y^2} \qquad 19.7$$

$$\arg z = \text{arc tan}\, y/x, \text{ i.e. the angle whose tangent is } y/x \qquad 19.8$$

In fact, by using the vector interpretation of a complex number, when we write:

$$z = r(\cos\theta + j\sin\theta) = [r, \theta]$$

we are using a notation for complex numbers corresponding to the magnitude and direction notation for a vector. When we write:

$$z = x + jy = (x, y)$$

we are using a notation for complex numbers corresponding to the components of a vector notation.

This modulus-argument notation enables us to illustrate multiplication and division of complex numbers on an Argand diagram. Note that we did not try to do this with the cartesian notation. In Fig. 19.2 P_1 and P_2 represent z_1 and z_2.

$$z_1 = r_1(\cos\theta_1 + j\sin\theta_1) \qquad 19.9$$

$$z_2 = r_2(\cos\theta_2 + j\sin\theta_2) \qquad 19.10$$

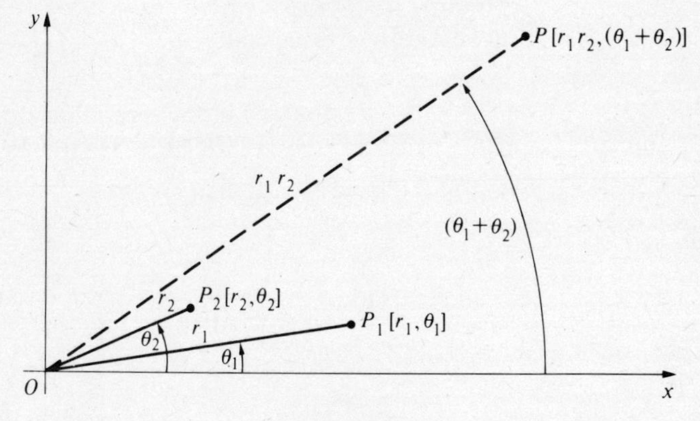

Figure 19.2

Multiply *19.9* by *19.10*:

$$\begin{aligned} z_1 z_2 &= r_1 r_2(\cos\theta_1 + j\sin\theta_1)(\cos\theta_2 + j\sin\theta_2) \\ &= r_1 r_2[(\cos\theta_1 . \cos\theta_2 - \sin\theta_1 . \sin\theta_2) \\ &\quad + j(\sin\theta_1 . \cos\theta_2 + \sin\theta_2 . \cos\theta_1)] \\ &= r_1 r_2[\cos(\theta_1 + \theta_2) + j\sin(\theta_1 + \theta_2)] \qquad 19.11 \end{aligned}$$

In other words:

$$|z_1 z_2| = r_1 r_2 = |z_1| \times |z_2| \qquad 19.12$$

and

$$\arg z_1 z_2 = \theta_1 + \theta_2 = \arg z_1 + \arg z_2 \qquad 19.13$$

Formulae *19.11* and *19.12* show how to construct a point (P) representing $z_1 z_2$ when the points P_1 and P_2, representing z_1 and z_2 respectively, are known.

Step 1 Draw OP such that it makes angle $X\hat{O}P = X\hat{O}P_1 + X\hat{O}P_2$.
Step 2 Measure $OP = OP_1 \times OP_2$.

Examples

Determine the modulus and argument of each of the following complex numbers.

1. $z = -8 + j.15$

$$|z| = \sqrt{(-8)^2 + 15^2} = 17$$

$$\arg z = \arctan(15/-8) = \arctan\left(-\frac{15}{8}\right)$$

Because $x = -8$ and $y = 15$ the point z is in the second quadrant. Therefore $\arg z = 180° - 61.927513° = 118.07249° \approx 118.07°$, i.e.

$$z = 17(\cos 118.07° + j \sin 118.07°)$$

2. $1/z$, where $z = -8 + j.15$

$$1/z = \frac{\bar{z}}{|z|^2} = \frac{-8 - j.15}{17^2} = \left(-\frac{8}{17^2}\right) + j.\left(-\frac{15}{17^2}\right)$$

$$\text{mod}\,(1/z) = \sqrt{\left(-\frac{8}{17^2}\right)^2 + \left(-\frac{15}{17^2}\right)^2} = \frac{17}{17^2} = \frac{1}{17}$$

$$\arg(1/z) = \arctan[(-15/17^2)/(-8/17^2)] = \arctan(15/8)$$

Because $x = -8/17^2$ and $y = -15/17^2$ the point $1/z$ is in the third quadrant.
Then $\arg(1/z) = 180° + 61.927513° = 241.927513° \approx 241.93°$. Therefore:

$$\frac{1}{z} = \frac{\bar{z}}{|z|^2} = \frac{1}{17}(\cos 241.93° + j.\sin 241.93°)$$

3. $1/z$ where $z = r(\cos\theta + j\sin\theta)$

$$\bar{z} = r(\cos\theta - j\sin\theta) = r[\cos(-\theta) + j\sin(-\theta)]$$
$$= r[\cos(2\pi - \theta) + j\sin(2\pi - \theta)]$$

$$\frac{1}{z} = \frac{\bar{z}}{z^2} = \frac{r[\cos(-\theta) + j\sin(-\theta)]}{r^2} \text{ or } \frac{r[\cos(2\pi - \theta) + j\sin(2\pi - \theta)]}{r^2}$$

$$= \frac{1}{r}[\cos(-\theta) + j\sin(-\theta)] \text{ or } \frac{1}{r}[\cos(2\pi - \theta) + j\sin(2\pi - \theta)]$$

Therefore $\text{mod}\,(1/z) = 1/r = 1/|z| = 1/\text{mod}\, z$.
And $\arg\,(1/z) = -\theta = -\arg z$, or $2\pi - \theta = 2\pi - \arg z$.

Exercise 19.2

Determine the modulus and argument of each of the following complex numbers.

1. $3 - j.4$
2. $-4 - j.3$
3. $5 + j.12$
4. $12 - j.5$
5. $\frac{1}{2} + j.\frac{1}{2}$
6. $-3 - j.2$
7. $1 + j$
8. $1 - j$
9. $3 + j.2$
10. $3 - j.2$
11. $\bar{z}$ where $z = a[\cos(\pi - \theta) + j\sin(\pi - \theta)]$
12. $\bar{z}$ where $z = r[\cos 2\theta + j\sin 2\theta]$
13. $1/z$ where $z = 2(\cos\pi/3 + j\sin\pi/3)$
14. $1/z$ where $z = \dfrac{1}{\sqrt{2}}\left(\cos\dfrac{5\pi}{6} + j\sin\dfrac{5\pi}{6}\right)$
15. $(1 + j)(1 - j)$
16. $(2 + j.3)(2 - j.3)$
17. $2(\cos\pi/6 + j\sin\pi/6) \times 5(\cos\pi/3 + j\sin\pi/3)$
    Hint: use questions 12 and 13.
18. $4(\cos 5\pi/6 + j\sin 5\pi/6) \times 5(\cos 4\pi/3 + j\sin 4\pi/3)$
    Hint: use questions 12 and 13.
19. $a(\cos\theta + j\sin\theta).b(\cos 2\theta + j\sin 2\theta)$
20. $\dfrac{a}{2}(\cos 2\theta + j\sin 2\theta).3b[\cos(2\pi - \theta) + j\sin(2\pi - \theta)]$
21. $r(\cos\theta + j\sin\theta).\dfrac{1}{r}[\cos(-\theta) + j\sin(-\theta)]$

19.4 Multiplication and division of complex numbers in modulus-argument form

In the previous section rule *19.11* dealt with multiplication. That is the rule to apply. Rules *19.12* and *19.13* provide the rule for writing down the modulus and argument of the product of two complex numbers. Examples 2 and 3 above and questions 11, 12, 13, 14 and 21 in Exercise 19.2 provide an introduction to discovering the corresponding rules for division.

Examples

1. Express z_1/z_2 in polar form. Write down the modulus and argument when $z_1 = r_1(\cos\theta_1 + j\sin\theta_1)$ and $z_2 = r_2(\cos\theta_2 + j\sin\theta_2)$.

$$z_1/z_2 = z_1 \times \frac{1}{z_2} = z_1 \times \frac{\bar{z}_2}{|z_2|^2}$$

$$= \frac{r_1(\cos\theta_1 + j\sin\theta_1) \times r_2[\cos(-\theta_2) + j\sin(-\theta_2)]}{r_2^2}$$

$$= \frac{r_1}{r_2}(\cos\theta_1 + j\sin\theta_1)[\cos(-\theta_2) + j\sin(-\theta_2)]$$

$$= \frac{r_1}{r_2}[\cos(\theta_1 - \theta_2) + j\sin(\theta_1 - \theta_2)] \text{ (by } \mathit{19.11}) \qquad \mathit{19.14}$$

From *19.14*:

$$\text{mod}(z_1/z_2) = r_1/r_2 = \frac{|z_1|}{|z_2|} \qquad \mathit{19.15}$$

and

$$\arg(z_1/z_2) = \theta_1 - \theta_2 = \arg z_1 - \arg z_2 \qquad \mathit{19.16}$$

19.14 provides the rule for the division of two complex numbers in polar form. *19.15* and *19.16* provide the rules for writing down the modulus and argument of a quotient of two complex numbers in polar form.

2. Express $4(\cos 5\pi/6 + j\sin 5\pi/6) \div 3(\cos \pi/3 + j\sin \pi/3)$ in polar form.

$$\frac{4}{3}\left[\cos\left(\frac{5\pi}{6} - \frac{\pi}{3}\right) + j\sin\left(\frac{5\pi}{6} - \frac{\pi}{3}\right)\right]$$

$$= \frac{4}{3}(\cos \pi/2 + j\sin \pi/2)$$

This reduces to:

$$\frac{4}{3}(0+j) = \frac{4}{3}j$$

3. Express $a(\cos 3\theta + j\sin 3\theta) \div \frac{1}{a}[\cos(-\theta) + j\sin(-\theta)]$

$$\left(a \div \frac{1}{a}\right)[\cos(3\theta+\theta) + j\sin(3\theta+\theta)] = a^2(\cos 4\theta + j\sin 4\theta)$$

Exercise 19.3

Express the following in polar form:

1. $3(\cos \pi/3 + j\sin \pi/3) \div 2(\cos \pi/6 + j\sin \pi/6)$
2. $4(\cos 2\pi/3 + j\sin 2\pi/3) \div 5(\cos \pi/2 + j\sin \pi/2)$
3. $7(\cos 7\pi/6 + j\sin 7\pi/6) \div 11(\cos 2\pi/3 + j\sin 2\pi/3)$
4. $6(\cos 3\pi/4 + j\sin 3\pi/4) \div 2(\cos 5\pi/4 + j\sin 5\pi/4)$
5. $9(\cos\theta + j\sin\theta) \div 3(\cos\theta/2 + j\sin\theta/2)$
6. $25(\cos 2\theta + j\sin 2\theta) \div 5[\cos(-\theta) + j\sin(-\theta)]$
7. Prove $a(\cos\theta + j\sin\theta) \times a(\cos\theta + j\sin\theta) = a^2(\cos 2\theta + j\sin 2\theta)$
8. Prove $r(\cos\phi/2 + j\sin\phi/2) \times r(\cos\phi/2 + j\sin\phi/2) = r^2(\cos\phi + j\sin\phi)$
9. Prove $[a(\cos\theta + j\sin\theta)]^2 = a^2(\cos 2\theta + j\sin 2\theta)$
   Hint: use question 7.
10. Prove $[r(\cos\phi/2 + j\sin\phi/2)]^2 = r^2(\cos\phi + j\sin\phi)$
    Hint: use question 8.
11. Prove $[\sqrt{r}(\cos\phi/2 + j\sin\phi/2)]^2 = r(\cos\phi + j\sin\phi)$
    Use question 10.

19.5 The square roots of a complex number

It might appear from question 11 in Exercise 19.3 that, by taking the square root of both sides, we obtain:

$$\sqrt{r(\cos\phi + j\sin\phi)} = \sqrt{r}(\cos\phi/2 + j\sin\phi/2) \qquad 19.17$$

This is only half the truth. Before we are able to proceed we must understand an important aspect of the polar form of a complex number. When a number may be written:

$$a(\cos\theta + j\sin\theta)$$

it may also be written:

$$a\cos(2\pi+\theta)+j\sin(2\pi+\theta)$$
$$\text{or}\quad a\cos(4\pi+\theta)+j\sin(4\pi+\theta)$$
$$\text{or}\quad a\cos(2n\pi+\theta)+j\sin(2n\pi+\theta)$$

where n is an integer, positive or negative. This follows from the Argand diagram. Whatever position P takes up to represent z it will take up the same position when rotated a further number of exact revolutions round O. To obtain the complete solution of the square root of $r(\cos\theta + j\sin\theta)$ we first convert the expression to:

$$r[\cos(2n\pi+\theta)+j\sin(2n\pi+\theta)]$$

The square root is then:

$$\sqrt{r}\,[\cos\tfrac{1}{2}(2n\pi+\theta)+j\sin\tfrac{1}{2}(2n\pi+\theta)]\ \text{(using } \mathit{19.17}\text{)}$$

(a) Now put $n = 0$. One answer is $\sqrt{r}\,[\cos(\theta/2)+j\sin(\theta/2)]$, by *19.17*
(b) put $n = 1$. Another answer is $\sqrt{r}\,[\cos(\pi+\theta/2)+j\sin(\pi+\theta/2)]$
(c) put $n = 2$, giving $\sqrt{r}\,[\cos(2\pi+\theta/2)+j\sin(2\pi+\theta/2)]$
(d) put $n = 3$, giving $\sqrt{r}\,[\cos(2\pi+\pi+\theta/2)+j\sin(2\pi+\pi+\theta/2)]$

(c) and (d) are repetitions of (a) and (b). By putting n = every conceivable integer, positive or negative, we shall merely repeat the answers in (a) and (b). *This means that a complex number has two square roots.* Fig. 19.3 gives the Argand diagram representation of a complex number and its two square roots. If P represents the complex number then P_1 and P_2 represent the square roots of that number.

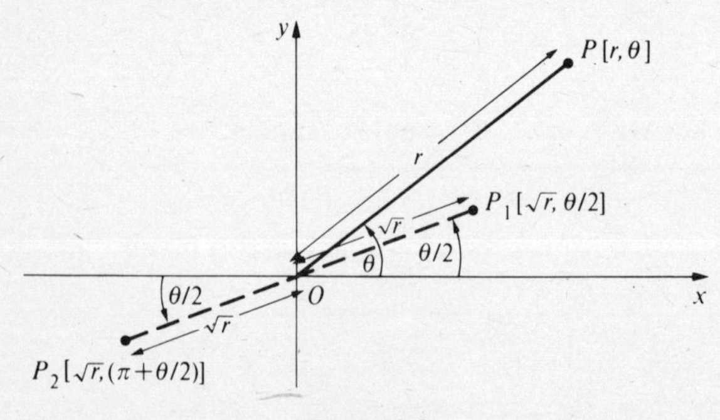

Figure 19.3

An interesting special case is provided when $\theta = 0$, i.e. when P lies on Ox. Fig. 19.4 represents this case. Here P_1 and P_2 also lie on Ox, one at each side of O. This case corresponds to the square roots of a real number. The square roots are $\pm\sqrt{r}$.

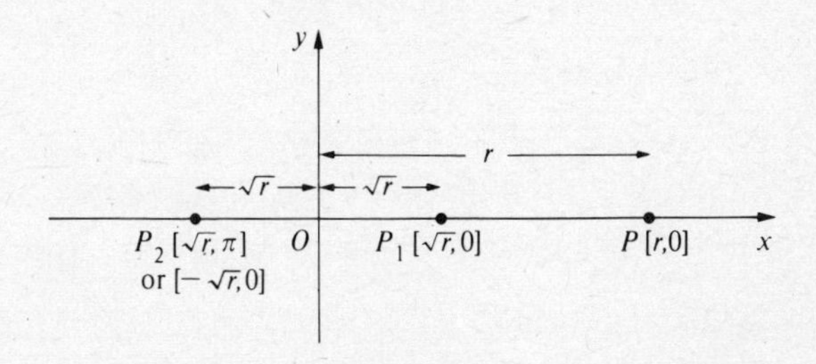

Figure 19.4

Examples

Find the square roots of the following.

1. $4(\cos \pi/3 + j\sin \pi/3)$
   Write:

$$\begin{aligned} z &= 4(\cos \pi/3 + j\sin \pi/3) \\ &= 4[\cos(2n\pi + \pi/3) + j\sin(2n\pi + \pi/3)] \end{aligned}$$

   Then:

$$\begin{aligned} \sqrt{z} &= \sqrt{4}[\cos\tfrac{1}{2}(2n\pi + \pi/3) + j\sin\tfrac{1}{2}(2n\pi + \pi/3)] \\ &= 2[\cos \pi/6 + j\sin \pi/6], \text{ when } n = 0 \\ &= 2(\cos 7\pi/6 + j\sin 7\pi/6), \text{ when } n = 1 \end{aligned}$$

   The roots may be written:

$$2(\sqrt{3}/2 + j.\tfrac{1}{2}) \text{ and } 2(-\sqrt{3}/2 - j.\tfrac{1}{2})$$

2. $8(\cos 3\pi/4 - j\sin 3\pi/4)$
   Put:

$$\begin{aligned} z &= 8[\cos(-3\pi/4) + j\sin(-3\pi/4)] \\ &= 8[\cos 5\pi/4 + j\sin 5\pi/4] \\ &= 8\left[\cos\left(2n\pi + \frac{5\pi}{4}\right) + j\sin\left(2n\pi + \frac{5\pi}{4}\right)\right] \end{aligned}$$

Then $\sqrt{z} = \sqrt{8}\left[\cos\frac{1}{2}\left(2n\pi + \frac{5\pi}{4}\right) + j\sin\frac{1}{2}\left(2n\pi + \frac{5\pi}{4}\right)\right]$

Put $n = 0$. One root $= 2\sqrt{2}(\cos 5\pi/8 + j\sin 5\pi/8)$

Put $n = 1$. Another root $= 2\sqrt{2}(\cos 13\pi/8 + j\sin 13\pi/8)$

3. -4

Write:

$$\begin{aligned} z = -4 &= 4(-1 + j.0) \\ &= 4(\cos\pi + j\sin\pi) \\ &= 4[\cos(2n+1)\pi + j\sin(2n+1)\pi] \\ \sqrt{z} &= 2[\cos\tfrac{1}{2}(2n+1)\pi + j\sin\tfrac{1}{2}(2n+1)\pi] \end{aligned}$$

Put $n = 0$: One root is $2(\cos\pi/2 + j\sin\pi/2) = 2(0 + j) = j.2$

Put $n = 1$: Another root is $2(\cos 3\pi/2 + j\sin 3\pi/2) = 2(0 - j) = -j.2$

The square roots of -4 are $\pm j.2$.

This is also obvious because $-4 = (-1)\times 4 = j^2.4$

Then $-4 = \pm j.2$

4. 8

Write $z = 8 = 8(1 + j.0) = 8(\cos 2n\pi + j\sin 2n\pi)$

Then $\sqrt{z} = \sqrt{8}(\cos n\pi + j\sin n\pi)$

Put $n = 0$: One root is $2\sqrt{2}(\cos 0 + j\sin 0) = 2\sqrt{2}$

Put $n = 1$: Another root is $2\sqrt{2}(\cos\pi + j\sin\pi) = -2\sqrt{2}$

Exercise 19.4

Determine the square roots of the following.

1. -1
2. 1
3. 25
4. -64
5. 6
6. -10
7. $25(\cos\pi/3 + j\sin\pi/3)$
8. $64(\cos 2\pi/3 + j\sin 2\pi/3)$
9. $125(\cos 2\pi/3 - j\sin 2\pi/3)$
10. $40(\cos 3\pi/2 - j\sin 3\pi/2)$
11. $a^2(\cos 2\theta + j\sin 2\theta)$
12. $b(\cos\theta - j\sin\theta)$
13. $a^2[\cos(\pi + \theta) + j\sin(\pi + \theta)]$

14. Determine z such that $z = 1/z$. (Hint: rearrange and then apply the method above.)
15. Determine $z = 4/z$
16. Determine z such that $z + 2/z = 0$
17. Determine z such that $z = \dfrac{3(\cos 5\pi/3 + j\sin 5\pi/3)}{z}$

20

Graphs of Sine and Cosine Functions

20.1 Approximations for sin x, cos x and tan x when x is small

In Fig. 20.1 O represents the centre of a circle of radius r. A and B are two points on the circle such that angle $A\hat{O}B = x$ radians. AT is the tangent to the circle at A. Then:

area triangle $OAB <$ area sector $OAB <$ area triangle OAT

i.e. $\frac{1}{2}OA.OB \sin x < \frac{1}{2}r^2.x < \frac{1}{2}OA.OT$

$\frac{1}{2}r^2 \sin x < \frac{1}{2}r^2 x < \frac{1}{2}r^2 \tan x$

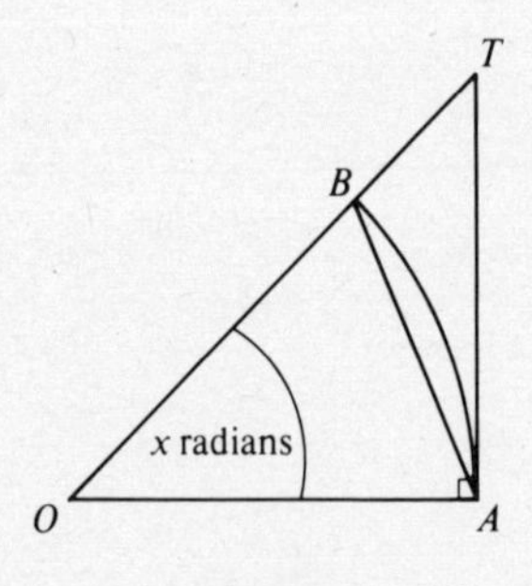

Figure 20.1

Divide throughout by $\frac{1}{2}r^2$:

$$\sin x < x < \tan x$$

Divide throughout by x (supposed positive):

$$\frac{\sin x}{x} < 1 < \frac{\tan x}{x}$$

Therefore $$\frac{\sin x}{x} < 1 \qquad 20.1$$

and $$\frac{\tan x}{x} > 1 \qquad 20.2$$

Multiply *20.2* by $\cos x$, which is positive for small positive x:

$$\frac{\sin x}{x} > \cos x \qquad 20.3$$

From *20.1* and *20.3*:

$$\cos x < \frac{\sin x}{x} < 1$$

As $x \to 0$, $\cos x \to 1$. Therefore:

$$\frac{\sin x}{x} \to 1$$

i.e. for small x, measured in radians,

$$\sin x \approx x \qquad 20.4$$

Divide *20.1* by $\cos x$:

$$\frac{\sin x/\cos x}{x} < \frac{1}{\cos x}$$

$$\text{i.e.} \quad \frac{\tan x}{x} < \sec x \qquad 20.5$$

As $x \to 0$, $\sec x \to 1$. From *20.2* and *20.5*:

$$1 < \frac{\tan x}{x} < \sec x$$

Therefore:

$$\frac{\tan x}{x} \to 1$$

i.e. $\tan x \to x$

or $\tan x \approx x$ $\qquad$ *20.6*

when x is small and measured in radians. In fact, for small x, from *20.1*, $\sin x$ is always slightly less than x. And, from *20.2*, $\tan x$ is always slightly greater than x. For all x and t measured in radians:

$$\int_0^t \sin x \, . \, dx = \Big[-\cos x\Big]_0^t \qquad \text{(i)}$$

when x and t are both small. From (i), using *20.4* as an approximation for $\sin x$:

$$\Big[-\cos x\Big]_0^t \approx \int_0^t x \, . \, dx$$

$$\text{i.e.} \quad -\cos t + \cos 0 \approx \left[\frac{1}{2}x^2\right]_0^t = \frac{1}{2}t^2$$

$$\text{i.e.} \quad -\cos t + 1 \approx \tfrac{1}{2}t^2$$

$$\text{or} \quad \cos t \approx 1 - \tfrac{1}{2}t^2$$

Therefore, for small x, $\cos x \approx 1 - \frac{1}{2}x^2$

20.2 The graphs of sines and cosines of x, $2x$ and $\frac{1}{2}x$ between $x = 0°$ and $x = 360°$

In Levels 1 and 2 graphs of $\sin x$, $\tan x$ and $\cos x$ were drawn or sketched between $x = 0°$ and $x = 360°$. Therefore we have some idea of the nature of the curves. As a preliminary to a closer look at two of those curves and others related to them, we shall construct a table of values obtained from trigonometric tables or by calculator (Table 20.1).

Fig. 20.2 represents sketches of the graphs of $y = \sin x$, $y = \sin 2x$ and $y = \sin \frac{1}{2}x$. The sketches also show the relative variations of the three graphs. The following points emerge:

1. All three graphs oscillate between the values $y = 1$ and $y = -1$. There is a central line $y = 0$ running through each one. The fluctuation between the central value and the peak value is called the amplitude for the curve. In each case the amplitude is 1.
2. Each curve goes through a certain cycle of values and then repeats that cycle over and over again. The cycle for $y = \sin x$ is 360°; the cycle for $y = \sin 2x$ is 180°; and the cycle for $y = \sin \frac{1}{2}x$ is 720°.

Table 20.1

x	$\sin x$	$\cos x$	$2x$	$\sin 2x$	$\cos 2x$	$\frac{1}{2}x$	$\sin \frac{1}{2}x$	$\cos \frac{1}{2}x$
0°	0.0000	1.0000	0°	0.0000	1.0000	0°	0.0000	1.0000
10°	0.1736	0.9848	20°	0.3420	0.9397	5°	0.0872	0.9962
20°	0.3420	0.9397	40°	0.6428	0.7660	10°	0.1736	0.9848
30°	0.5000	0.8660	60°	0.8660	0.5000	15°	0.2588	0.9659
40°	0.6428	0.7660	80°	0.9848	0.1736	20°	0.3420	0.9397
50°	0.7660	0.6428	100°	0.9848	−0.1736	25°	0.4226	0.9063
60°	0.8660	0.5000	120°	0.8660	−0.5000	30°	0.5000	0.8660
70°	0.9397	0.3420	140°	0.6428	−0.7660	35°	0.5736	0.8192
80°	0.9848	0.1736	160°	0.3420	−0.9397	40°	0.6428	0.7660
90°	1.0000	0.0000	180°	0.0000	−1.0000	45°	0.7071	0.7071
100°	0.9848	−0.1736	200°	−0.3420	−0.9397	50°	0.7660	0.6428
110°	0.9397	−0.3420	220°	−0.6428	−0.7660	55°	0.8191	0.5736
120°	0.8660	−0.5000	240°	−0.8660	−0.5000	60°	0.8660	0.5000
130°	0.7660	−0.6428	260°	−0.9848	−0.1736	65°	0.9063	0.4226
140°	0.6428	−0.7660	280°	−0.9848	0.1736	70°	0.9397	0.3420
150°	0.5000	−0.8660	300°	−0.8660	0.5000	75°	0.9660	0.2588
160°	0.3420	−0.9397	320°	−0.6428	0.7660	80°	0.9848	0.1736
170°	0.1736	−0.9848	340°	−0.3420	0.9397	85°	0.9962	0.0872
180°	0.0000	−1.0000	360°	0.0000	1.0000	90°	1.0000	0.0000
190°	−0.1736	−0.9848				95°	0.9962	−0.0872
200°	−0.3420	−0.9397				100°	0.9848	−0.1736
210°	−0.5000	−0.8660				105°	0.9662	−0.0872
220°	−0.6428	−0.7660				110°	0.9397	−0.3420
230°	−0.7660	−0.6428				115°	0.9063	−0.4226
240°	−0.8660	−0.5000				120°	0.8660	−0.5000
250°	−0.9397	−0.3420				125°	0.8192	−0.5736
260°	−0.9848	−0.1736				130°	0.7660	−0.6428
270°	−1.0000	0.0000				135°	0.7071	−0.7071
280°	−0.9848	0.1736				140°	0.6428	−0.7660
290°	−0.9397	0.3420				145°	0.5736	−0.8192
300°	−0.8660	0.5000				150°	0.5000	−0.8660
310°	−0.7660	0.6428				155°	0.4226	−0.9063
320°	−0.6428	0.7660				160°	0.3420	−0.9397
330°	−0.5000	0.8660				165°	0.2588	−0.9659
340°	−0.3420	0.9397				170°	0.1736	−0.9848
350°	−0.1736	0.9848				175°	0.0872	−0.9962
360°	0.0000	1.0000				180°	0.0000	−1.0000

3. Each cycle can be divided into four quadrants: 1, 2, 3 and 4. In quadrants 1 and 2 the curve is above the central line, while in quadrants 3 and 4 the curve is below the central line.
4. Note:

   The cycle for $y = \sin x$ is $\frac{360°}{1}$, where 1 is the multiple of x in $\sin x$.

   The cycle for $y = \sin 2x$ is $\frac{360°}{2}$, where 2 is the multiple of x in $\sin 2x$.

   The cycle for $y = \sin \frac{1}{2}x$ is $\frac{360°}{\frac{1}{2}}$, where $\frac{1}{2}$ is the multiple of x in $\sin \frac{1}{2}x$.

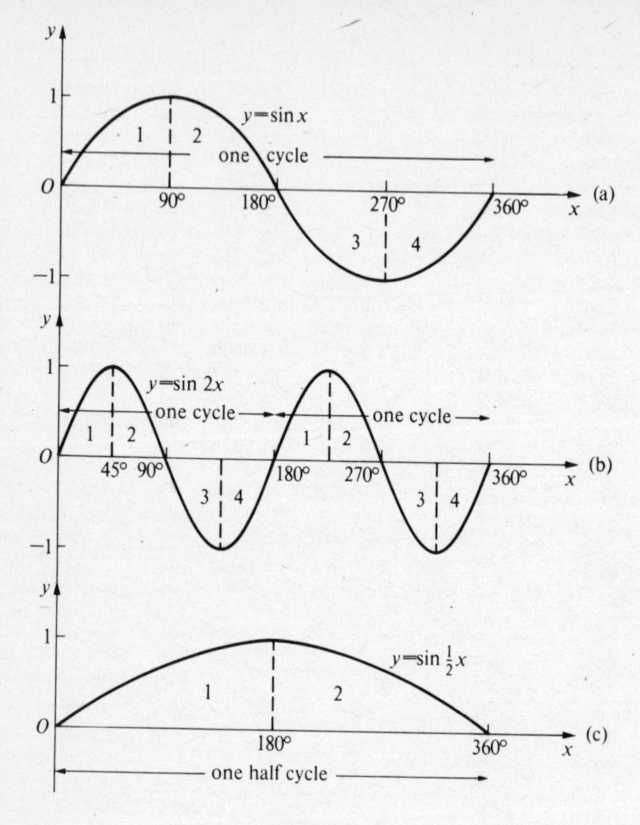

Figure 20.2

The inferences about cycles and amplitudes for Fig. 20.2 apply correspondingly to Fig. 20.3. From this it appears that the graphs of $y = \sin 3x$ and $y = \cos 3x$ must have cycles of $360°/3 = 120°$ and that their amplitudes must be 1.

Exercise 20.1

1. Check that the cycle of $\sin 3x$ is 120° by calculating the values of $\sin 3x$, using a calculator, for values of x from 0° to 360° in steps of 15°.
2. Check that the cycle of $\sin \frac{1}{3}x$ is 1080° by calculating the values of $\sin \frac{1}{3}x$, using a calculator, for values of x from 0° to 1080° in steps of 90°.
3. Check that the cycle of $\sin 4x$ is 90° by calculating the values of $\sin 4x$ for values of x from 0° to 90° in steps of 7.5°.
4. Check by calculation that the cycle of $\sin \frac{1}{4}x$ is 1440°.
5. In each of the cases above check that the amplitude is 1.

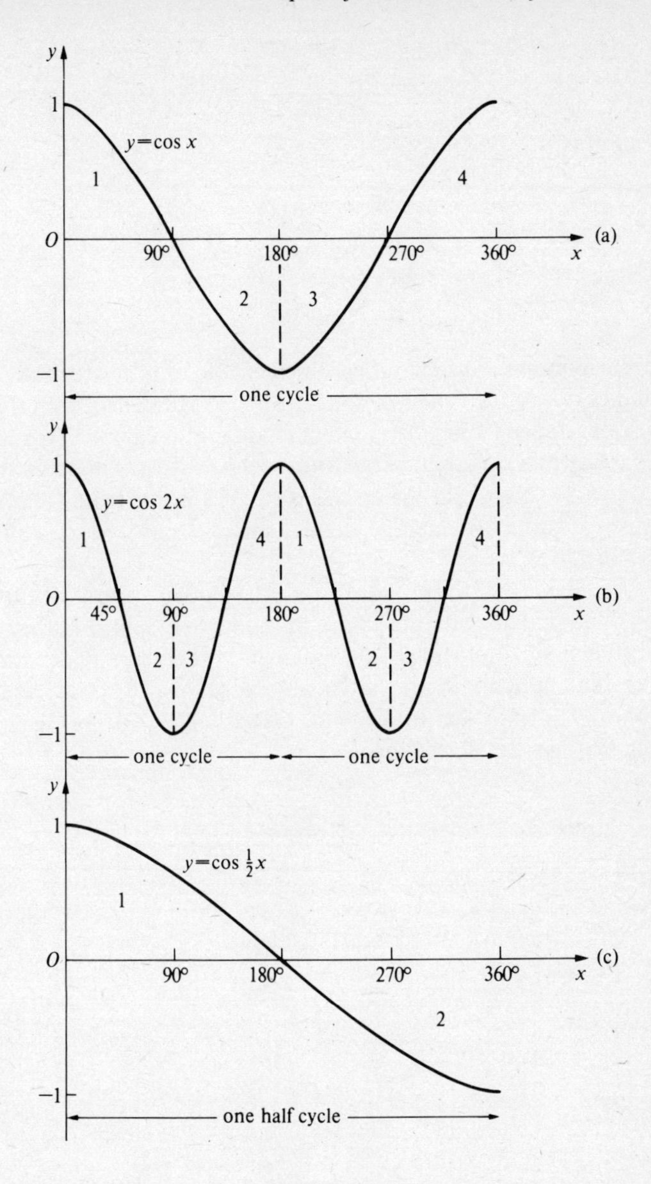

Figure 20.3

From the above it appears that the graph of $y = \sin nx$ will have a cycle equal to the value of x which makes:

$$nx = 360^\circ$$
$$\text{i.e.} \quad x = 360^\circ/n$$

Where x is measured in radians then

$$x = 2\pi/n$$

represents the cycle for the curve $y = \sin nx$.

20.3 The graphs of $\sin^2 x$ and $\cos^2 x$

There are trigonometrical identities which enable us to relate the values of $\sin^2 x$ and $\cos^2 x$ to the values of $\cos 2x$. They are obtained from the basic identities mentioned in the introductory chapter of the book. Alternatively, Table 20.2 provides a method of relating the graphs to one another and also to that of $\cos 2x$. Note that the entries for $\cos^2 x$ in the table are obtained from those for $\sin^2 x$ by using the identity $\sin^2 x + \cos^2 x = 1$, for all x; this appeared in Level 2.

Note that Table 20.2 stops short at $x = 180°$. For values of x between 180° and 360° $\sin x$ repeats the values in the table except that they are all negative. When these values are squared to give the column under $\sin^2 x$ the values in that column in the table are repeated. Fig. 20.4 represents the graph of $y = \sin^2 x$ between $x = 0°$ and $x = 360°$. From Fig. 20.4 the following features are determined:

Table 20.2

x	$2x$	$\sin x$	$\sin^2 x$	$\cos^2 x$	$\cos 2x$	$\frac{1}{2}(1 - \cos 2x)$
0°	0°	0.0000	0.0000	1.0000	1.0000	0.0000
10°	20°	0.1736	0.030137	0.9699	0.9397	0.0302
20°	40°	0.3420	0.1170	0.8830	0.7660	0.1170
30°	60°	0.5000	0.2500	0.7500	0.5000	0.2500
40°	80°	0.6428	0.4132	0.5868	0.1736	0.4132
50°	100°	0.7660	0.5868	0.4132	−0.1736	0.5868
60°	120°	0.8660	0.7500	0.2500	−0.5000	0.7500
70°	140°	0.9397	0.8830	0.1170	−0.7660	0.8830
80°	160°	0.9848	0.9698	0.0302	−0.9397	0.9698
90°	180°	1.0000	1.0000	0.0000	−1.0000	1.0000
100°	200°	0.9848	0.9698	0.0302	−0.9397	0.9698
110°	220°	0.9397	0.8830	0.1170	−0.7660	0.8830
120°	240°	0.8660	0.7500	0.2500	−0.5000	0.7500
130°	260°	0.7660	0.5868	0.4132	−0.1736	0.5868
140°	280°	0.6428	0.4132	0.5868	0.1736	0.4132
150°	300°	0.5000	0.2500	0.7500	0.5000	0.2500
160°	320°	0.3420	0.1170	0.8830	0.7660	0.1170
170°	340°	0.1736	0.0301	0.9699	0.9397	0.0302
180°	360°	0.0000	0.0000	1.0000	1.0000	0.0000

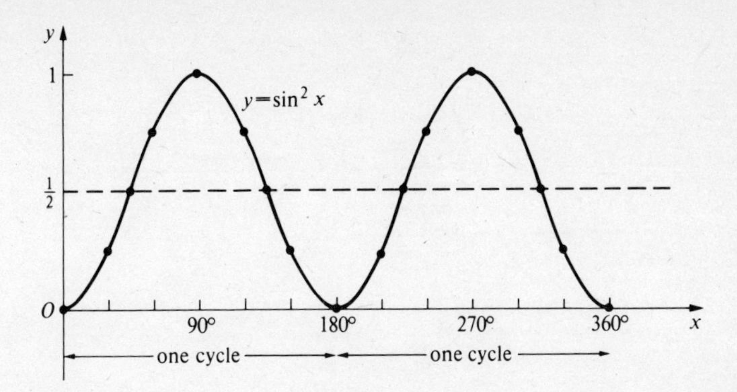

Figure 20.4

1. The central value of y is $\frac{1}{2}$.
2. The peak values of y are all 1.
3. The minimum values of y are all 0.
4. The curve oscillates between $y = 0$ and $y = 1$.
5. The amplitude is $\frac{1}{2}$.
6. The cycle is 180°.

Fig. 20.5 represents the graph of $y = \cos^2 x$. Since when $y = \cos^2 x$ it also equals $1 - \sin^2 x$ the curve might have been obtained in two stages.
Step 1 From Fig. 20.4 sketch the graph of $y = -\sin^2 x$, i.e. merely make all values of y in Fig. 20.4 negative. In other words, reflect the curve in Ox, Fig. 20.6.

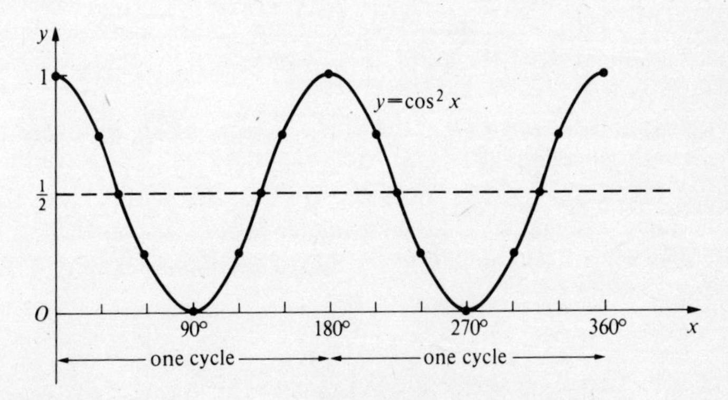

Figure 20.5

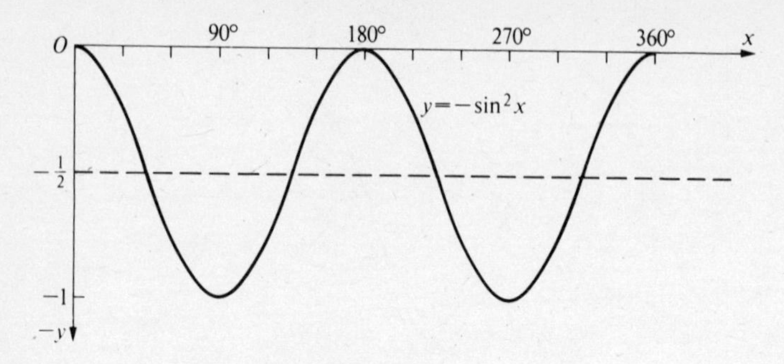

Figure 20.6

Step 2 From Fig. 20.6 obtain $y = 1 - \sin^2 x = -\sin^2 x + 1$ by adding 1 to every ordinate of the curve $y = -\sin^2 x$. In other words, shift the whole of the curve, Fig. 20.6, by moving every point one unit parallel to Oy.

Exercise 20.2

1. Using the values in the column under $\cos 2x$ in Table 20.2 determine another set of values, this time for $\frac{1}{2}(1 + \cos 2x)$. Compare the values with the corresponding values in the $\cos^2 x$ column.
2. By rearranging the equation $y = \cos^2 x$ in the form:

$$\begin{aligned} y &= \tfrac{1}{2}(1 + \cos 2x) \\ &= \tfrac{1}{2} + \tfrac{1}{2}\cos 2x \end{aligned}$$

   use the results of section 20.2 to determine: (a) the cycle; (b) the amplitude; (c) the central value of y; (d) the peak values of y; and (e) the minimum values of y; and check that the values agree with those obtained direct from the graph in Fig. 20.5.

Suppose now that the angles x in the previous sections are produced by a rotating vector such as that in Fig. 20.7.

Fig. 20.7 represents a circle, centre O, with OA a fixed initial line. OP is a rotating radius moving at a constant rate anticlockwise round the circle. Suppose that OP rotates round O at the rate of ω radians per second. Then the angle x rotated from OA is given by:

$$x = \omega . t$$

The value ω is the angular velocity of OP about O. When the cycle of values of x is 360° or 2π radians the corresponding values of t change by an amount

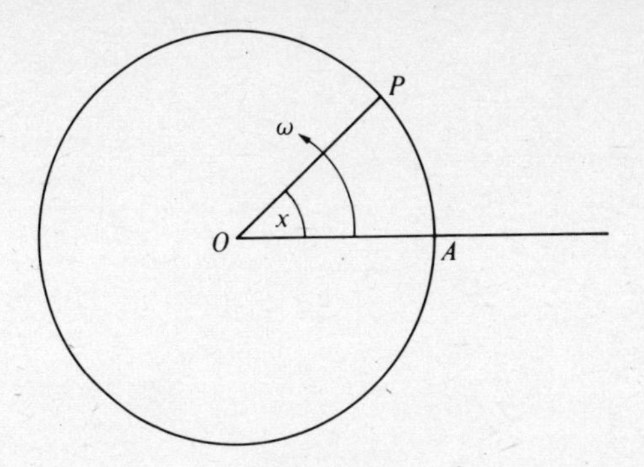

Figure 20.7

given by the relation:

$$2\pi = \omega . T$$

$$\text{i.e.} \quad T = \frac{2\pi}{\omega} \qquad \textit{20.7}$$

The value of T given in *20.7* is called the period of the function.

Examples

1. When the angular velocity is ω determine the periods of the following curves: (a) $y = \cos x$; (b) $y = \sin 2x$; (c) $y = \cos \frac{1}{2}x$; (d) $y = \sin 3x$; (e) $y = \cos ax$.

   (a) $T = \dfrac{2\pi}{\omega}$

   (b) $T = \dfrac{2\pi}{2\omega} = \dfrac{\pi}{\omega}$

   (c) $T = \dfrac{2\pi}{\frac{1}{2}\omega} = \dfrac{4\pi}{\omega}$

   (d) $T = \dfrac{2\pi}{3\omega}$

   (e) $T = \dfrac{2\pi}{a\omega}$

2. Determine the periods of the following curves where ω is constant: (a) $y = \cos 4\omega t$; (b) $y = \sin 3\omega t$; (c) $y = \sin \frac{1}{2}\omega t$; (d) $y = \sin a\omega t$.

(a) $T = \dfrac{2\pi}{4\omega} = \dfrac{\pi}{2\omega}$

(b) $T = \dfrac{2\pi}{3\omega}$

(c) $T = \dfrac{2\pi}{\frac{1}{2}\omega} = \dfrac{4\pi}{\omega}$

(d) $T = \dfrac{2\pi}{a\omega}$

3. Sketch the graph of $y = \sin \omega t$.
Fig. 20.8 is obtained from Fig. 20.2(a) by replacing the axis of x by the axis of t. The values along the t axis are obtained from those of the x axis by dividing by ω after converting the angle measure of degrees into radians. In particular, the value 360° becomes, along the t axis, $2\pi/\omega$ seconds. Note that whereas the unit along Ox is measured in degrees the unit along the t axis is measured in seconds.

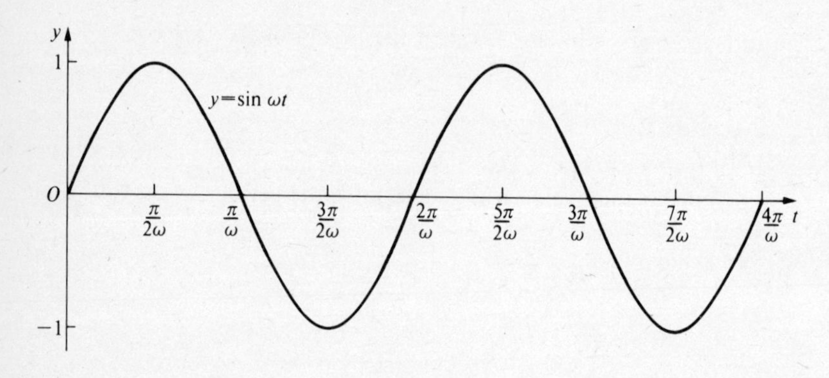

Figure 20.8

4. Sketch the graph of $y = \cos^2 \omega t$.
Fig. 20.9 represents the graph of $y = \cos^2 \omega t$. It is obtained from that of $y = \cos^2 x$ by similar principles to those used in example 3.

Exercise 20.3

Sketch the graphs of the following curves. In each case note the amplitude and the period.

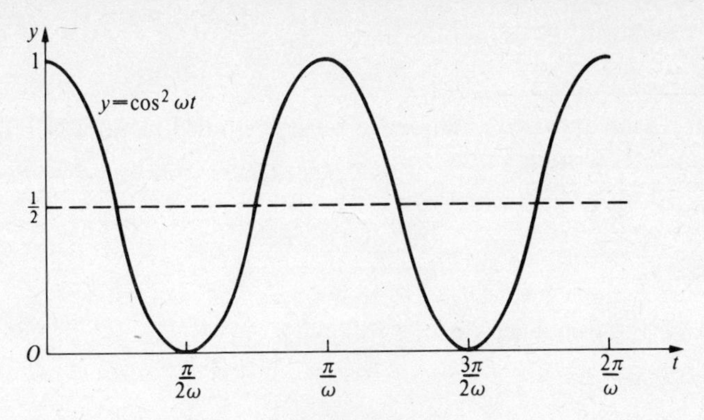

Figure 20.9

1. $y = \cos 2\omega t$
2. $y = \sin 3\omega t$
3. $y = \sin \frac{1}{2}\omega t$
4. $y = \cos \frac{1}{2}\omega t$
5. $y = \sin^2 \omega t$
6. $y = \sin 4\omega t$
7. $y = 2 \sin \omega t$
8. $y = 3 \cos \omega t$
9. $y = a \sin \frac{1}{2}\omega t$
10. $y = p \sin^2 \omega t$

Frequency

The curve $y = \sin \omega t$ has a period $T = 2\pi/\omega$ seconds. In other words, it takes $2\pi/\omega$ seconds to cover a whole cycle of values of y. In applications of this principle to wave motions, vibrations, and alternating currents the period will, more often than not, be less than a second. For instance, in the alternating current of power supply the period is 1/50 of a second. Consequently there are a number of periods in a second: in fact there are 50. The number of periods in a second, then, is:

$$\frac{1}{2\pi/\omega} = \frac{\omega}{2\pi} \qquad \textit{20.8}$$

The expression in *20.8* is called the frequency, usually denoted by the symbol f. With that notation *20.8* becomes:

$$f = \frac{\omega}{2\pi}$$

By rearrangement we obtain:

$$\omega = 2\pi . f$$

For that reason a revised notation for these periodic functions is frequently used. Sin ωt is written $\sin 2\pi ft$; $\cos 4\omega t$ is written $\cos 8\pi ft$; and so on.

Examples

Determine the frequencies of the following curves.

1. $y = 6 \sin 3\omega t.$
   $f = \frac{3\omega}{2\pi}$; the period $T = \frac{2\pi}{3\omega}$.
   Note that $f = 1/T$ and $T = 1/f$. Each of f and T is the reciprocal of the other.
2. $y = 5 \sin^2 \frac{1}{2}\omega t.$
   $f = \frac{\omega}{2\pi}$; $T = \frac{2\pi}{\omega}$.
   The amplitude is 5/2, since $y = 5\frac{1}{2}(1 - \cos \omega t)$
3. $y = \cos 6\pi t.$ (i)
   Rewrite the equation $y = \cos 2\pi . 3t.$ (ii)
   Then $f = 3$ from (ii), or $f = \frac{6\pi}{2\pi} = 3$ from (i).

Exercise 20.4

Determine the amplitudes, angular velocities, frequencies and periods of the following functions:

1. $y = \sin 2t$
2. $y = \cos \frac{1}{4}t$
3. $y = 3 \sin (1.5t)$
4. $y = \sin^2 (5t)$
5. $y = 3 \cos^2 4t + 8 \cos 8t$
6. $y = 4 \sin^2 \frac{1}{2}t + 5 \cos^2 \frac{1}{2}t$
7. $y = 4 + \sin 2t$
8. $y = 8 - \cos 3t$
9. $y = 5 - 6 \sin^2 \frac{1}{4}t$
10. $y = 11 + 12 \cos^2 (5.2t)$
11. $y = (\frac{5}{2} \cos \frac{3}{5} t)^2$
12. $y = a \sin^2 t + b \cos^2 t$

13. $y = a\sin^2 pt + b\cos^2 pt$
14. $y = a\sin^2 pt - b\cos^2 pt$
15. $y = a\sin^2 pt + 3a\cos 2pt$
16. $y = 4a + 6a\cos pt + \dfrac{3a}{4}\cos^2 \tfrac{1}{2}pt$

21

The Combination of Sine Waves

21.1 Rotating vectors

Fig. 21.1 represents a vector OP of fixed magnitude V which rotates about a fixed point O with a fixed rate of revolution in an anticlockwise sense about an axis through O perpendicular to the plane of the page.

In Fig. 21.1 OP represents an instantaneous position of the rotating vector when it has moved through an angle θ radians from OA. Such a rotating vector is called a phasor.

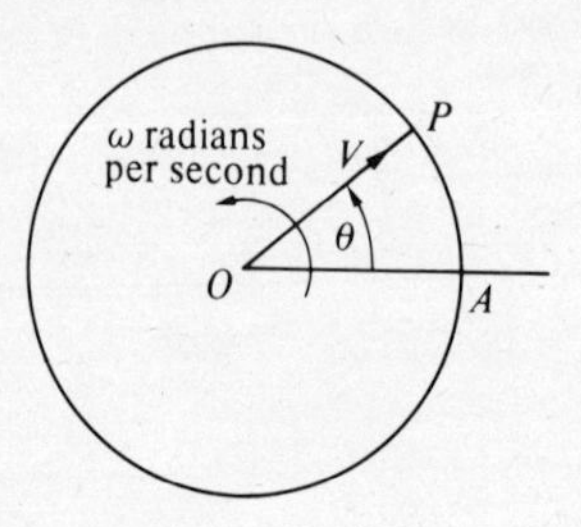

Figure 21.1

21.2 The relationship of a sine wave to a phasor

Fig. 21.2 is a sketch of the graph obtained by projecting the tip (P) of the phasor on the left onto axes on the right.

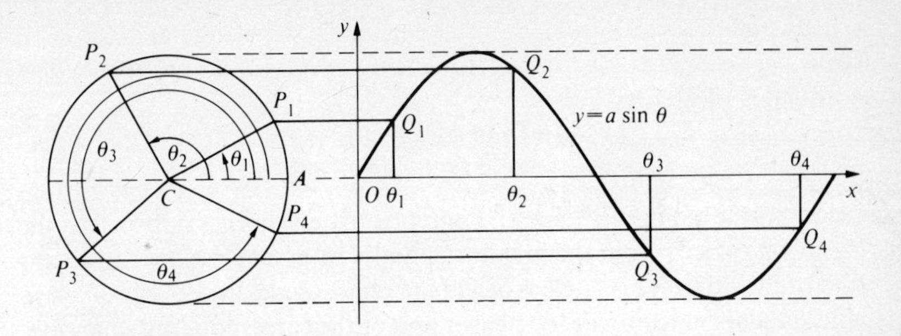

Figure 21.2

Suppose the radius of the circle, centre C, is a and the angular rotations from CA to CP_1, CP_2, CP_3 and CP_4 are θ_1, θ_2, θ_3 and θ_4 respectively. In relation to the axes on the right, Ox, Oy, Q_1 is the point $(\theta_1, a \sin \theta_1)$, Q_2 is the point $(\theta_2, a \sin \theta_2)$, Q_3 is the point $(\theta_3, a \sin \theta_3)$ and Q_4 is the point $(\theta_4, a \sin \theta_4)$. For the general point, P, on the circle, Q is $(\theta, a \sin \theta)$, or $(\omega t, a \sin \omega t)$, or $(2\pi ft, a \sin 2\pi ft)$, when OP rotates at ω radians per second, or f revolutions per second, i.e. f cycles per second, or f hertz.

Note:

1. Q is at O when P is at A.
2. The curve on the right is a complete sine wave.
3. The sine wave makes one complete cycle while P makes one complete revolution about O.
4. The amplitude of the sine wave $= a =$ the magnitude of the phasor.
5. The time taken to complete the cycle is called the period, i.e. T, the period, $= \dfrac{2\pi}{\omega} = \dfrac{2\pi}{2\pi f} = \dfrac{1}{f}$ seconds.

Fig. 21.3 represents a relation between the sine wave and the phasor when the cycle starts with P at B, i.e. ahead of A.

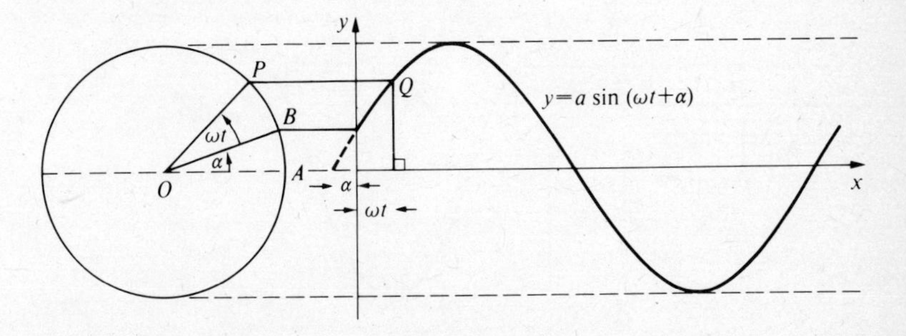

Figure 21.3

The two phasors in Figs 21.2 and 21.3 have the same angular velocity, ω radians per second, i.e. the same frequency, f revolutions per second. The co-ordinates of Q are:

$$[\omega t, a \sin(\omega t + \alpha)]$$
$$\text{or} \quad [2\pi f t, a \sin(2\pi f t + \alpha)]$$

The sine waves in Figs 21.2 and 21.3 differ from one another only in that the second is displaced from the first horizontally by an angle α. It is α radians ahead of the other. This angle is called the phase angle. Fig. 21.2 is taken to be the standard position of the phasor and of the corresponding sine wave. Any phasor different from that because of its displacement is said to have a phase difference.

21.3 The cosine wave and its phasor

Fig. 21.4 represents the relationship between a cosine wave and its associated phasor. The starting point of the cycle of the phasor is OC. The angular velocity of the phasor is ω radians per second or f revolutions per second. The co-ordinates of Q in relation to the axes of the graph are $(\omega t, a \cos \omega t)$ or $(2\pi f t, a \cos 2\pi f t)$. By Fig. 21.3 the phase angle in relation to $(\omega t, a \sin \omega t)$ is $\pi/2$ radians. Therefore an alternative notation for $(\omega t, a \cos \omega t)$, or $(2\pi f t, a \cos 2\pi f t)$, is:

$$\left[\omega t, a \sin\left(\omega t + \frac{\pi}{2}\right)\right] \quad \text{or} \quad \left[2\pi f t, a \sin\left(2\pi f t + \frac{\pi}{2}\right)\right]$$

The advantage of the latter notation is that all phasors are expressed in the standard form:

$$[\omega t, a \sin(\omega t + \alpha)]$$

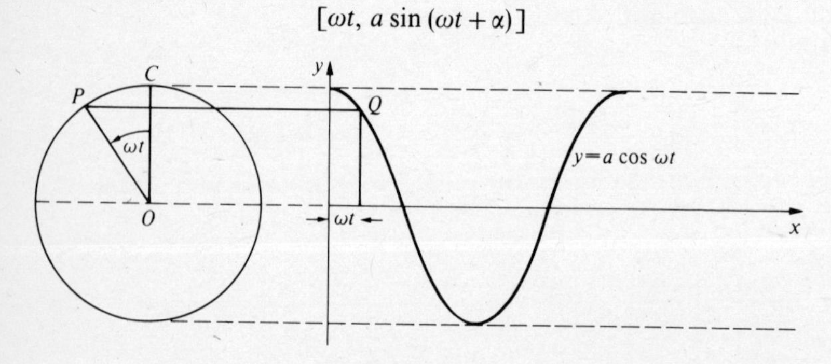

Figure 21.4

where α is the phase angle. When $0 < \alpha < \frac{\pi}{2}$ we usually say that the phasor is leading the standard phasor by α. When $0 > \alpha > \frac{-\pi}{2}$ we usually say that the phasor is lagging on the standard phasor by α. But the terms 'lagging' and 'leading' are purely relative. They might all be called leading or all be called lagging. For example, lagging by $\pi/4$ is equivalent to leading by $7\pi/4$, and leading by $\pi/3$ is equivalent to lagging by $5\pi/3$.

21.4 The addition of two phasors

The following examples illustrate the methods of adding two phasors.

1. Simplify $\sin \omega t + \sin (\omega t + \pi/2)$.
   In Fig. 21.5 the two terms are represented by the two phasors **OA** and **OB** respectively. In Fig. 21.5 $OA = OB = 1$ and angle $B\hat{O}A = \pi/2$ radians $= 90°$. The sum of the two phasors **OA** and **OB** is the phasor **OC**, which is obtained by drawing the diagonal of the rectangle produced by OA and OB. In that construction use is made of the parallelogram law of addition of two vectors. By Pythagoras the magnitude of the phasor **OC** is $\sqrt{2}$. The phase angle of the phasor **OC** is angle $C\hat{O}A = \pi/4$ radians $= 45°$.

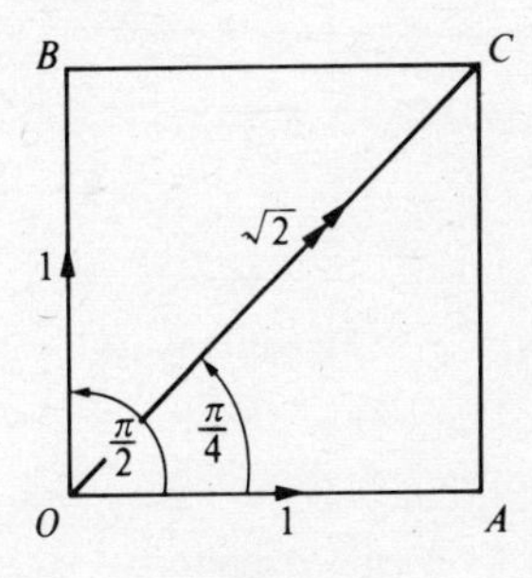

Figure 21.5

Therefore $\sin \omega t + \sin (\omega t + \pi/2) = \sqrt{2} \sin (\omega t + \pi/4)$.
This is a sinusoidal wave-form of amplitude $\sqrt{2}$ and phase angle $\pi/4$.

Alternatively, the result may be obtained by using the formulae in the introduction.

$$\sin \omega t + \sin(\omega t + \pi/2) = 2\sin\frac{1}{2}\left[\omega t + \left(\omega t + \frac{\pi}{2}\right)\right]\cos\frac{1}{2}\left[\omega t + \frac{\pi}{2} - \omega t\right]$$

$$= 2\sin\left(\omega t + \frac{\pi}{4}\right)\cos\frac{\pi}{4}$$

$$= 2\cos\frac{\pi}{4}\sin\left(\omega t + \frac{\pi}{4}\right)$$

$$= 2 \times \frac{1}{\sqrt{2}}\sin\left(\omega t + \frac{\pi}{4}\right)$$

$$= \sqrt{2}\sin\left(\omega t + \frac{\pi}{4}\right)$$

2. Simplify $\sin \omega t + \sin\left(\omega t + \frac{\pi}{4}\right)$.

   In Fig. 21.6 the phasor $\sin \omega t$ is represented by the vector **OA** and the phasor $\sin\left(\omega t + \frac{\pi}{4}\right)$ is represented by the vector **OB**, where $OA = OB = 1$ and angle $B\hat{O}A = \pi/4$ radians. The sum is represented by the vector **OC**, where angle $C\hat{O}A = \pi/8$ radians and the magnitude of OC = the amplitude of the sinusoidal wave-form $= 2\cos \pi/8$. Therefore the sum $= 2\cos(\pi/8).\ \sin(\omega t + \pi/8)$.

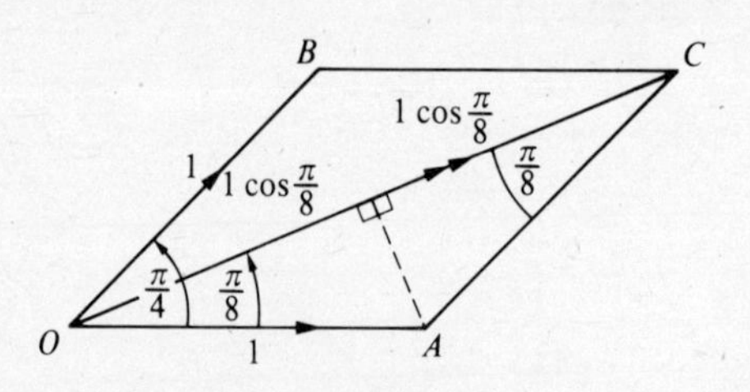

Figure 21.6

3. Simplify $4\sin \omega t + 3\cos \omega t$.

   This is equivalent to $4\sin \omega t + 3\sin(\omega t + \pi/2)$. Fig. 21.7 represents the phasor diagram. In Fig. 21.7 suppose that the vectors **OA** and **OB** represent the phasors $4\sin \omega t$ and $3\cos \omega t$ respectively and that **OC** represents the resultant sinusoidal wave-form. Suppose $OC = R$ and the angle $C\hat{O}A = \alpha$. Then, from right-angled triangle OAC:

$$4 = R\cos\alpha \qquad \text{(i)}$$

$$3 = R\sin\alpha \qquad \text{(ii)}$$

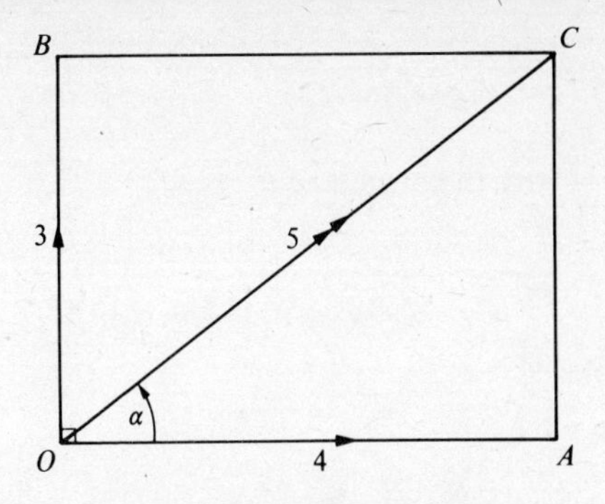

Figure 21.7

$(i)^2 + (ii)^2$ gives:

$$R^2(\cos^2\alpha + \sin^2\alpha) = 4^2 + 3^2 = 5^2$$

Therefore $R = 5$. Substitute for R in (i) and (ii):

$$\cos\alpha = 4/5, \ \sin\alpha = 3/5$$

$\cos\alpha$ and $\sin\alpha$ are both positive, therefore α is acute.

$$4\sin\omega t + 3\cos\omega t = 5\sin(\omega t + \alpha)$$

The amplitude of the resultant phasor is 5 and its phase angle is α where $\cos\alpha = 4/5$ and $\sin\alpha = 3/5$, i.e. α is acute.

4. Simplify $a\sin(\omega t + \alpha) + b\sin(\omega t + \beta)$.
   In Fig. 21.8 the vectors **OA**, **OB** and **OC** represent $a\sin(\omega t + \alpha)$, $b\sin(\omega t + \beta)$ and their sum respectively. $AOBC$ is a parallelogram. The

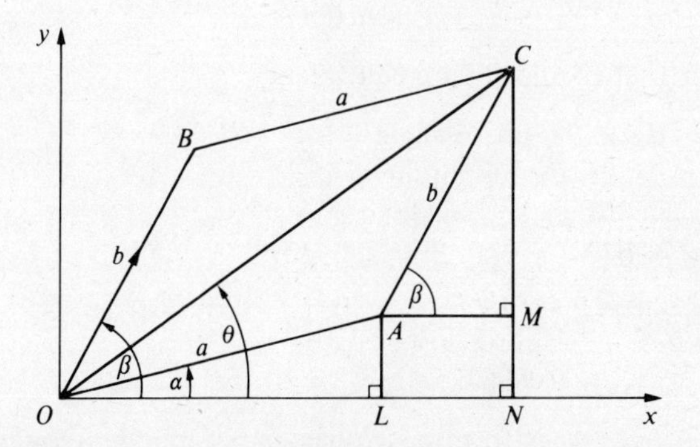

Figure 21.8

amplitude of the phasor which is the resultant of the two individual phasors is OC. The phase angle is θ.

To calculate OC

In triangle OAC, by the cosine rule:

$$OC^2 = a^2 + b^2 + 2ab\cos(\beta - \alpha)$$
$$OC = \sqrt{a^2 + b^2 + 2ab\cos(\beta - \alpha)}$$

To calculate θ

From triangle OCN:

$$\tan\theta = \frac{CN}{ON} = \frac{CM + MN}{OL + LN} = \frac{a\sin\alpha + b\sin\beta}{a\cos\alpha + b\cos\beta}$$

The amplitude of the resultant phasor $= \sqrt{a^2 + b^2 + 2ab\cos(\beta - \alpha)}$

The phase angle $= \text{arc tan}\left(\dfrac{a\sin\alpha + b\sin\beta}{a\cos\alpha + b\cos\beta}\right)$

5. An electric motor rotates at 1500 rev/min. Calculate its angular velocity.
   $f = 1500/60$ rev/s $= 25$ Hz
   $\omega = 2\pi f = 2\pi \times 25 = 50\pi$ rad/s
6. Calculate the period of a wave-form of 25 000 Hz, i.e. 25 kHz.
   $f = 25\,000$ rev/s
   $T = 1/f = \dfrac{1}{25\,000}\text{ s} = 0.00004\text{ s} = 4 \times 10^{-5}\text{ s}$
7. Calculate the amplitude, frequency, period and phase angle of the sine wave represented by:

$$48\sin 640t \qquad \text{(i)}$$

The standard sine wave-form is:

$$a\sin(\omega t + \alpha) \qquad \text{(ii)}$$

In order that (i) and (ii) are equivalent:

(a) $\alpha = 0$, i.e. the phase angle $= 0$
(b) $a = 48$, i.e. the amplitude $= 48$
(c) $\omega t = 640t$, i.e. $\omega = 640$ rad/s
(d) $2\pi f = 640$, $f = 640/2\pi = 320/\pi$ Hz
(e) period $T = 1/f = \dfrac{1}{320/\pi} = \dfrac{\pi}{320}\text{ s}$
$= 0.0098175\text{ s} \approx 9.8 \times 10^{-3}\text{ s}$

8. Determine the wave-form which has amplitude 28, frequency 50 and phase angle 30°.

The wave-form is:

$$a \sin(\omega t + \alpha)$$

where $a = 28$, $\omega = 2\pi f = 2\pi \times 50 = 100\pi$, and $\alpha = 30° \times \dfrac{\pi}{180°}$ $= \pi/6$ rad. Therefore the wave-form is:

$$28 \sin(100\pi t + \pi/6)$$

Exercise 21.1

1. An electric motor rotates at three different speeds: (a) 1500 rev/min, (b) 300 rev/min and (c) 12 000 rev/min. Calculate the angular velocity in each case.
2. Calculate the periods, in seconds, of the three speeds of the motor in question (1).
3. Determine the periods of wave-forms of frequencies 120 Hz, 120 kHz, 90 MHz, 395 Hz, 5.26×10^5 Hz.
4. Calculate the frequencies of wave-forms with periods 0.025 s, 3×10^{-4} s, 6.2×10^{-6} s, 3.5×10^{-8} s.

For the following wave-forms determine the amplitudes, angular velocities, frequencies, periods and phase angles (in radians).

5. $120 \sin 100\pi t$
6. $120 \sin(100\pi t + \pi/2)$
7. $120 \cos(100\pi t)$
8. $120 \cos(100\pi t + \pi/4)$

Hint: use the relation $\cos A = \sin(A + \pi/2)$.

9. $-120 \sin 100\pi t$ (Hint: use $-\sin A = \sin(A + \pi)$.)
10. $240 \sin(10^6 t + \pi/6)$
11. $6 \times 10^5 \sin(3 \times 10^9 t + \pi/3)$
12. $120 \sin(50\pi t - \pi/2)$
13. $120 \sin(550\pi t - \pi/4)$
14. $2400 \sin(100\pi t - 30°)$
15. $2.4 \times 10^3 \sin(200\pi t + 45°)$
16. $3650 \cos(50t + 30°)$
17. $4345 \cos(256t + 12°)$

Determine the wave-forms with the following characteristics:

18. amplitude 250, frequency 50 Hz, phase angle $= \pi/4$ leading
19. amplitude 100, period 2×10^{-3} s, phase angle $\pi/6$ lagging

20. amplitude 50, angular velocity 40 rad/s, phase angle 20° leading
21. amplitude 25, period 2×10^{-8} s, phase angle 40° lagging

Simplify the following to single wave-forms:

22. $10 \sin \omega t + 10 \cos \omega t$
23. $10 \sin \omega t - 10 \cos \omega t$
24. $120 \cos \omega t - 120 \sin \omega t$
25. $-120 \cos \omega t - 120 \sin \omega t$
26. $3 \sin \omega t + 4 \cos \omega t$
27. $4 \sin \omega t - 3 \cos \omega t$
28. $5 \sin \omega t + 12 \cos \omega t$
29. $12 \sin \omega t - 5 \cos \omega t$
30. $8 \cos \omega t + 15 \sin \omega t$
31. $8 \sin \omega t - 15 \cos \omega t$
32. $6 \sin \omega t + 5 \cos \omega t$
33. $10 \sin \omega t - 15 \cos \omega t$
34. $2 \sin (\omega t + \pi/6) + 3 \sin (\omega t + \pi/3)$
35. $5 \sin (\omega t + \pi/6) + 4 \sin (\omega t + \pi/4)$
36. $120 \sin (\omega t + 20°) + 250 \sin (\omega t + 40°)$
37. $180 \sin (\omega t + 10°) + 20 \sin (\omega t + 80°)$
38. $200 \sin (\omega t + 5°) + 40 \cos (\omega t - 5°)$

21.5 A graphical method of combining two waves of the same frequency

From Chapter 20 it is clear that the curve for $y = \cos x$ is basically the same as that for $y = \sin x$ except that the whole curve is shifted $\pi/2$ radians parallel to Ox. In fact the equation $y = \cos x$ might be written:

$$y = \sin (x + \pi/2)$$

In the examples below we shall study the graphical approach to the combination of two sine waves. These examples are:

1. $y = \sin x + \cos x$
2. $y = \sin 2x + \cos 2x$
3. $y = 3 \sin x + 4 \cos x$

To do that, data in Table 20.1 will be used to construct a new table, Table 21.1.

Table 21.1

x	$2x$	$\sin x + \cos x$	$\sin 2x + \cos 2x$	$3 \sin x + 4 \cos x$
0°	0°	1.0000	1.0000	4.0000
10°	20°	1.1585	1.2817	4.4602
20°	40°	1.2817	1.4088	4.7848
30°	60°	1.3660	1.3660	4.9641
40°	80°	1.4088	1.1585	4.9925
50°	100°	1.4088	0.8112	4.8693
60°	120°	1.3660	0.3660	4.5981
70°	140°	1.2817	−0.1233	4.1872
80°	160°	1.1585	−0.5977	3.6490
90°	180°	1.0000	−1.0000	3.0000
100°	200°	0.8112	−1.2817	2.3054
110°	220°	0.5977	−1.4088	1.4510
120°	240°	0.3660	−1.3660	0.5981
130°	260°	0.1233	−1.1585	−0.2730
140°	280°	−0.1233	−0.8112	−1.1358
150°	300°	−0.3660	−0.3660	−1.9641
160°	320°	−0.5977	0.1233	−2.7327
170°	340°	−0.8112	0.5977	−3.4183
180°	360°	−1.0000	1.0000	−4.0000
190°		−1.1585		−4.4602
200°		−1.2817		−4.7848
210°		−1.3660		−4.9641
220°		−1.4088		−4.9925
230°		−1.4088		−4.8693
240°		−1.3660		−4.5981
250°		−1.2817		−4.1872
260°		−1.1585		−3.6490
270°		−1.0000		−3.0000
280°		−0.8112		−2.3054
290°		−0.5977		−1.4510
300°		−0.3660		−0.5981
310°		−0.1233		0.2730
320°		0.1233		1.1358
330°		0.3660		1.9641
340°		0.5977		2.7327
350°		0.8112		3.4183
360°		1.0000		4.0000

Examples

1. From the data in Table 21.1 construct a graph of $y = \sin x + \cos x$ for values of x between 0° and 360°.
   Fig. 21.9 represents the curve. Fig. 21.9 is a sine wave displaced from $y = \sin x$ by 45° or $\pi/4$ radians parallel to Ox. It has an amplitude of approximately 1.41. In fact its amplitude is $\sqrt{2}$. The phase angle is 45° or $\pi/4$ radians.

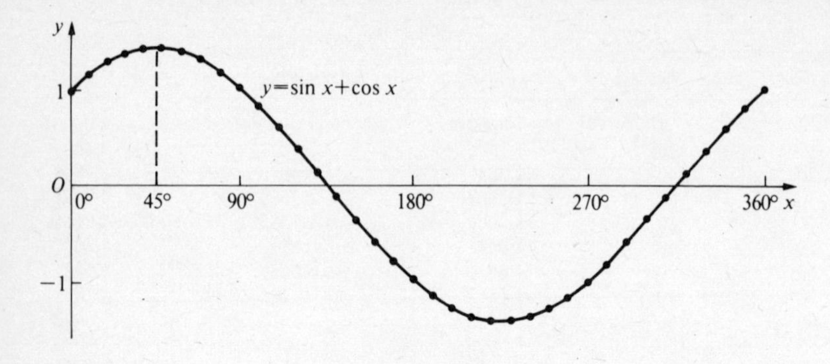

Figure 21.9

2. From the data in Table 21.1 construct a graph of the curve $y = \sin 2x + \cos 2x$.
   Fig. 21.10 represents the curve. Fig. 21.10 is a sine wave displaced from $y = \sin 2x$ by $22\frac{1}{2}°$ or $\pi/8$ radians parallel to Ox. It has an amplitude of approximately 1.41. In fact the amplitude is exactly $\sqrt{2}$. The phase angle is $\pi/8$ radians.

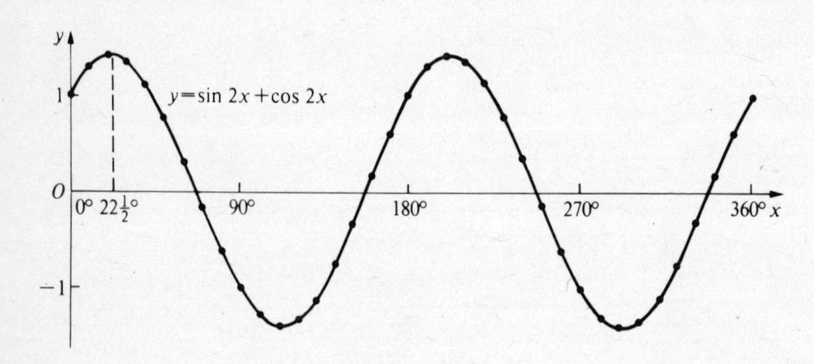

Figure 21.10

3. From the data in Table 21.1 construct a graph of the curve $y = 3 \sin x + 4 \cos x$.
   Fig. 21.11 represents the curve. Fig. 21.11 is a sine wave displaced from $y = \sin x$ by approximately 53° parallel to Ox. By calculator it is 53.130102°. From the graph the amplitude is approximately 5.0. In fact the amplitude is exactly 5. The phase angle is $\approx 53°$.

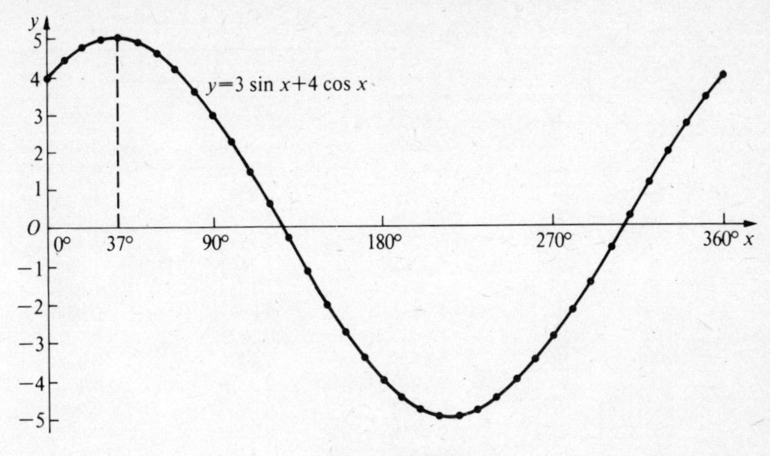

Figure 21.11

Exercise 21.2

Draw the graphs of the following functions between the values of the variable indicated. From the graphs determine the amplitudes and phase angles of the resulting sine waves.

1. $y = 4 \sin x + 3 \cos x$ between $x = 0°$ and $x = 360°$
2. $y = \cos x - \sin x$ between $x = 0°$ and $x = 360°$
3. $y = \sin 2x - \cos 2x$ between $x = 0°$ and $x = 180°$
4. $y = \sin \frac{1}{2}x + \cos \frac{1}{2}x$ between $x = 0°$ and $x = 720°$
5. $y = 3 \sin 2x + 4 \cos 2x$ between $x = 0°$ and $x = 180°$
6. $y = 2 \sin x + \cos x$ between $x = 0°$ and $x = 360°$
7. $y = 5 \sin x + 12 \cos x$ between $x = 0°$ and $x = 360°$
8. $y = 2 \sin x - \cos x$ between $x = 0°$ and $x = 360°$
9. $y = \sin^2 x + 3 \cos^2 x$ between $x = 0°$ and $x = 180°$
10. $y = 3 \sin^2 x + \cos^2 x$ between $x = 0°$ and $x = 180°$
11. $y = 2 \sin^2 x + 8 \cos^2 x$ between $x = 0°$ and $x = 180°$
12. $y = 2 \sin^2 x + 6 \cos 2x$ between $x = 0°$ and $x = 180°$

21.6 The combination of two sine waves of different frequencies

Table 21.2 is constructed from the data in Table 20.1.

Table 21.2

x	$2x$	$\frac{1}{2}x$	$\sin x + \sin 2x$	$\cos x + \cos \frac{1}{2}x$	$\sin 2x + \sin \frac{1}{2}x$
0°	0°	0°	0.0000	2.0000	0.0000
10°	20°	5°	0.5157	1.9810	0.4292
20°	40°	10°	0.9848	1.9245	0.8164
30°	60°	15°	1.3660	1.8320	1.1248
40°	80°	20°	1.6276	1.7057	1.3268
50°	100°	25°	1.7509	1.5491	1.4014
60°	120°	30°	1.7321	1.3660	1.3660
70°	140°	35°	1.5825	1.1612	1.2164
80°	160°	40°	1.3268	0.9397	0.9848
90°	180°	45°	1.0000	0.7071	0.7071
100°	200°	50°	0.6428	0.4691	0.4240
110°	220°	55°	0.2970	0.2316	0.1764
120°	240°	60°	0.0000	0.0000	0.0000
130°	260°	65°	−0.2188	−0.2202	−0.0785
140°	280°	70°	−0.3420	−0.4240	−0.0451
150°	300°	75°	−0.3660	−0.6072	0.0990
160°	320°	80°	−0.3008	−0.7660	0.3420
170°	340°	85°	−0.1684	−0.8977	0.6542
180°	360°	90°	0.0000	−1.0000	1.0000
190°	380°	95°	0.1684	−1.0720	1.3382
200°	400°	100°	0.3008	−1.1133	1.6276
210°	420°	105°	0.3660	−1.1248	1.8320
220°	440°	110°	0.3420	−1.1081	1.9245
230°	460°	115°	0.2188	−1.0654	1.8911
240°	480°	120°	0.0000	−1.0000	1.7321
250°	500°	125°	−0.2970	−0.9156	1.4619
260°	520°	130°	−0.6428	−0.8164	1.1081
270°	540°	135°	−1.0000	−0.7071	0.7071
280°	560°	140°	−1.3268	−0.5924	0.3008
290°	580°	145°	−1.5825	−0.4771	−0.0692
300°	600°	150°	−1.7321	−0.3660	−0.3660
310°	620°	155°	−1.7509	−0.2635	−0.5622
320°	640°	160°	−1.6276	−0.1736	−0.6428
330°	660°	165°	−1.3660	−0.0999	−0.6072
340°	680°	170°	−0.9848	−0.0451	−0.4691
350°	700°	175°	−0.5157	−0.0114	−0.2549
360°	720°	180°	0.0000	0.0000	0.0000

Examples

1. Construct the graph of $y = \sin x + \sin 2x$ between $x = 0°$ and $x = 360°$. From Table 21.2 Fig. 21.12 may be drawn. Note the curve is not sinusoidal.
2. Draw the graph of $y = \cos x + \cos \frac{1}{2}x$ between $x = 0°$ and $x = 360°$. Fig. 21.13 represents the curve. Again this is not sinusoidal.

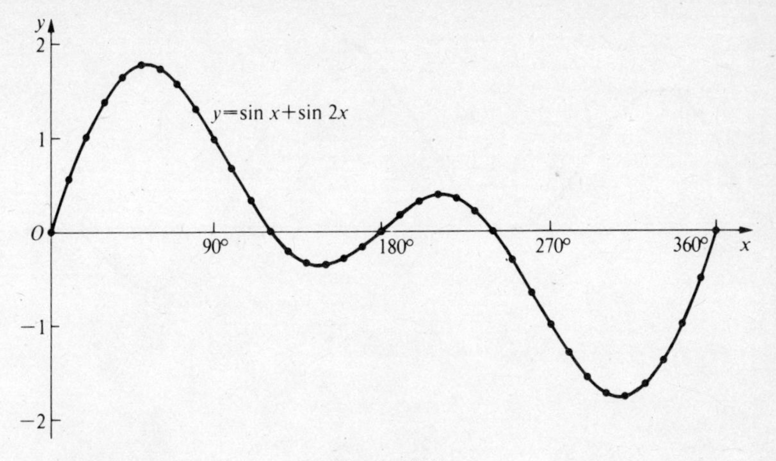

Figure 21.12

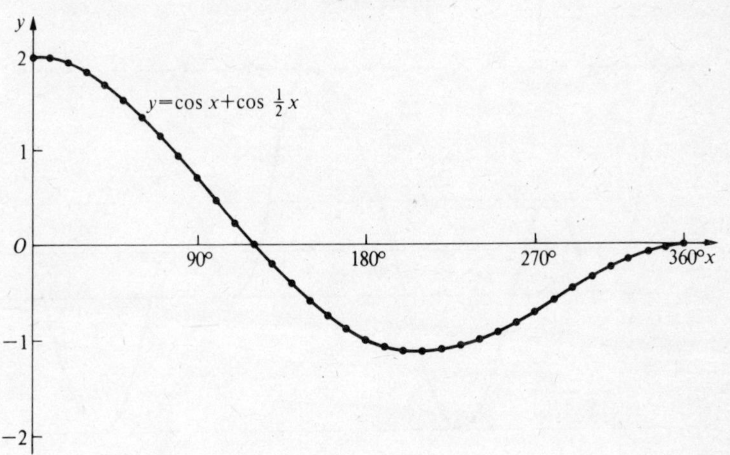

Figure 21.13

3. Draw the graph of $y = \sin 2x + \sin \frac{1}{2}x$ between $x = 0°$ and $x = 360°$. Fig. 21.14 represents the curve. Fig. 21.14 is not a sinusoidal curve.

Figs 21.15, 21.16 and 21.17 are sketches which represent extensions of Figs 21.12, 21.13 and 21.14 respectively. Note that all three curves have a complete cycle of values, that is, they are periodic.
The period of $y = \sin x + \sin 2x$ is 360°.
The period of $y = \cos x + \cos \frac{1}{2}x$ is 720°.
The period of $y = \sin 2x + \sin \frac{1}{2}x$ is 720°.

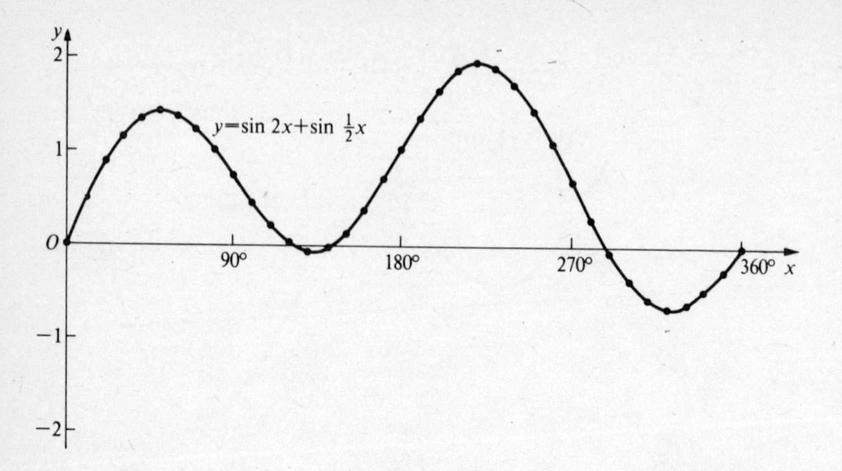

Figure 21.14

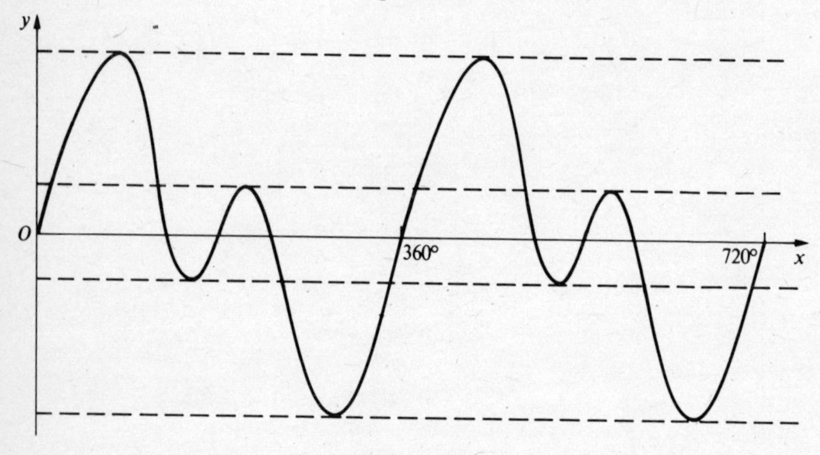

Figure 21.15

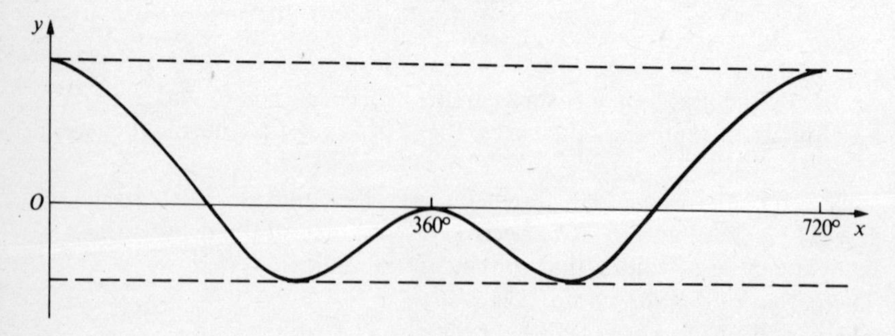

Figure 21.16

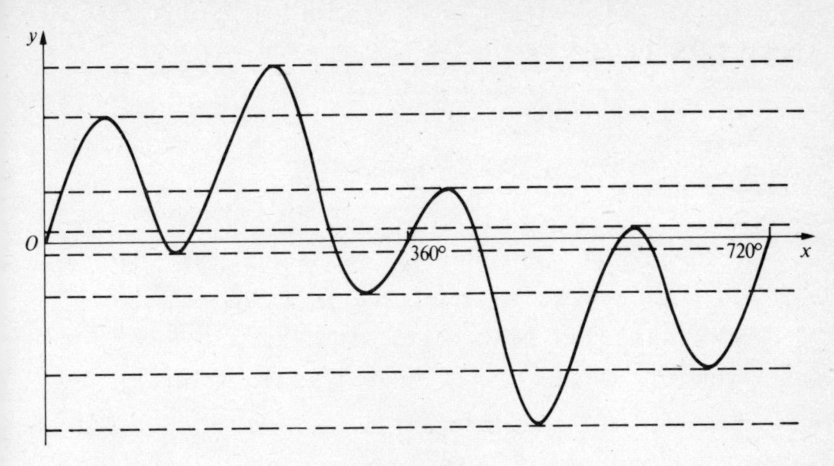

Figure 21.17

Exercise 21.3

Draw the graph of the following functions for the intervals stated. Determine the period of each function.

1. $y = \cos x + \cos 2x$: $x = 0°$ to $x = 360°$
2. $y = \sin x + \sin \frac{1}{2}x$: $x = 0°$ to $x = 720°$
3. $y = \cos 2x + \cos \frac{1}{2}x$: $x = 0°$ to $x = 720°$
4. $y = \sin x + \cos 2x$: $x = 0°$ to $x = 360°$
5. $y = \cos x + \sin \frac{1}{2}x$: $x = 0°$ to $x = 360°$
6. $y = \sin 2x + \cos \frac{1}{2}x$: $x = 0°$ to $x = 360°$
7. $y = \sin x + \frac{1}{10} \sin 2x$: $x = 0°$ to $x = 360°$
8. $y = \sin \frac{1}{2}x + \frac{1}{2} \sin x + \frac{1}{4} \sin 2x$: $x = 0°$ to $x = 360°$

ANSWERS TO EXERCISES

Exercise 1.1

1. $2e^{2x}$ **2.** $5e^{5x}$ **3.** $\frac{3}{5}e^{\frac{3x}{5}}$ **4.** $-\frac{1}{4}e^{-\frac{1}{4}x}$
5. $10e^{2x}$ **6.** $-2e^{-\frac{1}{3}x}$ **7.** $44e^{-4x}$ **8.** $-\frac{4}{3}e^{-2x}$
9. $-\frac{3}{16}e^{\frac{1}{4}x}$ **10.** $2\cos 2x$ **11.** $\frac{3}{5}\cos(3x/5)$ **12.** $\frac{\pi}{2}\cos(\pi x/2)$
13. $-4\sin 4x$ **14.** $3\sin(-3x) = -3\sin 3x$ **15.** $-\frac{1}{4}\sin\frac{1}{4}x$
16. $-\frac{\pi}{4}\sin(\pi x/4)$ **17.** $2\cos 3x$ **18.** $-\frac{16}{3}\sin 4x$ **19.** $-\frac{8}{3}\cos(3x/2)$
20. $-\frac{\pi}{9}\sin(\pi x/6)$ **21.** $1/x$ **22.** $1/x$ **23.** $2/x$ **24.** $2/x$
25. $3/x$ **26.** $3/x$ **27.** $1/2x$ **28.** $1/2x$
29. $3/x$ **30.** $3/x$ **31.** $8/x$ **32.** $3e^x + 10e^{2x} + 18e^{3x}$
33. e^x **34.** $12e^{2x}(2 - 5e^x)$ **35.** $6e^x(2 + 3e^x)$
36. $2\cos x - 6\cos 2x - 5\sin x - 16\sin 2x$ **37.** $-\frac{6}{25}\sin(3x/5) - \frac{8}{25}\cos(2x/5)$
38. $-4e^{-x} - \frac{7}{4}\sin\frac{1}{4}x - \frac{3}{10\sqrt{x}}$ **39.** $2\cos(3x/2) - \frac{1}{12}e^{-\frac{1}{3}x} + \frac{12}{5x^3}$
40. $\frac{37}{4x} + \frac{5}{9}e^{\frac{2x}{3}} - \frac{35}{2}x^2\sqrt{x}$

Exercise 1.2

1. $e^{-1} - 1 = -0.6321206 \approx -0.632$ **2.** $-3e^{-2} - 10e^{-4} \approx -0.589$
3. $\frac{1}{4}$ **4.** 6 **5.** $2\cos(\pi/4) = \sqrt{2}$ **6.** $4\sin\frac{\pi}{2} = 4$
7. $4\cos\frac{\pi}{2} - 6\cos\pi = 6$ **8.** $-3\sin\frac{\pi}{2} + 10\sin\pi = -3$
9. $-e^{-\frac{\pi}{6}} - 3\sin\frac{\pi}{6} = -e^{-\frac{\pi}{6}} - \frac{3}{2} = -2.0923848 \approx -2.092$
10. $\frac{4}{\pi} - \frac{4}{5\sqrt{2}} = 0.7075541 \approx 0.708.$
11. $1 - 6 - 4e^{-1} = -5 - 4e^{-1} = -6.4715178 \approx -6.472$
12. $6e^{\frac{1}{2}} = 9.8923276 \approx 9.892$ **13.** $-\frac{1}{26} - 0.85e^{-1.768} = -0.1835344 \approx -0.1835$
14. $3.0277778 \approx 3.028$ **15.** $-0.011334 \approx -0.011$

Exercise 2.1

1. $8x + 3$ **2.** $15x^2 - 6 + 11/x^2$ **3.** $14x + 6/x^3$ **4.** $2e^x - 4e^{-x}$
5. $14e^{2x} - 8e^x$ **6.** $10\cos x - 4\sin x - 2$ **7.** $6\cos 2x - 12\cos 3x - 2\sin 2x + 6x$
8. $e^{2x} + \frac{1}{4}e^x + 6/x^3$ **9.** $2/x + 3e^x - 6x^2$.
10. $20\cos 4x + 12\sin 2x - \frac{3}{4}e^{-x} - 3x/2$

Exercise 2.2

1. $4(2x + 3)$ **2.** $6(2x + 3)^2$ **3.** $10(5x + 1)$ **4.** $15(5x + 1)^2$

5. $-2(2x-1)^{-2}$ **6.** $-6(3x+2)^{-3}$ **7.** $9a(ax+b)^8$
8. $2(2x-3)(x^2-3x+5)$ **9.** $3(2x-3)(x^2-3x+6)^3$ **10.** $12e^{4x+1}$
11. $4xe^{\frac{1}{4}x^2}$ **12.** $\cos 2x \,.\, e^{-\sin 2x}$
13. $-15A \sin 3t \,.\, e^{5\cos 3t}$
14. $7(2x+5)/(x^2+5x-1)$ **15.** $-ap\tan pt$ **16.** $6e^x(3e^x+4)$

Exercise 2.3

1. $e^{ax}(a\cos bx - b\sin bx)$ **2.** $x+3/x+2x\ln x$ **3.** $2e^{2x}(x^2-x+2)$
4. $\dfrac{1}{x}\cos 2x - 2\sin 2x \,.\, \ln x$
5. $4(3\cos 3x \,.\, \cos 2x - 2\sin 3x \,.\, \sin 2x)$
6. $\left(1-\dfrac{1}{x^2}\right)\cos 4x - 4\left(x+\dfrac{1}{x}\right)\sin 4x$ **7.** $e^{-3x}(2\cos 2x - 3\sin 2x)$
8. $e^x[(2x+7)\cos x - (2x+5)\sin x]$
9. $\ln x[3\sin x + (3x+1)\cos x] + \left(3+\dfrac{1}{x}\right)\sin x$
10. $e^{-\frac{1}{4}x}\left[\dfrac{1}{x}\sin\dfrac{2}{3}x + \dfrac{2}{3}\cos\dfrac{2}{3}x \,.\, \ln x - \dfrac{1}{4}\sin\dfrac{2}{3}x \,.\, \ln x\right]$

Exercise 2.4

1. $-25/(4x-7)^2$ **2.** $\dfrac{(2x+1)\cos x - 2\sin x}{(2x+1)^2}$ **3.** $-2\dfrac{[\cos 2x + (x^2-4)\sin 2x]}{(x^2-4)^2}$
4. $-a\dfrac{[p\cos bx + b(px+q)\sin bx]}{(px+q)^2}$ **5.** $e^x \operatorname{cosec} 3x(1+3\cot 3x)$
6. $\dfrac{4}{x(x^2+16)} - \dfrac{8x\ln x}{(x^2+16)^2}$ **7.** $1/(x+2)^2$ **8.** $-4x/(x^2-1)^2$
9. $-2(5x+7)/(x^2+4x+2)$ **10.** $2e^x/(e^x+1)^2$
11. $2e^x/(e^x+1)^2$ **12.** $-e^{-2t} \,.\, (2t+1)/(t+1)^2$

Exercise 2.5

1. $a\sec(at) \,.\, \tan(at)$ **2.** $-a\operatorname{cosec}^2(at)$ **3.** $\sin(\frac{1}{2}x) \,.\, \cos(\frac{1}{2}x)$
4. $-\frac{3}{2}\cos(3x/4)\sin(3x/4)$
5. $2a\tan(at)\sec^2(at)$ **6.** $a\cos(at)e^{\sin(at)}$
7. $-a\sin(at)e^{\cos(at)}$ **8.** $-2\sin 2t/(1+\cos 2t)$
9. $2\sin 2t/(1-\cos 2t)$ **10.** $4t \,.\, \cos(2t^2+1)$ **11.** $\frac{1}{2}a\cos(2at) - \frac{1}{3}a\sin(at)$
12. $2(3x+4)\sec^2 2x + 3\tan 2x$ **13.** $\dfrac{\sec 3x\,[3(1+5x)\tan 3x - 5]}{(1+5x)^2}$
14. $\cos x - \sin x$ **15.** $\sec^2 x \,.\, e^{\tan x}$ **16.** $\cot x$
17. $-a\tan(ax)$ **18.** $a\sec ax \,.\, \operatorname{cosec} ax$ **19.** $-\cot x$ **20.** $-a\sec ax \,.\, \operatorname{cosec} ax$
21. $\tan x$ **22.** $(6x+5)/2\sqrt{3x^2+5x-1)}$ **23.** $\sec x\tan x$ **24.** $ae^{ax}/2\sqrt{(1+e^{ax})}$
25. $1/x\sqrt{\ln 3x}$ **26.** $-1/(1+x)\sqrt{(1-x^2)}$ **27.** $1/(1+\cos x)$
28. $e^t/(e^t+1)^{\frac{3}{2}}(e^t-1)^{\frac{1}{2}}$ **29.** $\frac{1}{2}(x+1)$ **30.** $6/(1+2x)$
31. $\dfrac{6}{\sqrt{x}}\cos 2x - \dfrac{3}{2x\sqrt{x}} \cdot \sin 2x$ **32.** $-1/(x+2)$

Exercise 2.6

1. -3; $0.0168849 \approx 0.0169$; $0.0269499 \approx 0.0269$
2. 4; $3.9376591 \approx 3.938$; $3.8762897 \approx 3.876$

3. $-\frac{16}{5}\sqrt{5}; -\frac{4}{3}\sqrt{3}$ **4.** $\frac{1}{2}; 1; \infty$ **5.** $0; \frac{1}{2}; \frac{4}{17}$
6. $2e(1+\ln\frac{1}{4}) = -2.1001139; e^2(1+2\ln\frac{1}{2}) = -2.8543507; e^8(\frac{1}{4}+2\ln 2) = 4.8777247$
7. $\frac{11}{4}e^{\frac{1}{2}} = 4.5339835 \approx 4.534; \frac{25}{8}e^{0.875} = 7.4964853 \approx 7.496$
8. $6\sin 1 = 5.0488259 \approx 5.049; 6\cos 2 . \ln 2 + 3\sin 2 = 0.9971862 \approx 0.997;$
$6\cos 4 . \ln 4 + \frac{3}{2}\sin 4 = -6.5720585 \approx -6.572$
9. $3; 0; -0.089856 \approx -0.09$ **10.** -1 **11.** π

Exercise 3.1

1. 6 **2.** $2/x^3$ **3.** $-6/x^3+42/x^4$ **4.** $8+24/x^4$
5. $8+54/x^4$ **6.** $12(2t^2+3t-1)(10t^2+15t+4)$ **7.** $-4a\sin 2t$
8. $-96\cos 3t$ **9.** $-a^2(A\cos at + B\sin at)$ **10.** $9Ae^{3x}$ **11.** $4Pe^{-2x}$
12. $\frac{1}{4}(Ae^{\frac{1}{2}x}+Be^{-\frac{1}{2}x})$
13. $-Ae^{-t}(3\sin 2t + 4\cos 2t)$
14. $Ae^{-t}(6\sin 3t - 8\cos 3t) - Be^{-2t}(5\sin 3t + 12\cos 3t)$
15. $8\,\mathrm{cosec}^2(2x+1)\cot(2x+1)$ **16.** $-3/x^2$ **17.** $-6/x^2$ **18.** $2/x+3/x^2$
19. $40(4x^2+1)^3(36x^2+1)$ **20.** $4(2x+3)\cos 2x - 2(2x^2+6x-1)\sin 2x$
21. $-\frac{1}{4}x^{-\frac{3}{2}}+\frac{3}{4}x^{-\frac{5}{2}}$
22. $2\sec^2 x[(3x-1)\tan x + 3]$
23. $18\sec 3x(1+2\tan^2 3x) + 12\,\mathrm{cosec}\,2x(1+2\cot^2 2x)$

Exercise 3.2

1. 10 **2.** 32 **3.** $-12, -3, -\frac{48}{49}$ **4.** $18, 320.4$ **5.** $205, 904.0192(904\frac{12}{625})$
6. $2, \frac{1}{4}, \frac{2}{125}$ **7.** $-1, 2e-3e^{-1} = 4.3329253, 2e^{10}-3e^{-10} = 44052.931$
8. $-11.5, 7.3e^{-1}-18.8e^{-2} = 0.1412166, 7.3e^{-2.5}-18.8e^{-5} = 0.4725471$
9. $\frac{3}{128}, 0.0121633$ **10.** $-6; 6\frac{3}{4}; -3.549731$
11. $-3.2927016; -1.2253424$ **12.** $12; 16e^{-\pi}; -12e^{-2\pi}$
13. $8; 4(\pi+24); \infty$ **14.** $-\frac{3}{25}; -0.0956633; -0.0780437$

Exercise 3.3

1. $v = -11; a = 28$ **2.** $8, 12; 20, 12$ **3.** $-9, -10; -29, -10$
4. $-4, 2; 12, 14; 52, 26$ **5.** $-8, 22; 2, -2; -12, -26; -50, -50$
6. $-6, 6; 14, 30; 178, 150$ **7.** $-6, 3; -4\frac{2}{3}, 0; -5\frac{1}{3}, -1; -6, 0$
8. $2, 2; 2e, 2e; 2e^2, 2e^2$ **9.** $-5, 5; -5e^{-1}, -5e^{-1}; -5e^{-2}, -5e^{-2}$
10. $-1, 7; (3e-4e^{-1}), (3e+4e^{-1}); (3e^2-4e^{-2}), (3e^2+4e^{-2})$
11. $4\pi, 0; 4\pi\cos\frac{\pi}{8}, -4\pi^2\sin\frac{\pi}{8}; \frac{4\pi}{\sqrt{2}}, -\frac{4\pi^2}{\sqrt{2}}$ **12.** $\frac{3\pi}{2}, \pi^2; \frac{7\pi}{2\sqrt{2}}, \frac{\pi^2}{4\sqrt{2}}; 2\pi, -\frac{3\pi^2}{4}$
13. $5\pi, -10\pi; \frac{5e^{-\frac{1}{2}}(\pi-2)}{\sqrt{2}}, \frac{5e^{-\frac{1}{2}}(4-4\pi-\pi^2)}{2\sqrt{2}}; -10e^{-1}, \frac{5e^{-1}(4-\pi^2)}{2}$
14. $2, 8; 5.5994573, 6.6998191; 18.456201, 6.5259341$

Exercise 4.1

1. $(1\frac{1}{2}, -\frac{1}{4})$ min. **2.** $(2\frac{1}{2}, -\frac{1}{4})$ min. **3.** $(6, -1)$ min. **4.** $(3, -1)$ min.
5. $(-2, -1)$ min. **6.** $(-4, -4)$ min. **7.** $(-1, 4)$ max. **8.** $(1, 36)$ max.
9. $(-\frac{7}{8}, -12\frac{1}{16})$ min. **10.** $(-\frac{3}{4}, -\frac{29}{32})$ min. **11.** $(-1\frac{5}{6}, 28\frac{1}{12})$ max.
12. $(6, -36)$ min. **13.** $(0, 5)$ min. **14.** $(0, -8)$ max.
15. $(-\frac{7}{8}, -9\frac{15}{16})$ max.

Exercise 4.2

1. $(1, 5)$ max.; $(3, 1)$ min. **2.** $(-1, 7)$ min.; $(-3, 11)$ max.
3. $(2, 20)$ max.; $(4, 16)$ min. **4.** $(2, -31)$ min.; $(-4, 77)$ max.

5. $(-5, 275)$ max.; $(2, -68)$ min. **6.** $(-\frac{1}{2}, -2\frac{1}{4})$ min.; $(-1\ -2)$ max.
7. $(\frac{1}{3}, 1\frac{13}{27})$ min.; $(-3, 20)$ max. **8.** $(1, 7)$ min.; $(3, 11)$ max.
9. $(2, 21)$ max.; $(-4, -33)$ min. **10.** $(\frac{1}{5}, \frac{23}{25})$ max.; $(-3, -81)$ min.

Exercise 4.4

1. $(\ln\sqrt{2}, 2\sqrt{2})$ min. **2.** $(-\ln\sqrt{2}, 2\sqrt{2})$ min.

3. $(\overline{n\pi + \frac{1}{4}\pi}, 5)$ max.; $(\overline{n\pi - \frac{1}{4}\pi}, -5)$ min.

4. $\left(2n\frac{\pi}{3}, 4\right)$ max.; $\left[\frac{(2n+1)\pi}{3}, -4\right]$ min.

5. $\left[\left(\frac{n\pi}{2} + \frac{\pi}{8}\right), -4\right]$ min.; $\left[\left(\frac{n\pi}{2} - \frac{\pi}{8}\right), 4\right]$ max.

6. $[(2n\pi + \frac{1}{4}\pi), \sqrt{2}]$ max.; $[(\overline{2n+1}\pi + \frac{1}{4}\pi), -\sqrt{2}]$ min.

7. $[(2n\pi - \alpha/2), -5]$ min.; $[(\overline{2n+1}\,.\,\pi - \alpha/2), 5]$ max. where $\alpha \approx 0.6435011$ radians

8. $(0, \frac{1}{2})$ max.

9. $(2n\pi + \frac{1}{4}\pi), \frac{1}{\sqrt{2}}e^{-(2n\pi + \frac{1}{4}\pi)}$ max.; $\left[\overline{(2n+1)}\pi + \frac{1}{4}\pi, -\frac{1}{\sqrt{2}}e^{-(2n+\frac{5}{4})\pi}\right]$ min.

10. $\left[(2n\pi - \frac{1}{4}\pi), \frac{1}{\sqrt{2}}e^{-(2n\pi - \frac{1}{4}\pi)}\right]$ max.; $\left[(\overline{2n+1}\,.\,\pi - \frac{1}{4}\pi), -\frac{1}{\sqrt{2}}e^{-(2n+\frac{3}{4})\pi}\right]$ min.

11. $\left[(2n\pi + \alpha), \frac{1}{\sqrt{5}}e^{-2(2n\pi + \alpha)}\right]$ max.; $\left[(\overline{2n+1}\,.\,\pi + \alpha), -\frac{1}{\sqrt{5}}e^{-2(\overline{2n+1}\,.\,\pi + \alpha)}\right.$ min., where
$\alpha \approx 0.4636476$ radians

12. $\left[\left(\frac{2n\pi}{4} - \frac{\beta}{4}\right), \frac{2}{\sqrt{5}}e^{-(n\pi - \frac{1}{2}\beta)}\right]$ max.; $\left[\left(\frac{\overline{2n+1}\,.\,\pi}{4} - \frac{\beta}{4}\right); -\frac{2}{\sqrt{5}}e^{-(\overline{\frac{2n+1}{2}}\ \pi - \frac{1}{2}\beta)}\right]$ min.,
where $\beta \approx 0.4636476$ radians

Exercise 4.5

1. (a) $t = 1, 5$ (b) $-4, -36$ (c) $-12, 12$ (d) $t = 3$
(e) -20 (f) -32 (g) 25

2. (a) 3 (b) $v = 11, a = 19$ (c) 59 (d) 244
(e) 13.221326 (f) $6\frac{1}{3}$

3. $\sqrt{50}$ m $\times \sqrt{50}$ m **4.** 100 m $\times$ 100 m

5. 400 m $\times$ 200 m; fencing 800 m **6.** 400 m $\times$ 200 m

7. $r = 5\sqrt{\frac{10}{\pi}}$ cm; $h = 10\sqrt{\frac{10}{\pi}}$ cm **8.** $r = 5\sqrt[3]{\frac{4}{\pi}}$ cm; $h = 10\sqrt[3]{\frac{4}{\pi}}$ cm

9. $r = h = 35/(4+\pi)$ **10.** $\frac{45}{2}\sqrt{\frac{4+\pi}{2}}$ m ≈ 42.517239 m

11. At $x = \frac{5}{2}$ **12.** $\sqrt[3]{60}$; $2\sqrt[3]{60}$; $\frac{4}{3}\sqrt[3]{60}$

13. $\frac{1}{2}$ **14.** (a) $R = r - 50$ (b) $R = r + 50$ **15.** $R = rA/(A - r)$

16. For integral values of n, $t = \frac{2n\pi}{p}$ for maximum current, $t = \frac{(2n-1)\pi}{p}$ for minimum current

17. $t = \frac{1}{2\pi f}\left(2n\pi + \frac{\pi}{4}\right)$ for integral values of n; max. current $\frac{5}{2}\sqrt{2}\,A$

18. Percentage loss of power = 53.19 per cent

Exercise 5.1

1. $2\sin 2x$ 2. $12\sin\frac{1}{2}x$ 3. $\frac{6}{5}\sin\frac{2}{3}x$ 4. $-\frac{5}{2}\cos 2x$
5. $-16\cos\frac{1}{2}x$ 6. $-\frac{9}{8}\cos\frac{2}{3}x$ 7. $-\frac{1}{2}(\cos 2x-\sin 2x)$
8. $40\sin\frac{1}{4}x+18\cos\frac{2}{3}x$ 9. $\frac{p}{a}\sin ax$ 10. $-\frac{q}{a}\cos ax$
11. $-\frac{A}{a}\cos ax+\frac{B}{b}\sin bx$ 12. $\frac{1}{2}x^2+\sin x$
13. $x+\cos x$ 14. $x-\frac{1}{2}\sin 2x$ 15. $\frac{1}{2}x+\frac{1}{8}\sin 4x$ 16. $2e^x+2x\frac{1}{2}e^{2x}$
17. $2x-3e^{-x}+2e^{-2x}$ 18. $\frac{1}{2}x+\frac{1}{4}\sin 2x$ 19. $\frac{1}{2}x-\frac{1}{8}\sin 4x$ 20. $\frac{5}{8}x+\frac{15}{32}\sin\frac{4}{3}x$
21. $\frac{1}{2}x+\frac{1}{12}\sin 6x$ 22. $\frac{1}{2}\cos 2x-\frac{1}{6}\cos 6x$ 23. $\sin x+\frac{1}{3}\sin 3x$
24. $-\frac{1}{12}\cos 6x-\frac{1}{4}\cos 2x$ 25. $\frac{1}{4}\sin 2x-\frac{1}{16}\sin 8x$
26. $2e^{3x}+2x$ 27. $\frac{2}{5}e^{2x}+\frac{1}{5}e^{-3x}$

Exercise 5.2

1. 1 2. $1/\sqrt{2}$ 3. $\sqrt{3}/2$ 4. $\frac{1}{2}$ 5. $\sqrt{3}$ 6. $\frac{11}{2}-2\sqrt{3}$
7. $\sqrt{2}+\frac{3}{4}\sqrt{3}-\frac{5}{2}$ 8. $\frac{1}{2}(e^2-1)$ 9. $4(1-e^{-1})$ 10. $\frac{5}{2}(e^2-e^{-2})$
11. $\frac{3}{8}(1-e^{-2})$ 12. $3e^4-3e^2+8$ 13. $\frac{1}{3}-\frac{1}{4}e^{-1}$ 14. $13\frac{1}{2}+2e-\frac{15}{2}e^2$
15. $\frac{\pi}{2}$ 16. $\frac{\pi}{8}-\frac{1}{4}$ 17. $\frac{\pi}{16}$ 18. $\frac{\pi}{12}$
19. 1 20. $\sqrt{3}-\frac{4}{3}$ 21. $\frac{\pi}{8}+\frac{1}{4}+\frac{1}{\sqrt{2}}$

Exercise 5.3

1. $\frac{2}{\pi}$ 2. $\frac{4}{\pi}$ 3. $\frac{14}{\pi}$ 4. $\frac{2(a+b)}{\pi}$ 5. $\frac{3(\sqrt{3}-1)(a+b)}{\pi}$
6. $10\frac{5}{6}$ 7. $\frac{4}{3}$ 8. $\frac{1}{9}\ln 10$ 9. $3\ln 2$ 10. $\frac{1}{5}\ln 2$
11. $\frac{1}{2}(e^2-1)$ 12. $\frac{1}{2}(1-e^{-2})$ 13. $\frac{1}{2}(e-e^{-1})$ 14. $\frac{2}{\pi}$ 15. $\frac{2\sqrt{2}.a}{\pi}$
16. 0 17. $3(\sqrt{3}-1)\frac{p}{\pi}$

Exercise 5.4

1. $\frac{1}{\sqrt{2}}$ 2. $\frac{1}{\sqrt{2}}$ 3. $\frac{3}{\sqrt{2}}$ 4. $\frac{5}{\sqrt{2}}$ 5. $3\sqrt{2}$ 6. $\sqrt{2}$
7. $\frac{I_m}{\sqrt{2}}$ 8. $\frac{I_m}{\sqrt{2}}$ 9. $\frac{I_m}{\sqrt{2}}$ 10. $\frac{V_m}{\sqrt{2}}$

Exercise 6.1

1. 0.6420728; 0.4957211 per cent error
2. (a) 10.022799; 0.2279876 per cent error
   (b) 10.001346; 0.0134582 per cent error
3. (a) 20.045598; 0.2279876 per cent error
   (b) 20.008632; 0.0431594 per cent error
4. (a) 284.35556; 0.0279517 per cent error
   (b) 284.28889; 0.0044991 per cent error
   (c) 284.27751; 0.000496 per cent error

5. (a) 7.6829276; 5.9170568 per cent error
(b) 7.2896787; 0.4957169 per cent error
6. (a) 0.9928056; −0.7194432 per cent error
(b) 0.9995912; −0.0408775 per cent error
7. 3.1429486; 0.0431611 per cent error
8. (a) 8.5504132; 1.9234234 per cent error
(b) 8.4003755; 0.1349306 per cent error
9. (a) 7.0159592; 0.2279886 per cent error
(b) 7.0009421; 0.0134586 per cent error
10. (a) 23.52042; −0.1701477 per cent error
(b) 23.548976; −0.0489468 per cent error
(c) 23.559432; −0.004567 per cent error
11. 5.0239626 (16 strips); 0.4792642 per cent error
12. 10.001346; 0.0134582 per cent error (four strips)
13. 525.38757; 0.0013413 per cent error (eight strips)
14. 1.1071401; −0.0007768 per cent error (eight strips)

Exercise 7.1

1. $y = x^2 + 3x + c;\ y = x^2 + 3x + 2$ **2.** $y = a + 5x - 2x^2;\ y = -1 + 5x - 2x^2$
3. $y = x^3 + 2x^2 - 3x + c;\ y = x^3 + 2x^2 - 3x - 79$ **4.** $y = \ln(ax);\ y = 2 + \ln(\frac{1}{5}x)$
5. $y = \dfrac{1}{6}\sin 2x + c;\ y = \dfrac{1}{6}\sin 2x - \dfrac{3}{2} - \dfrac{1}{6\sqrt{2}}$
6. $y = -6e^{-x} + c;\ y = 6(e^{-3} - e^{-x}) - 10$
7. $y = \frac{1}{3}e^{2x} + \frac{4}{15}e^{-3x} + c;\ y = \frac{1}{3}(e^{2x} - e^2) + \frac{4}{15}(e^{-3x} - e^{-3}) + 5$
8. $y = 2x^2 + \dfrac{8}{x} + c;\ y = 2x^2 + \dfrac{8}{x} - 193\frac{4}{5}$
9. $v = \frac{1}{3}t\sqrt{t} - \frac{2}{3}\sqrt{t} + c;\ v = \frac{1}{3}t\sqrt{t} - \frac{2}{3}\sqrt{t} + \frac{5}{3}$
10. $i = \frac{3}{10}t^2 + 8t + c;\ i = \frac{3}{10}t^2 + 8t - \frac{254}{5}$
11. $A = \frac{5}{2}x^2 + 2\tan x + c;\ A = \frac{5}{2}x^2 + 2\tan x - 9 - \dfrac{5\pi^2}{32}$
12. $y = -\frac{5}{2}e^{-2x} + 3e^{-x} + c;\ \ y = -\frac{5}{2}e^{-2x} + 3e^{-x} + 2 + \frac{5}{2}e^{-2} - 3e^{-1}$
13. $i = 5t + 3\ln t + c;\ \ i = 5t - 9 + 3\ln(t/2)$
14. $i = \frac{3}{25}\sin(\frac{5}{2}t) + c;\ i = \frac{3}{25}[\sin(\frac{5}{2}t) - \sin 5] + 8$
15. $y = \dfrac{7}{2}\sin 2x + \cos 3x + c;\ y = \dfrac{7}{2}\sin 2x + \cos 3x + \dfrac{3}{2} - \dfrac{1}{\sqrt{2}}$

Exercise 7.2

1. $y = 2x^2 + 2x + c;\ y = 2(x^2 + x - 1)$
2. $y = 2x^3 - 4x^2 + 3x + c;\ y = 2x^3 - 4x^2 + 3x - 11$
3. $y = x^2 + \frac{1}{2}x^3 - \frac{5}{8}x^4 + c;\ y = x^2 + \frac{1}{2}x^3 - \frac{5}{8}x^4 - 3\frac{7}{8}$
4. $s = -5e^{-t} + 6t + c;\ s = -5e^{-t} + 6t + 5$
5. $s = -\frac{3}{2}e^{-2t} + 4e^{-t} - 4t + c;\ s = -\frac{3}{2}(e^{-2t} - e^{-2}) + 4(e^{-t} - e^{-1}) - 4(t - 1)$
6. $y = 2\sin x - 3\cos x + c;\ y = 2\sin x - 3\cos x - \frac{5}{2} - \sqrt{3}$
7. $y = 3\tan x + c;\ y = 3\tan x - 1$
8. $y = x + \frac{1}{2}x^2 + c;\ y = x + \frac{1}{2}x^2 - \frac{29}{2}$ **9.** $y = \frac{3}{2}x^2 - 3x + c;\ y = \frac{3}{2}x^2 - 3x - \frac{1}{96}$
10. $y = 7x - 3x^2 + c;\ y = 7x - 3x^2 + 207$

Exercise 8.2

1. $y = ae^x;\ y = 2e^x$ **2.** $y = ae^{2x};\ y = -5e^{2x}$ **3.** $y = ae^{-3x};\ y = 20e^{3(1-x)}$
4. $y = ae^{-3x/4};\ y = 10e^{-\frac{3}{4}(x+1)}$ **5.** $y = ae^{-5x/8};\ y = 3e^{\frac{5}{8}(2-x)}$

6. $y = Ae^{-2x/a}$; $y = 8e^{2(1-x/a)}$ **7.** $x = Ae^{-qt/p}$; $x = 5ae^{-(2+qt/p)}$
8. $T = T_0e^{-\frac{1}{4}\theta}$; $T = 450e^{-\frac{1}{4}\theta}$ **9.** $i = Ie^{-3t/2}$; $i = 0.25e^{-3t/2}$
10. $i = Ie^{-Rt/L}$; $i = 0.35e^{-5t/9}$ **11.** $i = Ie^{-t/CR}$; $i = 0.85e^{250t}$
12. $M = M_0e^{-at}$; $M = 285e^{-1.56 \times 10^{-2}t}$; 301.37 kg; 50.67 years
13. $80.91918 \approx 80.9°C$; $24.356961 \approx 24.36$ minutes
14. $2.3199478 \approx 2.320$ hours; $4.3125623 \times 10^{11} \approx 4.31 \times 10^{11}$
15. $76.221692 \approx 76$ N; where $\theta = 4.95021 \approx 4.950$ radians
16. $0.486549 \approx 0.49$ A; $7.374494 \times 10^{-1} \approx 7.4 \times 10^{-1}$ seconds

Exercise 9.1

1. 1.2214028 **2.** 2.117 **3.** 4.4816891 **4.** 148.41316
5. 5.1847055×10^{21} **6.** 2.6881171×10^{43} **7.** 1.3937096×10^{65}
8. 7.2259738×10^{86} **9.** 3.505791×10^{95} **10.** 7.7220185×10^{99}

Exercise 9.2

1. 0.4723666 **2.** 0.1353353 **3.** 2.0611536×10^{-9}
4. $1.9287498 \times 10^{-22}$ **5.** 3.720076×10^{-44} **6.** $1.3838965 \times 10^{-87}$
7. $2.8524233 \times 10^{-96}$ **8.** $2.6010734 \times 10^{-99}$

Exercise 9.3

1. 3.490; 12.182; 268 337.29; 7.2004898×10^{10} **2.** 1.916; 7.029; 181.272; 5 956 538
3. 135.86; 210.96; 3911.40; 3 547 131.3 **4.** 258.70; 137.78; 22.55; 0.07
5. 3.244; 3.232; 3.067; 2.856
6. 12.057; 15.621; 436.426; 28 878 842; 1.2644955×10^{17}
7. 0.111; 2.212; 46.575

Exercise 9.4

1. 0.3365 **2.** 1.0296 **3.** 1.2809 **4.** 1.5686 **5.** 1.8245
6. 2.1633 **7.** 0.3988 **8.** 1.0613 **9.** 1.2947 **10.** 1.5707
11. 1.8278 **12.** 2.1702 **13.** 0.3584 **14.** 1.0627 **15.** 1.2952
16. 1.5721 **17.** 1.8288 **18.** 2.1712 **19.** 2.0009 **20.** 2.2954

Exercise 9.5

By calculator:
1. 2.6610 **2.** 3.3652 **3.** 3.5979 **4.** 3.8747
5. 4.1313 **6.** 4.4738 **7.** 4.3035 **8.** 4.5979
9. 4.6386 **10.** −1.2399 **11.** −1.0073 **12.** −0.7304
13. −0.1314 **14.** −0.3016 **15.** −0.007226 **16.** 5.7605
17. 8.8896 **18.** −5.9721 **19.** 15.7830 **20.** −9.6530

Exercise 9.6

By calculator:
1. 1.0534 **2.** 2.4449 **3.** 0.4090 **4.** 0.9493
5. 0.2393 **6.** 0.0556 **7.** 0.0305 **8.** 1.0531
9. 2.4468 **10.** 49.9989 **11.** 86.5048 **12.** 515.997
13. 2987.22 **14.** 9 376 340 **15.** 0.1205 **16.** 0.1932
17. 0.1336 **18.** 0.1255 **19.** 0.2800 **20.** 0.7570
21. 0.0757 **22.** 0.0005432 **23.** 0.0000924 **24.** 0.0000028

Exercise 9.7

By calculator:

1. 0.3206 **2.** 0.9014 **3.** 3.6336 **4.** 4.0992
5. 5.4583 **6.** 5.9937 **7.** 11.2923 **8.** -0.4695
9. -0.6507 **10.** -3.1543 **11.** -4.5338 **12.** -6.0727

Exercise 9.8

1. 12.05; 17.97; 48.85; 1617.7; 240088 **2.** 8.05; 1.09; 0.0006; 2.56×10^{-21}
3. 2.4995; 2.495; 2.450 **4.** 103.07; 106.46; 147.20; 3758.6
5. 0.1795; 1.789; 17.30; 126.32; 232.86 **6.** 0.106; 0.003; 0; 0
7. 0.037; 0.00046; 0 **8.** 43.2; 53.9; 129.9; 552
9. 26.33; 54.90; 57.49; 57.62 **10.** 0.171; 107.2; 107.2; 107.2; 107.2
11. 3.00; -0.85; 0.57; 0.08; -0.0004 **12.** 0.058; 0.191; 0.779; 8.087

Exercise 9.9

1. $\ln 5$ **2.** $\ln 3.5$ **3.** no solution **4.** no solution
5. $\frac{1}{2}\ln 1.5$ **6.** $\ln 1.25$ **7.** no solution **8.** $\frac{1}{2}\ln(4/7)$
9. $\frac{3}{2}\ln 1.5$ **10.** $\ln 4$ **11.** 0 or $\ln 3$ **12.** $\ln 2$ or $\ln 4$
13. $\ln 3$ **14.** $\ln\left(\frac{9+\sqrt{21}}{6}\right)$ or $\ln\left(\frac{9-\sqrt{21}}{6}\right)$ **15.** $-\ln 3$ or $-\ln 5$ **16.** $\ln 3$
17. no solution **18.** no solution **19.** no solution **20.** e^3
21. $e^{\frac{7}{3}}$ **22.** e^{-5} **23.** $e^{-\frac{11}{4}}$ **24.** e^{-1}
25. e or e^{-2} **26.** $e^{\frac{1}{2}}$ or $e^{-\frac{2}{3}}$ **27.** $2+e^{\frac{1}{2}}$ **28.** $\frac{5 \pm \sqrt{5}}{2}$
29. 2 **30.** $\frac{1 \pm \sqrt{73}}{12}$ **31.** 12 **32.** $\frac{1}{3}(e^{-2}-1)$
33. $2+e^5$ **34.** $3+e^{-\frac{1}{4}}$ **35.** $\frac{1}{4}(e^{10.5}-1)$ **36.** e
37. $\frac{1}{4}(e-1)$ **38.** $\frac{1}{p}(e^{cb/a}-q)$ **39.** $\frac{1}{p}[e^{c/(a+b)}-q]$ **40.** e^{-2}

Exercise 10.1

1. $2+\frac{3}{x-1}$ **2.** $2-\frac{1}{x+1}$ **3.** $3+\frac{4}{x-2}$ **4.** $4-\frac{9}{x+3}$
5. $\frac{5}{2}+\frac{27}{2(2x-3)}$ **6.** $2-\frac{3}{5x-2}$ **7.** $x+\frac{2}{x+1}$ **8.** $x-3+\frac{4}{x-1}$
9. $x+4+\frac{6}{x-1}$ **10.** $x+8+\frac{29}{x-3}$ **11.** $x+2-\frac{2}{2x-1}$
12. $x^2+2x+3+\frac{10}{2x-1}$ **13.** $x^2+3x+7+\frac{15}{x-2}$ **14.** x^2-x+2
15. $2x^2-x+3+\frac{2}{3x-2}$ **16.** $1+\frac{7x+9}{x^2-2x-3}$
17. $3+\frac{4(5x+1)}{2x^2-3x+5}$ **18.** $2x-\frac{7}{3}+\frac{69x-62}{3(3x^2+12x-5)}$
19. $6x+11+\frac{10x+57}{x^2+x-6}$ **20.** $6+\frac{17x^2+27x+27}{x^3-7x-6}$

Exercise 10.2

1. $\frac{7}{x-2}-\frac{5}{x-1}$ 2. $\frac{11}{x-2}-\frac{8}{x-1}$ 3. $\frac{3}{x-1}+\frac{1}{x-2}$ 4. $\frac{7}{3(x-1)}-\frac{7}{3(x+2)}$

5. $\frac{8}{x+1}-\frac{8}{x+2}$ 6. $\frac{3}{x+2}-\frac{3}{x+3}$ 7. $\frac{4}{5x}+\frac{21}{5(x+5)}$ 8. $\frac{12}{7(x-2)}+\frac{65}{7(x+5)}$

9. $-\frac{3}{2(x-2)}+\frac{9}{2(x-6)}$ 10. $\frac{11}{3x}-\frac{11}{3(x+3)}$

Exercise 10.3

1. $-\frac{3}{x}+\frac{3}{2(x-1)}+\frac{3}{2(x+1)}$ 2. $\frac{2}{x}-\frac{4}{x-1}+\frac{2}{x-2}$

3. $\frac{17}{2(x-1)}-\frac{17}{x-2}+\frac{17}{2(x-3)}$ 4. $\frac{11}{15(x+1)}-\frac{11}{6(x+4)}+\frac{11}{10(x+6)}$

5. $-\frac{2}{x}+\frac{3}{2(x-1)}+\frac{1}{2(x+1)}$ 6. $\frac{4}{9(x-1)}-\frac{17}{9(x-4)}+\frac{13}{9(x-7)}$

7. $\frac{11}{16x}+\frac{1}{12(x-2)}-\frac{37}{48(x-8)}$ 8. $\frac{37}{24(x+3)}+\frac{19}{30(x-9)}-\frac{87}{40(x+11)}$

9. $-\frac{2}{x}+\frac{3}{2(x-1)}+\frac{3}{2(x+1)}$ 10. $\frac{1}{6(x-1)}+\frac{8}{15(x+2)}+\frac{3}{10(x-3)}$

11. $-\frac{1}{x}+\frac{3}{2(x-1)}+\frac{1}{2(x+1)}$ 12. $-\frac{5}{4(x-2)}-\frac{1}{44(x+6)}+\frac{25}{11(x-5)}$

13. $-\frac{1}{2x}-\frac{7}{6(x+2)}+\frac{14}{3(x+5)}$ 14. $-\frac{107}{20(x-5)}+\frac{107}{120(x+5)}+\frac{203}{24(x-7)}$

15. $\frac{19}{40x}+\frac{15}{8(x-4)}+\frac{53}{20(x+10)}$ 16. $-\frac{85}{18(x+11)}-\frac{27}{14(x-5)}-\frac{22}{63(x+2)}$

Exercise 10.4

1. $\frac{1}{x-1}+\frac{2}{(x-1)^2}$ 2. $\frac{1}{x+1}-\frac{2}{(x+1)^2}$ 3. $\frac{2}{x+1}+\frac{1}{(x+1)^2}$

4. $\frac{5}{x-2}+\frac{16}{(x-2)^2}$ 5. $\frac{11}{x+3}-\frac{45}{(x+3)^2}$ 6. $\frac{2}{x-1}-\frac{1}{(x-1)^2}$

7. $\frac{15}{x-6}+\frac{134}{(x-6)^2}$ 8. $\frac{3}{2(2x-1)}+\frac{7}{2(2x-1)^2}$

9. $\frac{2}{2x-3}+\frac{1}{(2x-3)^2}$ 10. $\frac{2}{3(3x-2)}+\frac{13}{3(3x-2)^2}$

11. $\frac{10}{x-2}-\frac{10}{x-1}-\frac{10}{(x-1)^2}$ 12. $\frac{9}{x-1}-\frac{9}{x-2}+\frac{9}{(x-2)^2}$

13. $\frac{16}{x+2}-\frac{16}{x+1}+\frac{16}{(x+1)^2}$ 14. $\frac{8}{x-1}-\frac{8}{x}-\frac{8}{x^2}$

15. $\frac{5}{9x}-\frac{20}{9(4x-3)}+\frac{29}{3(4x-3)^2}$ 16. $-\frac{62}{121(2x+3)}+\frac{31}{121(x-4)}+\frac{12}{11(x-4)^2}$

17. $-\frac{3}{1444(3x-7)}+\frac{5}{1444(5x+1)}-\frac{49}{38(5x-1)^2}$ 18. $\frac{9}{x-2}-\frac{8}{x-1}-\frac{5}{(x-1)^2}$

19. $\frac{34}{25(x-3)}+\frac{141}{25(x+2)}-\frac{54}{5(x+2)^2}$ 20. $\frac{113}{1089(6x+1)}-\frac{281}{363(3x+5)}-\frac{442}{99(3x-5)^2}$

Exercise 10.5

1. $\frac{3}{x}-\frac{3x}{x^2+1}$ 2. $\frac{4}{5x}-\frac{5x}{4(x^2+4)}$ 3. $-\frac{7}{25x}+\frac{7x}{25(x^2+25)}$
4. $\frac{11}{2(x-1)}-\frac{11(x+1)}{2(x^2+1)}$ 5. $\frac{9}{x}-\frac{36x}{(4x^2+1)}$ 6. $\frac{2}{x}-\frac{2x}{x^2+1}$
7. $\frac{3}{x}-\frac{6x}{2x^2+1}$ 8. $\frac{5}{7(x+2)}-\frac{5(x-2)}{7(x^2+3)}$
9. $\frac{1}{x}-\frac{(x-1)}{x^2+1}$ 10. $\frac{3}{x}-\frac{(3x-2)}{x^2+1}$
11. $-\frac{4}{41(x-4)}+\frac{5(4x-25)}{41(5x^2+2)}$ 12. $-\frac{45}{32(x+5)}+\frac{9(5x+7)}{32(x^2+7)}$
13. $\frac{26}{23(2x+1)}-\frac{(39x-146)}{23(3x^2+5)}$ 14. $\frac{93}{131(3x-5)}-\frac{(775x+2041)}{655(5x^2+2)}$
15. $\frac{3}{2(x-1)}-\frac{(x-1)}{2(x^2+1)}$ 16. $-\frac{1}{5(x+1)}-\frac{(4x-9)}{5(x^2+4)}$
17. $\frac{11}{4(x-2)}-\frac{(3x-14)}{4(x^2+4)}$ 18. $\frac{23}{25(x-3)}+\frac{77x+181}{25(x^2+16)}$
19. $-\frac{17}{9(x+2)}-\frac{(19x-65)}{9(x^2+5)}$ 20. $-\frac{6}{7(2x+3)}+\frac{(37x-17)}{7(3x^2+2)}$

Exercise 10.6

1. $1+\frac{4}{x-1}$ 2. $1-\frac{6}{x+1}$ 3. $2+\frac{11}{x-2}$ 4. $4-\frac{21}{x+3}$
5. $\frac{7}{2}+\frac{5}{2(2x+1)}$ 6. $-2+\frac{13}{x+5}$ 7. $1+\frac{1}{x}-\frac{2}{x+1}$ 8. $1+\frac{1}{x}$
9. $1+\frac{2}{x+1}-\frac{5}{x+2}$ 10. $1-\frac{4}{x-1}+\frac{7}{x-2}$ 11. $2+\frac{13}{x+2}-\frac{23}{x+3}$
12. $1+\frac{1}{x}-\frac{4}{3x+1}$ 13. $-1+\frac{2}{x^2+1}$ 14. $1+\frac{1}{x+1}-\frac{3}{x+2}$
15. $1+\frac{1}{3(x-1)}-\frac{7}{3(x+2)}$ 16. $-1+\frac{1}{1-x}+\frac{1}{2-x}$
17. $4+\frac{1}{x+1}-\frac{14}{x+3}$ 18. $6+\frac{177}{7(x-5)}-\frac{16}{7(x+2)}$
19. $-4+\frac{19}{2x^2+3}$ 20. $1-\frac{6}{x-1}+\frac{12}{x-2}$

Exercise 11.1

1. $a^6+6a^5x+15a^4x^2+20a^3x^3+15a^2x^4+6ax^5+a^6$
2. $a^7+7a^6x+21a^5x^2+35a^4x^3+35a^3x^4+21a^2x^5+7ax^6+a^7$
3. $a^8+8a^7b+28a^6b^2+56a^5b^3+70a^4b^4+56a^3b^5+28a^2b^6+8ab^7+b^8$
4. $a^{10}+10a^9x+45a^8x^2+120a^7x^3+210a^6x^4+252a^5x^5+210a^4x^6+120a^3x^7+45a^2x^8$ $+10ax^9+x^{10}$
5. $x^{11}+11x^{10}y+55x^9y^2+165x^8y^3+330x^7y^4+462x^6y^5+462x^5y^6+330x^4y^7$ $+165x^3y^8+55x^2y^9+11xy^{10}+y^{11}$
6. $a^{12}+12a^{11}b+66a^{10}b^2+220a^9b^3+495a^8b^4+792a^7b^5+924a^6b^6+792a^5b^7$ $+495a^4b^8+220a^3b^9+66a^2b^{10}+12ab^{11}+b^{12}$

Exercise 11.2

1. $x^6+6x^5y+15x^4y^2+20x^3y^3+15x^2y^4+6xy^5+y^6$
2. $x^4+8x^3y+24x^2y^2+8xy^3+y^4$
3. $16x^4+32x^3y+24x^2y^2+8xy^3+y^4$
4. $x^5+15x^4y+90x^3y^2+270x^2y^3+405xy^4+243y^5$
5. $x^4-4x^3y+6x^2y^2-4xy^3+y^4$
6. $x^4-8x^3y+24x^2y^2-32xy^3+16y^4$
7. $16x^4-32x^3y+24x^2y^2-8xy^3+y^4$
8. $x^6-18x^5y+135x^4y^2-540x^3y^3+1215x^2y^4-1458xy^5+729y^6$
9. $x^8+4x^7y+7x^6y^2+7x^5y^3+\frac{35}{8}x^4y^4+\frac{7}{4}x^3y^5+\frac{7}{16}x^2y^6+\frac{1}{16}xy^7+\frac{1}{256}y^8$
10. $x^6-3x^5y+\frac{15}{4}x^4y^2-\frac{5}{2}x^3y^3+\frac{15}{16}x^2y^4-\frac{3}{16}xy^5+\frac{1}{64}y^6$
11. $16x^4+16x^3y+6x^2y^2+xy^3+\frac{1}{16}y^4$
12. $729x^6+2916x^5+4860x^4+4320x^3+2160x^2+576x+64$
13. $16-160x+600x^2-1000x^3+625x^4$
14. $32+80x+80x^2+40x^3+10x^4+x^5$
15. $1+8x+24x^2+32x^3+16x^4$
16. $1-10x+40x^2-80x^3+80x^4-32x^5$
17. $126x^5y^4$ **18.** $-462x^6y^5$ **19.** $560x^4y^3$ **20.** $1792x^2$
21. $59\,136x^6$ **22.** $-960x^3$ **23.** $22\,680x^3$ **24.** 1020.6
25. 114 688 **26.** -0.54

Exercise 11.3

1. $1-x+x^2-x^3$; $(-1)^rx^r$; $|x|<1$
2. $1+x+x^2+x^3$; x^r; $|x|<1$
3. $1-2x+3x^2-4x^3$; $(-1)^r(r+1)x^r$; $|x|<1$
4. $1+\frac{1}{2}x-\frac{1}{8}x^2+\frac{1}{16}x^3$; $\frac{\frac{1}{2}.-\frac{1}{2}.-\frac{3}{2}\,.\,.\,.\,.\,-(r-\frac{1}{2})}{r!}$; $|x|<1$
5. $1-\frac{1}{2}x+\frac{3}{8}x^2-\frac{5}{16}x^3$; $(-1)^r\frac{\frac{1}{2}.\frac{3}{2}.\frac{5}{2}\,.\,.\,.\,.\,(r-\frac{1}{2})}{r!}$; $|x|<1$
6. $1-4x+12x^2-32x^3$; $(-1)^r(r+1)(2x)^r$; $|x|<\frac{1}{2}$
7. $1+x-\frac{1}{2}x^2+\frac{1}{2}x^3$; $\frac{\frac{1}{2}.-\frac{1}{2}.-\frac{3}{2}\,.\,.\,.\,.-(r-\frac{3}{2})}{r!}(2x)^r$; $|x|<\frac{1}{2}$
8. $1-x+\frac{3}{2}x^2+\frac{5}{2}x^3$; $(-1)^r\frac{1.3.5\,.\,.\,.\,.(2r-1)}{r!}x^r$; $|x|<\frac{1}{2}$
9. $1+4x+12x^2+32x^3$; $(r+1)2^rx^r$; $|x|<\frac{1}{2}$
10. $1+x+\frac{3}{2}x^2+\frac{5}{2}x^3$; $\frac{1.3.5.\,.\,.\,.\,(2r-1)}{r!}x^r$; $|x|<\frac{1}{2}$
11. $1-x-\frac{1}{2}x^2-\frac{1}{2}x^3$; $-\frac{1.1.3.4.\,.\,.\,.\,(2r-3)}{r!}x^r$; $|x|<\frac{1}{2}$
12. $1-x+\frac{3}{4}x^2-\frac{1}{2}x^3$; $(-1)^r(r+1)\left(\frac{x}{2}\right)^r$; $|x|<2$
13. $1+\frac{1}{4}x-\frac{1}{32}x^2-\frac{1}{64}x^3$; $(-1)^{r-1}\frac{1.1.3.5.\,.\,.\,.\,(2r-3)}{r!}\left(\frac{x}{4}\right)^r$; $|x|<2$
14. $1-\frac{1}{4}x+\frac{3}{3x}x^2-\frac{5}{128}x^3$; $(-1)^r\frac{1.3.5.\,.\,.\,.\,(2r-1)}{r!}\left(\frac{x}{4}\right)^r$; $|x|<2$

15. $1+x+\frac{3}{4}x^2+\frac{1}{2}x^3$; $(r+1)\left(\frac{x}{2}\right)^r$; $|x|<2$

16. $1-\frac{1}{4}x-\frac{1}{32}x^2-\frac{1}{128}x^3$; $-\frac{1.1.3.5.\ldots.(2r-3)}{r!}\left(\frac{x}{4}\right)^r$; $|x|<2$

17. $1+\frac{1}{4}x+\frac{3}{32}x^2+\frac{5}{128}x^3$; $\frac{1.3.5.\ldots.(2r-1)}{r!}\left(\frac{x}{4}\right)^r$; $|x|<2$

18. $\frac{1}{4}-\frac{1}{4}x+\frac{3}{16}x^2-\frac{1}{8}x^3$; $(-1)^r\frac{1}{4}(r+1)\left(\frac{x}{2}\right)^r$; $|x|<2$

19. $\frac{1}{4}+\frac{1}{4}x+\frac{3}{16}x^2+\frac{1}{8}x^3$; $(r+1)\cdot\frac{1}{4}\left(\frac{x}{2}\right)^r$; $|x|<2$

20. $2+\frac{2}{4}x-\frac{2}{32}x^2+\frac{2}{128}x^3$; $(-1)^{r-1}\frac{1.1.3.\ldots.(2r-3)}{r!}\left(\frac{x}{4}\right)^r$; $|x|<\frac{1}{2}$

21. $2+\frac{3\times 2}{4}x-\frac{9\times 2}{32}x^2+\frac{27\times 2}{128}x^3$; $(-1)^{r-1}2\frac{1.1.3.\ldots.(2r-3)}{r!}\left(\frac{3x}{4}\right)^r$; $|x|<\frac{2}{3}$

22. $\frac{1}{4}+\frac{3}{4}x+\frac{27}{16}x^2+\frac{27}{8}x^3$; $\frac{1}{4}(r+2)\left(\frac{3x}{2}\right)^r$; $|x|<\frac{2}{3}$

23. $1+5x+\frac{15}{2}x^2+\frac{5}{2}x^3$; $(-1)^{r-1}\frac{5.3.1.1.\ldots.(2r-7)}{r!}x^r$; $|x|<\frac{1}{2}$

24. $1-x^2+x^4-x^6$; $(-1)^r x^{2r}$; $|x|<1$

25. $1-2x^2+3x^4-4x^6$; $(-1)^r(r+1)x^r$; $|x|<1$

26. $1+\frac{1}{2}x^2-\frac{1}{8}x^4+\frac{1}{16}x^6$; $(-1)^{r-1}\frac{1.1.3.\ldots.(2r-3)}{r!}\left(\frac{x^2}{2}\right)^r$; $|x|<1$

27. $1-2x^2+4x^4-8x^6$; $(-1)^r(2x^2)^r$; $|x|<\frac{1}{\sqrt{2}}$

28. $1+4x^2+16x^4+64x^6$; $(4x^2)^r$; $|x|<\frac{1}{2}$

29. $1+2x^2+4x^4+8x^6$; $(2x^2)^r$; $|x|<\frac{1}{2}$

30. $1+\frac{1}{10}x^2+\frac{1}{100}x^4+\frac{1}{1000}x^6$; $(\frac{1}{10}x^2)^r$; $|x|<\sqrt{10}$

31. $1+\frac{1}{2}x-\frac{1}{8}x^2$ **32.** $1-\frac{1}{2}x-\frac{1}{8}x^2$

33. $1+\frac{1}{2}x+\frac{3}{8}x^2$ **34.** $1-\frac{1}{2}x^2$ **35.** $1-\frac{1}{2}x^2$ **36.** $1+x+\frac{1}{2}x^2$

37. $1+x+\frac{1}{2}x^2$ **38.** $1+\frac{3}{2}x-\frac{1}{8}x^2$ **39.** $1+\frac{1}{2}x-\frac{5}{8}x^2$ **40.** $1-x+x^2$

41. $\sqrt{5}-\sqrt{5}.x+\sqrt{5}.x^2$ **42.** $1-\frac{1}{10}x+\frac{3}{200}x^2$

Exercise 11.4

1. $3x^2h$ **2.** 0.03 **3.** $\frac{1}{50}x^2$ **4.** 0.44

5. $-\frac{1}{10}$ **6.** 9 per cent

Exercise 12.1

1. $y\approx 1+x+\frac{3}{2}x^2+\frac{7}{6}x^3$ **2.** $y\approx 1-x+\frac{3}{2}x^2-\frac{7}{6}x^3+\frac{13}{24}x^4$

3. $x-\frac{1}{2}x^2-\frac{1}{3}x^3+\frac{5}{6}x^4$ **4.** $1+x+\frac{1}{2}x^2+\frac{1}{8}x^3+\frac{1}{48}x^4$

5. $\frac{1}{x}+\frac{1}{x^2}+\frac{1}{2x^3}$ **6.** $\frac{1}{x}-\frac{1}{x^2}+\frac{1}{x^3}-\frac{1}{3x^4}$

7. 1.010 **8.** 0.954 **9.** 1.131 **10.** 0.0101 **11.** 0.0005

Exercise 12.2

1. 0.368 **2.** 1.1052 **3.** 0.9048 **4.** 1.6487 **5.** 0.6065

6. 1.0101 **7.** 0.9900 **8.** 7.389 **9.** 20.086 **10.** 0.135
11. 0.050 **12.** 16.487 **13.** −0.6550 **14.** 0.0735

Exercise 13.1

15. 99.7; 1.95 **16.** 99; 3 **17.** 148; 2.9 **18.** 203; 3.1
19. 89; 1.85 **20.** 248; 0.93 **21.** 93; 2.43 **22.** 0.8; 6.2

Exercise 13.2

1. $m_0 = 1$; $p = 1.35 \times 10^{-11}$; 5.14×10^{10} s; 7.78×10^{10} s; 0.48 kg
2. $T_0 \approx 50$; $\mu \approx 0.4$; 144 N; 1.73 rad **3.** $Q_0 \approx 2.5 \times 10^{-4}$; $p \approx 2$
4. $I_0 \approx 0.21$; $a \approx 42$; 0.199 A; 0.03 s **5.** $H_0 \approx 10$; $k \approx 0.07$; 36.5 cm; 17.5 days
6. $a \approx 50$; $b \approx 2$; 800; 2.32

Exercise 14.1

1. $y = 3.1$; $x = \pm 3.3$ **2.** $y = 37.4$; $x = 2.52$ or 0.48
3. $y = 0.04$; $x = 3.9$ or -0.9 **4.** $y = 1.2$; $x = \pm 1$
5. $y = -0.25$; $x = 3.8$ or 1.2 **6.** Crosses Oy where $y = 1$; $x = -0.27$ or -3.73.
7. Axis of symmetry $x = 3/4$, vertex (3/4, 7/8)
8. Axis of symmetry $x = 1$, vertex $(1, \frac{1}{2})$
9. vertex (0, 0), $y = x^2$ magnified parallel to Oy by 3
10. vertex (0, 0), $y = x^2$ magnified parallel to Oy by $-4/5$
11. vertex (1, 0), $y = (x-1)^2$ magnified parallel to Oy by 2
12. vertex $(-2, 0)$, $y = -(x+2)^2$ magnified parallel to Oy by $-\frac{1}{3}$
13. vertex (4, 0), $y = (x-4)^2$ magnified parallel to Oy by 2
14. vertex (0, 3), $y = x^2$ translated 3 units along Oy
15. vertex $(0, -6)$, $y = x^2$ translated -6 units along Oy
16. vertex (0, 5/8), $y = x^2$ translated 5/8 units along Oy then magnified by $-3/4$
17. vertex (2, 5), $y = x^2$ translated 5 units along Oy then 2 units parallel to Ox
18. vertex $(-1, -4/3)$, $y = x^2$ translated $-4/3$ units along Oy then -1 unit parallel to Ox, then magnified by $-2/3$
19. vertex (2/5, 11/6), $y = x^2$ translated 11/6 units along Oy then 2/5 units parallel to Ox, then magnified by 4/3
20. vertex (q, r), $y = x^2$ translated r units along Oy then q units along Ox, then magnified by p
21. vertex (1, 2), $y = x^2$ translated 2 units along Oy, then 1 unit along Ox
22. vertex $(-3/2, -12)$, $y = x^2$ translated -12 units along Oy then $-3/2$ units parallel to Ox, then magnified by 4
23. vertex $(\frac{1}{2}, \frac{3}{4})$, $y = x^2$ translated $\frac{3}{4}$ unit along Oy then $\frac{1}{2}$ unit parallel to Ox
24. vertex (3, 1), $x = y^2$ translated 3 units along Ox then 1 unit parallel to Oy, then magnified by 2

Exercise 14.2

1. $y = 1.5$; $x = 0.5$ **2.** $y = 0.4$; $x = 0.6$ **3.** $y = 5.5$; $x = 2.25$
4. $y = 7.5$; $x = -5.9$ **5.** $y = 7.3$; $x = 2.4$
6. $y = -5.2$; $x = 56.6$ **7.** $y = 8.05$; $x = 4.5$

Exercise 14.3

1. $y = 18.0$; $x = 6.2$ **2.** $y = 58.7$; $x = 8.7$ **3.** $y = 60.8$; $x = 8.5$
4. $y = 7.52$; $x = 4.5$ **5.** $y = 12.2$; $x = 0.27$ **6.** $y = 19$; $x = 0.244$
7. $y = 48.8$; $x = 0.156$ **8.** $y = 58.7$; $x = 1.52$

Exercise 15.1

1. $y = 138\,000$; $x = 10.63$ **2.** $y = 20.41$; $x = 0.224$

3. $y = 87.39$; $x = 4.27$ **4.** $k = 0.000125$; $W = 14.45$ kg; $R = 518$ mm
5. $a = 1.4$; $V = 8.52$ m/s; $h = 77.2$ m
6. $I = 0.004$ V; $I = 0.242$ A; $V = 162.5$ V
7. $T = 2.L^{\frac{1}{2}}$; $L = 6.25$ m; $T = 3.162$ s
8. $t = 42.V^{-1}$; $t = 1.87$ s; $V = 14$ m/s
9. $y = 8.95x^{0.423}$; $y = 11.6$; $x = 0.62$
10. $N \approx 8300 \times d^{-1.07}$ N: 260.6, 168.8; d: 25.45, 36.85
11. $a \approx 200$; $n \approx -2$ F: 8, 4.08; d: 2.5, 6.2
12. $P \approx 42000 \times v^{-1.015}$ P: 3320, 4730; v: 13.0, 8.3
13. $a = 0.52$; $n = -3$ W: 0.005; $V = 7.5$

Exercise 16.1

1. $\begin{pmatrix} 3 & 6 \\ 6 & 5 \end{pmatrix}$ **2.** $\begin{pmatrix} 15 & 24 \\ 21 & 6 \end{pmatrix}$ **3.** $\begin{pmatrix} 9 & 7 \\ 5 & 8 \end{pmatrix}$ **4.** $\begin{pmatrix} 8 & 14 \\ 5 & -2 \end{pmatrix}$

5. $\begin{pmatrix} 3a & 11a \\ 9a & 17a \end{pmatrix}$ **6.** $\begin{pmatrix} 6b & 5b \\ -2b & -16b \end{pmatrix}$ **7.** $\begin{pmatrix} \overline{2a-b} & \overline{3a+2b} \\ \overline{5a-b} & \overline{a+5b} \end{pmatrix}$

8. $\begin{pmatrix} 1 & 1 \\ 1 & 1 \end{pmatrix}$ **9.** $\begin{pmatrix} 6 & 1 \\ 3 & 6 \end{pmatrix}$ **10.** $\begin{pmatrix} 19 & -5 \\ -13 & -13 \end{pmatrix}$

11. $\begin{pmatrix} 62 & 7 \\ 31 & 80 \end{pmatrix}$ **12.** $\begin{pmatrix} 73 & -54 \\ 12 & -72 \end{pmatrix}$

13. $\begin{pmatrix} 11a & 11a \\ -22a & 22a \end{pmatrix}$ **14.** $\begin{pmatrix} \overline{-a+9b} & \overline{-2a+5b} \\ \overline{4a-b} & \overline{-3b-5a} \end{pmatrix}$

Exercise 16.2

1. $\begin{pmatrix} 7 & 8 \\ 18 & 17 \end{pmatrix}$ **2.** $\begin{pmatrix} -2 & 38 \\ -6 & 54 \end{pmatrix}$ **3.** $\begin{pmatrix} 13 & -4 \\ 4 & -1 \end{pmatrix}$

4. $\begin{pmatrix} 5a & -b \\ -c & 5d \end{pmatrix}$ **5.** $\begin{pmatrix} \overline{pa+r} & \overline{-qa+r} \\ \overline{qa+r} & pa \end{pmatrix}$ **6.** $\begin{pmatrix} 3ax & 7bx \\ -7ay & 3by \end{pmatrix}$

7. $\begin{pmatrix} 1 & 0 \\ 0 & 1 \end{pmatrix}$ **8.** $\begin{pmatrix} 1 & 0 \\ 0 & 1 \end{pmatrix}$ **9.** $\begin{pmatrix} 1 & 0 \\ 0 & 1 \end{pmatrix}$ **10.** $\begin{pmatrix} 1 & 0 \\ 0 & 1 \end{pmatrix}$

Exercise 16.3

1. $\begin{pmatrix} 4 \\ 10 \end{pmatrix}$ **2.** $\begin{pmatrix} 23 \\ 3 \end{pmatrix}$ **3.** $\begin{pmatrix} -14 \\ 1 \end{pmatrix}$ **4.** $\begin{pmatrix} 9 \\ -29 \end{pmatrix}$ **5.** $\begin{pmatrix} a \\ b \end{pmatrix}$

6. $\begin{pmatrix} -a \\ -3a \end{pmatrix}$ **7.** $\begin{pmatrix} \overline{-3a^2+4b^2} \\ \overline{-5a^2-7b^2} \end{pmatrix}$ **8.** $\begin{pmatrix} \overline{aA+bB} \\ \overline{cA+dB} \end{pmatrix}$ **9.** $\begin{pmatrix} \overline{aC+bD} \\ \overline{cC+dD} \end{pmatrix}$

Exercise 16.4

1. $\begin{pmatrix} 5 & 11 \\ 11 & 25 \end{pmatrix}$ **2.** $\begin{pmatrix} 5 & 4 \\ 4 & 5 \end{pmatrix}$ **3.** $\begin{pmatrix} 5 & 4 \\ 4 & 5 \end{pmatrix}$ **4.** $\begin{pmatrix} 50 & 82 \\ -7 & -5 \end{pmatrix}$

5. $\begin{pmatrix} 36 & 0 \\ 39 & 9 \end{pmatrix}$ **6.** $\begin{pmatrix} -28 & 13 \\ 21 & -6 \end{pmatrix}$ **7.** $\begin{pmatrix} -34 & 7 \\ 15 & 0 \end{pmatrix}$ **8.** $\begin{pmatrix} \overline{ac+bd} & \overline{ad+bc} \\ \overline{bc+ad} & \overline{bd+ac} \end{pmatrix}$

9. $(A \times B) \times C = A \times (B \times C) = \begin{pmatrix} 59 & 27 \\ 133 & 61 \end{pmatrix}$

10. $\begin{pmatrix} -93 & 141 \\ 224 & -192 \end{pmatrix}$ **11.** $\begin{pmatrix} -351 & -26 \\ 363 & 66 \end{pmatrix}$

12. $\begin{pmatrix} \overline{ac+bd} & \overline{ad+bc} \\ \overline{bc+ad} & \overline{bd+ac} \end{pmatrix}$ **13.** $\begin{pmatrix} \overline{ca+db} & \overline{cb+da} \\ \overline{da+cb} & \overline{db+ca} \end{pmatrix}$

14. $\begin{pmatrix} \overline{aA+bB} & \overline{aC+bD} \\ \overline{cA+dB} & \overline{cC+dD} \end{pmatrix}$ **15.** $\begin{pmatrix} \overline{Aa+Cc} & \overline{Ab+Cd} \\ \overline{Ba+Dc} & \overline{Bb+Dd} \end{pmatrix}$

16. $\begin{pmatrix} 3 & 5 \\ -2 & 7 \end{pmatrix}$ **17.** $\begin{pmatrix} -18 & -35 \\ 29 & -17 \end{pmatrix}$ **18.** $\begin{pmatrix} 3 & 5 \\ -2 & 7 \end{pmatrix}$

19. $\begin{pmatrix} -18 & -35 \\ 29 & -17 \end{pmatrix}$ **20.** $\begin{pmatrix} a & b \\ c & d \end{pmatrix}$ **21.** $\begin{pmatrix} a & b \\ c & d \end{pmatrix}$

Exercise 16.5

1. $\begin{pmatrix} 1 & 0 \\ 0 & 1 \end{pmatrix}$ **2.** $\begin{pmatrix} 1 & 0 \\ 0 & 1 \end{pmatrix}$ **3.** $\begin{pmatrix} 1 & 0 \\ 0 & 1 \end{pmatrix}$ **4.** $\begin{pmatrix} 1 & 0 \\ 0 & 1 \end{pmatrix}$

5. $\begin{pmatrix} 1 & 0 \\ 0 & 1 \end{pmatrix}$ **6.** $\begin{pmatrix} 1 & 0 \\ 0 & 1 \end{pmatrix}$ **7.** $\begin{pmatrix} 5 & 0 \\ 0 & 5 \end{pmatrix}$ **8.** $\begin{pmatrix} 5 & 0 \\ 0 & 5 \end{pmatrix}$

9. $\begin{pmatrix} 8 & 0 \\ 0 & 8 \end{pmatrix}$ **10.** $\begin{pmatrix} 8 & 0 \\ 0 & 8 \end{pmatrix}$ **11.** $\begin{pmatrix} 11 & 0 \\ 0 & 11 \end{pmatrix}$ **12.** $\begin{pmatrix} 11 & 0 \\ 0 & 11 \end{pmatrix}$

Exercise 16.6

1. 1 **2.** 1 **3.** 1 **4.** 5 **5.** 5 **6.** 1 **7.** 8 **8.** 8
9. -44 **10.** 32 **11.** $2a^2$ **12.** a^2 **13.** 2 **14.** a^2 **15.** $3a$

Exercise 17.3

1. $\begin{pmatrix} 2 & -1 \\ -5/2 & 3/2 \end{pmatrix}$ **2.** $\begin{pmatrix} 2/5 & -1/5 \\ -3/5 & 4/5 \end{pmatrix}$ **3.** $\begin{pmatrix} 2/5 & -3/5 \\ -3/5 & 7/5 \end{pmatrix}$

4. $\begin{pmatrix} 4/21 & -1/7 \\ -5/21 & 3/7 \end{pmatrix}$ **5.** $\begin{pmatrix} 3/22 & -5/22 \\ 1/11 & 2/11 \end{pmatrix}$ **6.** $\begin{pmatrix} 1 & -3/2 \\ 2 & -5/2 \end{pmatrix}$ **7.** $\begin{pmatrix} 3/5 & -2/5 \\ 4/5 & -1/5 \end{pmatrix}$

8. $\begin{pmatrix} -3/23 & -2/23 \\ 7/23 & -3/23 \end{pmatrix}$ **9.** $\begin{pmatrix} 1/2a & -1/2a \\ 1/2b & 1/2b \end{pmatrix}$ **10.** $\begin{pmatrix} -5/97 & 12/97 \\ 11/97 & -7/97 \end{pmatrix}$

Exercise 17.4

1. $x = 7, y = -2$ **2.** $x = 3, y = 1$ **3.** $x = -3\frac{4}{15}, y = \frac{3}{5}$ **4.** $a = 3, b = 1$
5. $x = 13, y = -20$ **6.** $p = -23/39, q = 5/13$ **7.** $x = 2\frac{3}{5}, y = 1\frac{19}{20}$

Exercise 17.5

8. $x = 8, y = 3$ **9** $x = 3/2, y = 3/4$ **10.** $x = 2/7, y = 1$
No solutions for 11, 12, 13, 14, 15, 16

Exercise 17.6

The following are singular matrices: 1, 3, 4, 5, 7.

9. $p = \pm 1$ **10.** $q = 3\frac{3}{4}$ **11.** $p = -8\frac{3}{4}$
12. $p = \pm 9$ **13.** $a = 5$ **14.** $a = -5$ **15.** $x + 3y = 0$
16. $p + 6 \neq 0$ **17.** $5a + 6 \neq 0$ **18.** $a \neq 0$ and $b \neq 0$

Exercise 17.7

1. $i_1 = 1\frac{32}{35};\ i_2 = 1\frac{31}{70}$ **2.** $V_1 = 46;\ V_2 = 34$
3. $i_1 = -9/46;\ i_2 = 2\frac{47}{184}$ **4.** $a = 1/15;\ b = 83\frac{1}{3}$
5. $p = 2894\frac{14}{19};\ q = 5/19$ **6.** $V_1 = 108\frac{1}{29};\ V_2 = 35\frac{1}{29}$
7. $r_1 = 5\frac{325}{384};\ r_2 = 5\frac{17}{192}$

Exercise 18.1

1. $\pm j.3$ **2.** $\pm j.5$ **3.** $\pm j.11$ **4.** $\pm j.a$ **5.** $\pm j.\sqrt{2}$
6. $\pm j.\sqrt{7}$ **7.** $\pm j.\sqrt{2.5}$ **8.** $\pm j$ **9.** $\pm j.\sqrt{2}$ **10.** $\pm j.2\sqrt{\frac{2}{3}}$
11. $\pm j.\dfrac{q}{p}$ **12.** $\pm j.\sqrt{\dfrac{q}{p}}$

Exercise 18.2

1. $-1 \pm j.3$ **2.** $1 \pm j.4$ **3.** $-2 \pm j.2$ **4.** $2 \pm j.2$
5. $-\dfrac{1}{2} \pm j.\dfrac{\sqrt{3}}{2}$ **6.** $\dfrac{1}{2} - j.\dfrac{\sqrt{3}}{2}$ **7.** $-\dfrac{1}{2} \pm j.\dfrac{\sqrt{11}}{2}$ **8.** $-\dfrac{5}{8} \pm j.\dfrac{\sqrt{23}}{8}$

Exercise 18.4

1. $\begin{pmatrix} 1 \\ 3 \end{pmatrix}$ **2.** $\begin{pmatrix} -1 \\ 2 \end{pmatrix}$ **3.** $\begin{pmatrix} -2 \\ -3 \end{pmatrix}$ **4.** $\begin{pmatrix} -\frac{3}{2} \\ -\frac{5}{2} \end{pmatrix}$ **5.** $\begin{pmatrix} -2 \\ 3 \end{pmatrix}$ **6.** $\begin{pmatrix} a \\ b \end{pmatrix}$

7. $\begin{pmatrix} 2a \\ 2b \end{pmatrix}$ **8.** $\begin{pmatrix} 3a \\ 3b \end{pmatrix}$ **9.** $\begin{pmatrix} -a \\ -b \end{pmatrix}$ **10.** $\begin{pmatrix} pa \\ pb \end{pmatrix}$ **11.** $\begin{pmatrix} a+c \\ b+d \end{pmatrix}$

Exercise 18.5

1. $3 + j.3$ **2.** $6 + j.6$ **3.** $7 + j.9$ **4.** $-1 + j.3$
5. $-1 - j.1$ **6.** $-3 - j.1$ **7.** $2 - j.4$ **8.** $1 + j.5$
9. $(2a + 3c) + j(2b + 3d)$ **10.** $ax + j.ay$ **11.** $-x - j.y$ **12.** $0 + j.0 = 0$
13. $(x_1 - x_2) + j(y - y_2)$

Exercise 18.6

1. $2 + j.2$ **2.** $1 + j.4$ **3.** $6 + j.2$ **4.** $6 + j.4$
5. $1 - j.10$ **6.** $-6 - j.8$ **7.** $-6 + j.8$ **8.** $-20 - j.1$
9. $13 - j.2$ **10.** $6 + j.1$ **11.** $27 + j.20$

Exercise 18.7

1. $-6 + j.17$ **2.** $6 + j.17$ **3.** $20 - j.21$ **4.** $-1 - j.18$
5. $2 - j.59$ **6.** $0 - j.2$ **7.** 2 **8.** 5 **9.** 2
10. 1 **11.** 1 **12.** -10 **13.** 10 **14.** $b(a^2 + b^2) + j.a(a^2 + b^2)$

Exercise 18.8

1. $\frac{1}{5} - j.\frac{2}{5}$ **2.** $\frac{1}{2} + j.\frac{1}{2}$ **3.** $\frac{3}{13} - j.\frac{2}{13}$ **4.** $\frac{5}{41} + j.\frac{4}{41}$

5. $\frac{5}{61}+j.\frac{6}{61}$ 6. $-\frac{6}{61}+j.\frac{5}{61}$ 7. $\frac{3}{17}-j.\frac{12}{17}$ 8. $\frac{96}{169}+j.\frac{40}{169}$
9. $\frac{3}{25}-j.\frac{4}{25}$ 10. $\frac{15}{17}+j.\frac{8}{17}$ 11. $-\frac{7}{25}+j.\frac{24}{25}$ 12. $\frac{40}{41}-j.\frac{9}{41}$

Exercise 18.9

1. $\frac{3}{5}+j.\frac{1}{5}$ 2. $\frac{5}{13}+j.\frac{1}{13}$ 3. $\frac{1}{13}-j.\frac{8}{13}$ 4. $\frac{24}{25}+j.\frac{7}{25}$
5. $0-j$ 6. $\frac{66}{53}-j.\frac{19}{53}$ 7. $\frac{7}{13}-j.\frac{9}{13}$ 8. $\frac{347}{169}-j.\frac{46}{169}$
9. $\frac{7}{26}+j.\frac{17}{26}$ 10. $0+j.2$ 11. $-5+j.12$ 12. $-2+j.2$
13. $1+j.3$ 14. $3+j.7$ 15. $\frac{13}{5}-j.\frac{14}{5}$ 16. $\frac{6}{5}-j.\frac{8}{5}$
17. -1.6 18. $-117-j.44$

Exercise 19.1

1. $\left[2\sqrt{2}, \frac{\pi}{4}\right]$ or $2\sqrt{2}\left(\cos\frac{\pi}{4}+j\sin\frac{\pi}{4}\right)$
2. $[5, \theta]$ or $5(\cos\theta+j\sin\theta)$ where $\tan\theta = \frac{3}{4}$ and θ is acute
3. $\left[2, \frac{\pi}{3}\right]$ or $2\left(\cos\frac{\pi}{3}+j\sin\frac{\pi}{3}\right)$
4. $[\sqrt{34}, (2\pi-\alpha)]$ where $\tan\alpha = \frac{5}{3}$
5. $[13, \beta]$ where $\tan\beta = 2.4$
6. $[25, (2\pi-\theta)]$ where $\tan\theta = \frac{7}{24}$
7. $[17, (\pi+\gamma)]$ where $\tan\gamma = \frac{15}{8}$
8. $[\sqrt{85}, \alpha]$ where $\tan\alpha = \frac{7}{6}$
9. $\left[a\sqrt{2}, \frac{\pi}{4}\right]$ 10. $[1, (\pi-\theta)]$ where $\tan\theta = \frac{3}{4}$ 11. $2\sqrt{2}+j.2\sqrt{2}$
12. $\frac{5\sqrt{3}}{2}-j.\frac{5}{2}$ 13. $5\sqrt{2}-j.5\sqrt{2}$ 14. $-5\sqrt{2}-j.5\sqrt{2}$
15. $-\sqrt{3}+j$ 16. $-4+j.4\sqrt{3}$ 17. $8-j.8\sqrt{3}$
18. $1.9318517+j(0.5176381)$ 19. $-1.1247553+j(4.8718503)$
20. $-2.0914323-j(5.6236919)$ 21. $9.9923493-j(13.753289)$

Exercise 19.2

1. mod 5; arg $(\pi-\theta)$ where $\tan\theta = \frac{4}{3}$
2. mod 5; arg $(\pi+\theta)$ where $\tan\theta = \frac{3}{4}$
3. mod 13; arg α where $\tan\alpha = \frac{12}{5}$
4. mod 13; arg $(\pi-\theta)$ where $\tan\theta = \frac{5}{12}$
5. mod $\frac{1}{\sqrt{2}}$; arg $\frac{\pi}{4}$
6. mod $\sqrt{13}$; arg $(\pi+\alpha)$ where $\tan\alpha = \frac{2}{3}$
7. mod $\sqrt{2}$; arg $\frac{\pi}{4}$ 8. mod $\sqrt{2}$; arg $\frac{3\pi}{4}$
9. mod $\sqrt{13}$; arg θ where $\tan\theta = \frac{2}{3}$
10. mod $\sqrt{13}$; arg $(\pi+\theta)$ where $\tan\theta = \frac{2}{3}$
11. mod a; arg $(\pi+\theta)$ 12. mod r; arg $2(\pi-\theta)$
13. mod $\frac{1}{2}$; arg $\left(\frac{5\pi}{3}\right)$ 14. mod $\sqrt{2}$; arg $\left(\frac{7\pi}{6}\right)$
15. mod 2; arg 0 16. mod 13; arg 0 17. mod 12; arg $\frac{\pi}{2}$ 18. mod 20; arg $\frac{\pi}{6}$
19. mod ab; arg 3θ 20. mod $3ab/2$; arg $(2\pi+\theta)$ or θ 21. mod 1; arg 0

Exercise 19.3

1. $\left[\frac{3}{2}, \frac{\pi}{6}\right]$ **2.** $\left[\frac{4}{5}, \frac{\pi}{6}\right]$ **3.** $\left[\frac{7}{11}, \frac{\pi}{2}\right]$ **4.** $\left[3, \frac{3\pi}{2}\right]$

5. $\left[3, \frac{\theta}{2}\right]$ **6.** $[5, 3\theta]$

Exercise 19.4

1. $\pm j$ **2.** ± 1 **3.** ± 5 **4.** $\pm j.8$ **5.** $\pm\sqrt{6}$

6. $\pm j.\sqrt{10}$ **7.** $\pm\left(\frac{5\sqrt{3}}{2} + j.\frac{5}{2}\right)$ **8.** $\pm(4 + j.4\sqrt{3})$ **9.** $\pm\left(\frac{5\sqrt{5}}{2} - j.\frac{5\sqrt{5}}{2}\right)$

10. $\pm(2\sqrt{5} + j.2\sqrt{5})$ **11.** $\pm(a\cos\theta + j.a\sin\theta)$ **12.** $\pm\left(b\cos\frac{\theta}{2} - j.b\sin\frac{\theta}{2}\right)$

13. $\pm\left(a\sin\frac{\theta}{2} - j.a\cos\frac{\theta}{2}\right)$ **14.** $z = 1$ or -1 **15.** $z = 2$ or -2

16. $z = \pm(1 + j)$ **17.** $\pm\left(\frac{3}{2} - j.\frac{\sqrt{3}}{2}\right)$

Exercise 20.3

1. $1; \frac{\pi}{\omega}$ **2.** $1; \frac{2\pi}{3\omega}$ **3.** $1; \frac{4\pi}{\omega}$ **4.** $1; \frac{4\pi}{\omega}$ **5.** $\frac{1}{2}; \frac{\pi}{\omega}$

6. $1; \frac{\pi}{2\omega}$ **7.** $2; \frac{2\pi}{\omega}$ **8.** $3; \frac{2\pi}{\omega}$ **9.** $a; \frac{4\pi}{\omega}$ **10.** $\frac{p}{2}; \frac{\pi}{\omega}$

Exercise 20.4

1. 1; 2 rad/s; $\frac{1}{\pi}$; π seconds

2. 1; $\frac{1}{4}$ rad/s; $\frac{1}{8\pi}$; 4π seconds

3. 3; 1.5 rad/s; $\frac{3}{4\pi}$; $\frac{4\pi}{3}$ seconds

4. $\frac{1}{2}$; 10 rad/s; $\frac{5}{\pi}$; $\frac{\pi}{5}$ seconds

5. $\frac{19}{2}$; 8 rad/s; $\frac{4}{\pi}$; $\frac{\pi}{4}$ seconds

6. $\frac{1}{2}$; 1 rad/s; $\frac{1}{2\pi}$; 2π seconds

7. 1; 2 rad/s; $\frac{1}{\pi}$; π seconds

8. 1; 3 rad/s; $\frac{3}{2\pi}$; $\frac{2\pi}{3}$ seconds

9. 3; $\frac{1}{2}$ rad/s; $\frac{1}{4\pi}$; 4π seconds

10. 6; 10.4 rad/s; $\dfrac{5.2}{\pi}$; $\dfrac{\pi}{5.2}$ seconds

11. $\dfrac{25}{8}$; 1.2 rad/s; $\dfrac{0.6}{\pi}$; $\dfrac{5\pi}{3}$ seconds

12. $\left|\dfrac{b-a}{2}\right|$; 2 rad/s; $\dfrac{1}{\pi}$, π seconds

13. $\left|\dfrac{b-a}{2}\right|$; $2p$ rad/s; $\dfrac{p}{\pi}$; $\dfrac{\pi}{p}$ seconds

14. $\dfrac{(a+b)}{2}$; $2p$ rad/s; $\dfrac{p}{\pi}$; $\dfrac{\pi}{p}$ seconds

15. $\dfrac{5a}{2}$; $2p$ rad/s; $\dfrac{p}{\pi}$; $\dfrac{\pi}{p}$ seconds

16. $\dfrac{15a}{2}$; p rad/s; $\dfrac{p}{2\pi}$; $\dfrac{2\pi}{p}$ seconds

Exercise 21.1

1. (a) 50π (b) 10π (c) 400π rad/s

2. (a) $\frac{1}{25}$ (b) $\frac{1}{5}$ (c) $\frac{1}{200}$

3. 8.3×10^{-3}; 8.3×10^{-4}; 1.1×10^{-8}; 2.53×10^{-3}; 1.9×10^{-6}

4. 40 Hz; 3333.3 Hz = $3\frac{1}{3}$ kHz; 1.6129×10^5 Hz; 28.6 MHz

5. 120; 100π rad/s; 50 Hz; $\frac{1}{50}$ s; zero phase angle

6. 120; 100π rad/s; 50 Hz; $\dfrac{1}{50}$ s; $\dfrac{\pi}{2}$ leading

7. 120; 100π rad/s; 50 Hz; $\dfrac{1}{50}$ s; $\dfrac{\pi}{2}$ leading

8. 120; 100π rad/s; 50 Hz; $\dfrac{1}{50}$ s; $\dfrac{3\pi}{4}$ leading

9. 120; 100π rad/s; 50 Hz; $\dfrac{1}{50}$ s; π leading

10. 240; 10^6 rad/s; $\dfrac{2 \times 10^5}{\pi}$ Hz; $\dfrac{\pi}{2} \times 10^{-5}$ s; $\dfrac{\pi}{6}$ leading

11. 6×10^5; 3×10^9 rad/s; $\dfrac{15 \times 10^8}{\pi}$ Hz; $\dfrac{\pi}{15} \times 10^{-8}$ s; $\dfrac{\pi}{3}$ leading

12. 120; 50π rad/s; 25 Hz; $\dfrac{1}{25}$ s; $\dfrac{\pi}{2}$ lagging

13. 120; $550\,\pi$ rad/s; 275 Hz; $\dfrac{1}{275}$ s; $\dfrac{\pi}{4}$ lagging

14. 2400; 100π rad/s; 50 Hz; $\dfrac{1}{50}$ s; $\dfrac{\pi}{6}$ lagging

15. 2.4×10^3; 200π rad/s; 100 Hz; $\dfrac{1}{100}$ s; $\dfrac{\pi}{4}$ leading

16. 3650; 50 rad/s; $\dfrac{25}{\pi}$ Hz; $\dfrac{\pi}{25}$ s; $\dfrac{2\pi}{3}$ leading

17. 4345; 256 rad/s; $\frac{128}{\pi}$ Hz; $\frac{\pi}{128}$ s; $\frac{17\pi}{30}$ leading

18. $250\sin\left(100\pi t+\frac{\pi}{4}\right)$

19. $100\sin\left(1000\pi t-\frac{\pi}{6}\right)$

20. $50\sin\left(40t+\frac{\pi}{18}\right)$

21. $25\sin\left(10^8\times\pi t-\frac{\pi}{9}\right)$

22. $10\sqrt{2}\sin\left(\omega t+\frac{\pi}{4}\right)$

23. $10\sqrt{2}\sin\left(\omega t-\frac{\pi}{4}\right)$

24. $120\sqrt{2}\sin\left(\omega t+\frac{3\pi}{4}\right)$

25. $120\sqrt{2}\sin\left(\omega t-\frac{3\pi}{4}\right)$

26. $5\sin(\omega t+\alpha)$ where $\tan\alpha=\frac{4}{3}$
27. $5\sin(\omega t-\alpha)$ where $\tan\alpha=\frac{3}{4}$
28. $13\sin(\omega t+\theta)$ where $\tan\theta=\frac{12}{5}$
29. $13\sin(\omega t-\beta)$ where $\tan\beta=\frac{5}{12}$
30. $17\sin(\omega t+\gamma)$ where $\tan\gamma=\frac{8}{15}$
31. $17\sin(\omega t-\theta)$ where $\tan\theta=\frac{15}{8}$
32. $\sqrt{61}\sin(\omega t+\alpha)$ where $\tan\alpha=\frac{5}{6}$
33. $5\sqrt{13}\sin(\omega t-\beta)$ where $\tan\beta=\frac{3}{2}$
34. $4.8\sin(\omega t+48^\circ)$ **35.** $8.9\sin(\omega t+37^\circ)$
36. $365\sin(\omega t+33.5^\circ)$ **37.** $188\sin(\omega t+15.75^\circ)$
38. $211\sin(\omega t+15.75^\circ)$

Exercise 21.2

1. 5; 37° leading **2.** $\sqrt{2}$; $\frac{3\pi}{4}$ leading **3.** $\sqrt{2}$; $\frac{\pi}{4}$ lagging **4.** $\sqrt{2}$; $\frac{\pi}{4}$ leading

5. 5; 53° leading **6.** $\sqrt{5}$; 26.5° leading **7.** 13; 67° leading

8. $\sqrt{5}$; 26.5° lagging **9.** 1; $\frac{\pi}{2}$ leading **10.** 2; $\frac{\pi}{2}$ lagging

11. 2; $\frac{\pi}{2}$ leading **12.** 5; $\frac{\pi}{2}$ leading

Exercise 21.3

1. 2π **2.** 4π **3.** 4π **4.** 2π
5. 4π **6.** 4π **7.** 2π **8.** 4π

Index